Manufacturing Strategies and Systems

Advancements in manufacturing strategies and systems have sparked a profound transformation, ushering in a new era of efficiency, precision, and sustainability, driven by the integration of automation, artificial intelligence, and advanced materials, reshaping industries, boosting productivity, reducing costs, and improving the overall quality of products. This book focuses on practical applications of manufacturing technologies, providing case studies and real-world examples of how these advancements in manufacturing are being implemented to solve manufacturing challenges and improve efficiency.

Manufacturing Strategies and Systems: Technologies, Processes, and Machine Tools presents numerical, experimental, and computational approaches for various methods of manufacturing and offers different concepts from cross-disciplinary fields, including discussions from mechanical engineering, production engineering, and industrial engineering, and acts as a guide on the modeling and optimization of various manufacturing methods. The book explores key emerging trends in manufacturing technologies, such as Industry 4.0, additive manufacturing, robotics and automation, advanced materials, digital twins, augmented reality/virtual reality, edge computing, sustainable manufacturing, and cyber security. Key chapters on micro- and nano-manufacturing and cellular manufacturing are included and details on the advances in machining, joining, forming, powder metallurgy, casting, and molding science are discussed. Included are original theoretical, experimental, and modeling results of advancements in manufacturing techniques along with recent developments, outlook, and advanced and analytical modeling techniques of manufacturing with examples backed by experimental and numerical data.

This reference title provides logical, technical, and analytical solutions and ideas to complex problems faced by researchers and professionals in the field of advancements in manufacturing. Academicians and students will get a comprehensive update on the state of the arts in this area and ample ideas for further research and innovation in manufacturing strategies.

Advancements in Intelligent and Sustainable Technologies and Systems

Series Editor: Ajay Kumar

This book series aims to provide a platform for academicians, researchers, professionals, and individuals to participate and to provide novel systematic theoretical, experimental, and computational work in the form of edited text, reference books, or monographs in the area of intelligent and sustainable technologies and systems from engineering, management, applied science, and healthcare domains. This book series will educate and inform the readers with a comprehensive overview of advancements in intelligent and sustainable techniques and systems with novel intelligent tools and algorithms to move industries from different domains from a data-centric community to a sustainable world. The book series covers ideas and innovations to help the research community and professionals to understand fundamentals, opportunities, challenges, future outlook, layout, lifecycle, and framework of intelligent and sustainable technologies and systems for different sectors. It serves as a guide for computer science, mechanical, manufacturing, electrical, electronics, civil, automobile, industrial engineering, biomedical, healthcare, and management professionals.

If you are interested in writing or editing a book for the series or would like more information, please contact Cindy Carelli, cindy.carelli@taylorandfrancis.com

5G-Based Smart Hospitals and Healthcare Systems
Evaluation, Integration, and Deployment
Edited by Arun Kumar, Sumit Chakravarty, Aravinda K., and Mohit Kumar Sharma

Handbook of Intelligent and Sustainable Smart Dentistry
Nature and Bio-Inspired Approaches, Processes, Materials, and Manufacturing
Edited by Ajay Kumar, Namrata Dogra, Sarita, Surbhi Bhatia, M S Sidhu

Handbook of Intelligent and Sustainable Manufacturing
Tools, Principles, and Strategies
Edited by Ajay Kumar, Parveen Kumar, Yang Liu, and Rakesh Kumar

Advances in Sustainable Biomaterials
Bioprocessing 4.0, Characterizations, and Applications
Edited by Ajay Kumar, D.K. Rajak, Parveen Kumar, and Ashwini Kumar

Manufacturing Strategies and Systems
Technologies, Processes, and Machine Tools
Edited by Ajay Kumar, Parveen Kumar, Ashish Kumar Srivastava, and Lokesh Saharan

Handbook of AI in Engineering Applications
Tools, Techniques, and Algorithms
Edited by Ajay Kumar, Sangeeta Rani, Krishna Dev Kumar, and Manish Jain

Handbook of Deep Learning Models for Healthcare Data Processing
Disease Prediction, Analysis, and Applications
Edited by Ajay Kumar, Deepak Dembla, Seema Tinker, and Surbhi Bhatia Khan

Manufacturing Strategies and Systems

Technologies, Processes, and Machine Tools

Edited by
Ajay Kumar, Parveen Kumar,
Ashish Kumar Srivastava, and Lokesh Saharan

CRC Press is an imprint of the
Taylor & Francis Group, an informa business

Designed cover image: Shutterstock - EtiAmmos

First edition published 2025
by CRC Press
2385 NW Executive Center Drive, Suite 320, Boca Raton FL 33431

and by CRC Press
4 Park Square, Milton Park, Abingdon, Oxon, OX14 4RN

CRC Press is an imprint of Taylor & Francis Group, LLC

ISBN: 978-1-032-71541-4 (hbk)
ISBN: 978-1-032-72506-2 (pbk)
ISBN: 978-1-032-72508-6 (ebk)

DOI: 10.1201/9781032725086

Typeset in Times
by KnowledgeWorks Global Ltd.

Contents

Preface

Advancements in manufacturing strategies and systems refer to the continuous improvement and innovation in technologies, processes, and machine tools used in the manufacturing industry. These advancements aim to enhance productivity, efficiency, flexibility, and overall performance of manufacturing operations enabling to meet evolving market demands and drive economic growth. The technological advancements in the manufacturing techniques have always been at the forefront in societal progress and in economic growth of any country. The manufacturing landscape is undergoing another profound evolution, fueled by the convergence of various emerging technologies like Industry 4.0 strategies in manufacturing sector to model, design, and improve production. The advancements in manufacturing technologies have matured and become more accessible, empowering manufacturers with new capabilities and opportunities to revolutionize their processes and products by increasing automation and personal production. In recent years, development of different strategies and techniques of manufacturing has been dramatic. Now, it is essential to combine and implement advanced manufacturing techniques in industries to divert the world toward green production.

The book *Manufacturing Strategies and Systems: Technologies, Processes, and Machine Tools* is likely to cater to a wide-ranging audience such as engineers, production managers, and technicians seeking to stay updated on the latest trends and advancements in manufacturing to improve processes and product quality in their respective industries: scholars, researchers, and educators in the intersection of manufacturing and technology. This book has the following key features:

- It presents numerical, experimental, and computational approach for various methods of manufacturing.
- It presents different concepts from the cross-disciplinary field of manufacturing, including discussions from mechanical engineering, production engineering, and industrial engineering.
- It provides a detailed guide on modeling and optimization of various manufacturing methods.
- It includes original theoretical, experimental, and modeling results of advancements in manufacturing techniques.
- The book would explore a range of emerging trends in manufacturing technologies, such as Industry 4.0, additive manufacturing, robotics and automation, advanced materials, digital twins, augmented reality/virtual reality, edge computing, sustainable manufacturing, and cyber security.

The book consists of 16 chapters that describe perspectives of *Manufacturing Strategies and Systems: Technologies, Processes, and Machine Tools.*

- Chapter 1 investigates the densification characteristics of Fe-0.8%C steel alloy with varying Mo content under cold upsetting conditions, using surface plots and analysis of variance (ANOVA) to analyze the influence of Mo on densification behavior across different strains, contributing valuable insights for optimizing powder metallurgy processes in high-density steel component production.
- Chapter 2 summarizes the outcomes of electrical discharge machining using sustainable bio-dielectric fluids and provides an overview of different dielectric fluids and their varieties.
- Chapter 3 aims to design, analyze, and discuss the observation of properties such as von Mises stress, equivalent elastic strain, and total deformation for hardness and tensile properties.
- Chapter 4 highlights the need for identifying novel methods for composite filament processing and developing new print nozzles that can mitigate clogging issues and enhance better printability.
- Chapter 5 covers the difficulties in parameter optimization, such as the parts quality and speed, and the necessity for customized solutions based on particular application needs.
- Chapter 6 creates a framework for comprehending the potential advantages of smart manufacturing in the defense industry by drawing on existing research.
- Chapter 7 focuses on design and analysis of indexing jig using CAD software, in which design of all parts of a drill jig, their assembly for real-time applications, and the jig are fabricated using hardened steel having hardness of 58–68 HRC.
- Chapter 8 applies Oyane's fracture criteria to establish workability characteristics and safe zones for artificial intelligence (AI) composites during cold forging under two different lubricant conditions and utilizing ANOVA to assess parameter effects on strain at fracture.
- Chapter 9 employs the Industrial Internet of Things (IIoT) in the manufacturing processes so as to increase the efficiency, productivity, and competitiveness of the manufacturing organizations in the dynamic market.
- Chapter 10 delves into how augmented and virtual reality technologies are revolutionizing modern manufacturing by improving training, digitizing workflows, enhancing data accessibility, and fostering innovation within the context of Industry 4.0.
- Chapter 11 highlights the diverse computer-aided design and computer-aided manufacturing (CAD-CAM) applications in orthodontics which is an advanced step toward the digital and customized orthodontics.
- Chapter 12 offers an exploration of the application of machine learning and deep learning in manufacturing. Beginning with the fundamentals, it covers machine learning (ML) and deep learning (DL) concepts, delving into their respective applications in manufacturing processes.
- Chapter 13 shows that the mechanical properties of fused deposition modeling (FDM) are relatively low when compared to the processes of selective laser melting (SLM) and selective laser sintering (SLS).

- Chapter 14 examines the process variables such as burnishing feed, speed, number of passes, force, lubricants, and ball material and its influence on process responses such as surface roughness, hardness, residual stress, corrosion and wear resistance, and fatigue strength of various materials after burnishing.
- Chapter 15 sheds light on the potential of LPBF (Laser Powder Bed Fusion) technology in revolutionizing metal part manufacturing by offering the ability to create intricate designs with high precision and efficiency.
- Chapter 16 explores advanced techniques for integrating reinforcements like SiC and B_4C into aluminum matrices, highlighting enhanced material properties and application potential.

This book is intended for both the academia and the industry. The postgraduate students, Ph.D. students, and researchers in universities and institutions, who are involved in the areas of manufacturing strategies and systems, will find this compilation useful.

The editors acknowledge the professional support received from CRC Press and express their gratitude for this opportunity.

Readers' observations, suggestions, and queries are welcome.

Editors
Ajay Kumar
Parveen Kumar
Ashish Kumar Srivastava
Lokesh Saharan

About the Editors

Ajay Kumar is currently serving as Professor in the Department of Mechanical Engineering in School of Engineering and Technology, JECRC University, Jaipur, Rajasthan, India. He received his Ph.D. in the field of Advanced Manufacturing and Automation from Guru Jambheshwar University of Science & Technology, Hisar, India, after B.Tech. (Hons.) in mechanical engineering and M.Tech. (Distinction) in Manufacturing and Automation. His areas of research include Biomedical Engineering, Incremental Sheet Forming, Artificial Intelligence, Sustainable Materials, Robotics and Automation, Additive Manufacturing, Mechatronics, Smart Manufacturing, Industry 4.0, Industrial Engineering Systems, Waste Management, and Optimization Techniques. He has over 100 publications in international journals of repute, including *SCOPUS*, *Web of Science*, and *SCI* indexed database and refereed international conferences. He has organized various national and international events, including an international conference on Mechatronics and Artificial Intelligence (ICMAI-2021) as conference chair. He is currently organizing international conference on Artificial Intelligence, Advanced Materials, and Mechatronics Systems (AIAMMS-2023) as conference chair. He has more than 20 national and international patents to his credit. He has supervised more than eight M.Tech., Ph.D. scholars and numerous undergraduate projects/theses. He has a total of 15 years of experience in teaching and research. He has served as a guest editor for many reputed journals. He has contributed to many international conferences/symposiums as a session chair, expert speaker, and member of the editorial board. He has won several proficiency awards during the course of his career, including merit awards and best teacher awards. He has also co-authored and co-edited more than 15 books and proceedings. He has also authored many in-house course notes, lab manuals, monographs, and invited chapters in books. He has organized a series of Faculty Development Programs, International Conferences, workshops, and seminars for researchers, Ph.D., UG, and PG level students. He is associated with many research, academic, and professional societies in various capacities.

Parveen Kumar is currently serving as Assistant Professor and Head in the Department of Mechanical Engineering at Rawal Institute of Engineering and Technology, Faridabad, Haryana, India. Currently, he is pursuing a Ph.D. from National Institute of Technology, Kurukshetra, Haryana, India. He completed his B.Tech. (Hons.) from Kurukshetra University, Kurukshetra, India, and M.Tech. (Distinction) in Manufacturing and Automation from Maharshi Dayanand University, Rohtak, India. His areas of research include Intelligent Manufacturing Systems, Materials, Mechatronics Systems, Advanced Materials, Industry 4.0, Die Less Forming, Design of Automotive Systems, Additive Manufacturing, CAD/CAM and Artificial Intelligence, Machine Learning and Internet of Things in Manufacturing, and Multi-Objective Optimization Techniques. He has over 50 publications in international journals of repute including *SCOPUS*, *Web of Science*, and *SCI* indexed database and refereed international conferences. He has eight national and international patents in his credit. He has supervised several M.Tech. scholars and numerous undergraduate projects/theses. He has a total of 13 years of experience in teaching and research. He has co-authored/co-edited several books. He has organized a series of Faculty Development Programs, workshops, and seminars for researchers and UG level students. He is the Editor of book series on "Industrial and Manufacturing Systems and Technologies: Sustainable and Intelligent Perspectives," CRC Press, Taylor & Francis.

Ashish Kumar Srivastava completed his Ph.D. from Indian Institute of Technology (ISM), Dhanbad, Jharkhand, India, in February 2018 and is currently working as Associate Professor in the Department of Mechanical Engineering, Muzaffarpur Institute of Technology (MIT), Muzaffarpur, Department of Science, Technology and Technical Education, Government of Bihar, India. He is a good learner and listener and contributed a lot to society through his academics, administration, and research work.

As a researcher, he has published more than 120 research papers in various journals of repute, indexed in SCI/SCIE/SCOPUS. He has also presented more than 20 research papers at various international and national conferences in India and abroad. Dr. Srivastava is among the Top 2% Scientists of the world listed by Stanford University, USA, and Elsevier B.V. in 2022 and 2023. He has successfully completed two research-funded projects under the collaborative research scheme of TEQIP-III. He presently received a grant of Rs 10.19 lakhs from the Council of Science and Technology, Uttar Pradesh (CST-UP). Recently he has published nine Indian patents which are likely to be granted in the upcoming years. Dr. Srivastava has also worked with the Institute of the Geonics of the CAS, Ostrava-Poruba, Czech Republic, and Slovak Research and Development Agency (Slovak Republic) under contract No. APVV-207-12 to carry out his Ph.D. research work related to Abrasive Waterjet Cutting and Turning. The current research areas of Dr. Srivastava are 3D Printing (Metal and Fiber), Friction Stir Additive Manufacturing (Through Robotic Milling, CNC Milling, and Conventional VMC), Wire Arc Additive Manufacturing (Through Robotic Welding), Friction Stir welding, Friction Stir Processing, Mechanical and Electromagnetic Stir Casting, Abrasive Waterjet Cutting and Turning, and Wire Electric Discharge Cutting and Turning.

Lokesh Saharan is working as Assistant Professor in the Department of Mechanical Engineering, College of Engineering at the University of Texas at Dallas, USA, and has more than ten year of teaching and research experience. He received his Ph.D. from the University of Texas at Dallas in December 2017. His primary research interests include rehabilitation robotics, smart materials, soft robotics, and additive manufacturing. Dr. Saharan has authored several research articles in peer-reviewed journals and conferences along with a book chapter. He recently received a UT System Rising STARs grant with a matching UTPB school of engineering grant worth $470,000. Currently, Dr. Saharan is also leading the Advanced Manufacturing Center in the College of Engineering. This center is funded by $1.1M from the

Odessa Development Corporation. Moreover, he has served as a reviewer for several peer-reviewed conferences and journals.

Before joining UTPB, he was an assistant teaching professor at Penn State Behrend, where he developed new courses for the biomedical minor in the mechanical engineering department. Besides, Dr. Saharan worked as a postdoctoral fellow at Miami University of Ohio for one year, where he contributed to the Miami University Center for Assistive Technology. Before moving to the United States, he also worked as Assistant Professor (Instruction) of mechanical engineering at the National Institute of Technology. His research interests are Additive Manufacturing, Rehabilitation Robotics, Smart Materials and Structures, Engineering Education, Biomechanics, and Soft Robotics.

Contributors

Saiyathibrahim Abdul Pari
Department of Mechanical Engineering, Karpagam Institute of Technology
Coimbatore, India

Shalini Anand
Department of Mechanical Engineering, Muzaffarpur Institute of Technology (MIT)
Muzaffarpur, Bihar, India

Rajeshkannan Ananthanarayanan
Mechanical Engineering, STEMP, The University of the South Pacific
Suva, Fiji

Firdoos Afzal Bhat
Department of Mechanical Engineering, National Institute of Technology
Srinagar, India

Loubna Bouhachlaf
Faculty of Sciences, Mohammed V University in Rabat
Morocco

Khalid Bouiti
LS3MN2E, CERNE2D, ENSAM, Mohammed V University in Rabat
Morocco

Samson Jerold Samuel Chelladurai
Department of Mechanical Engineering, Sri Krishna College of Engineering Technology
Coimbatore, India

Vikas Choudhary
Department of Mechanical Engineering, School of Engineering and Technology, JECRC University
Jaipur, Rajasthan, India

Mamta Dahiya
Computer Science and Engineering, School of Engineering and Technology, Manav Rachna International Institute of Research and Studies
Faridabad, Haryana, India

Mohamed Dalimi
LS3MN2E, CERNE2D, ENSAM, Mohammed V University in Rabat
Morocco

Venkata Krishna Devara
Department of Mechanical Engineering, Indian Institute of Technology Tirupati
Andhra Pradesh, India

Sunil Dhull
Mechanical Engineering Department, Panipat Institute of Engineering and Technology
Samalkha, India

Namrata Dogra
Department of Orthodontics, Faculty of Dental Sciences, SGT University
Gurugram, Haryana, India

Gyander Ghangas
Mechanical Engineering Department, Panipat Institute of Engineering and Technology
Samalkha, India

Surendra C. Ghorpade
Defence Institute of Advanced Technology
Pune, India

Anuj Gupta
Department of Mechanical Engineering, School of Engineering and Technology, JECRC University
Jaipur, Rajasthan, India

Souad El Hajjaji
LS3MN2E, CERNE2D, Faculty of Sciences, Mohammed V University in Rabat
Morocco

Infant Jegan Rakesh Antony John Bosco
Department of Mechanical Engineering, Karpagam Institute of Technology
Coimbatore, India

Archana Jaglan
Department of Orthodontics, Faculty of Dental Sciences, SGT University
Gurugram, Haryana, India

A. K. Jeevanantham
School of Mechanical Engineering, Vellore Institute of Technology
Vellore, India

Kaoutar Kara
LS3MN2E, CERNE2D, ENSAM, Mohammed V University in Rabat
Morocco

Jahangir Khan
Department of Mechanical Engineering, School of Engineering and Technology, JECRC University
Jaipur, Rajasthan, India

Mohd Yunus Khan
National Institute of Technical Teachers Training and Research
Chandigarh, India

Ajay Kumar
Department of Mechanical Engineering, School of Engineering and Technology, JECRC University
Jaipur, Rajasthan, India

Ashish Kumar
Department of Mechanical Engineering, Muzaffarpur Institute of Technology (MIT)
Muzaffarpur, Bihar, India

Manoj Kumar
Department of Mechanical Engineering, Dr. B. R. Ambedkar National Institute of Technology
Jalandhar, India

Pardeep Kumar
Mechanical Engineering Department, SRM Institute of Science and Technology
Delhi NCR Campus, Ghaziabad, India

Parveen Kumar
Department of Mechanical Engineering, Rawal Institute of Engineering and Technology
Faridabad, Haryana, India

Pramod Kumar
Department of Mechanical Engineering, Dr. B. R. Ambedkar National Institute of Technology
Jalandhar, India

Ravichandran Rajeeth Kumar
School of Mechanical Engineering, Vellore Institute of Technology
Vellore, India

Sanjay Kumar
Department of Mechanical Engineering, Dr. B. R. Ambedkar National Institute of Technology
Jalandhar, India

Sharda Kumari
Department of Mechanical Engineering, Muzaffarpur Institute of Technology (MIT)
Muzaffarpur, Bihar, India

Dhivya Priya Erode Loganathan
Department of Electronics and Communication Engineering, Erode Sengunthar Engineering College
Erode, Tamil Nadu, India

Houda Labjar
Faculty of Sciences and Technologies of Mohammedia, Hassan II University
Casablanca, Morocco

Najoua Labjar
Faculty of Sciences and Technologies of Mohammedia, Hassan II University
Casablanca, Morocco

Makeshkumar Mani
Department of Mechanical Engineering, KPR Institute of Engineering and Technology
Coimbatore, India

Shanmuka Srinivas Maddula
Department of Mechanical Engineering, Indian Institute of Technology Tirupati
Andhra Pradesh, India

Ravi Sankar Mamilla
Department of Mechanical Engineering, Indian Institute of Technology Tirupati
Andhra Pradesh, India

Vishwa Ratan Mishra
G.L. Bajaj Institute of Technology and Management
Greater Noida, India

Vijayan Singarajan Nagammal
Department of Mechanical Engineering, Karpagam Institute of Technology
Coimbatore, India

Hamid Nasrellah
Higher School of Education and Training, Chouaib Doukkali University
El Jadida, Morocco

Elango Natarajan
Faculty of Engineering, Technology and Built Environment, UCSI University
Kuala Lumpur, Malaysia

Bahadur Singh Pabla
National Institute of Technical Teachers Training and Research
Chandigarh, India

Bhargav Chandan Palivela
Department of Mechanical Engineering, Indian Institute of Technology Tirupati
Andhra Pradesh, India

Saad Parvez
Department of Mechanical Engineering, National Institute of Technology
Srinagar, India

Elayaraja Radhakrishnan
School of Mechanical Engineering, Vellore Institute of Technology
Vellore, Tamil Nadu, India

Rajkumar Shashank Rajkumar
School of Mechanical Engineering, Vellore Institute of Technology
Vellore, India

Sudhakar Rao Patange
National Institute of Technical Teachers Training and Research
Chandigarh, India

Boopathi Sampath
Department of Mechanical Engineering, Muthayammal Engineering College
Rasipuram, Tamil Nadu, India

Suresh Subramanian
Department of Mechanical Engineering, Erode Sengunthar Engineering College
Erode, Tamil Nadu, India

Jai Aultrin Kamalan Sundarabai
Department of Marine Engineering, Noorul Islam Centre for Higher Education
Kanyakumari, India

Devi Seenivasagam
School of Management, CBHTS, Fiji National University
Nasinu, Fiji

Kirti Sehrawat
Department of Orthodontics, Faculty of Dental Sciences, SGT University
Gurugram, Haryana, India

Pankaj Shakkarwal
Department of Mechanical Engineering, Manav Rachna International Institute of Research and Studies
Faridabad, India

Sumati Sidharth
Defence Institute of Advanced Technology
Pune, India

Ashish Kumar Srivastava
Department of Mechanical Engineering, Muzaffarpur Institute of Technology (MIT)
Muzaffarpur, Bihar, India

Siva Prasad Tadi
Department of Mechanical Engineering, Indian Institute of Technology Tirupati
Andhra Pradesh, India

Ravindra Tajane
Department of Automation and Robotics Engineering, Amrutvahini College of Engineering
Sangamner, Maharashtra, India

Nitin Jalindar Varpe
Department of Automation and Robotics Engineering, Amrutvahini College of Engineering
Sangamner, Maharashtra, India

Nitin Kumar Waghmare
Department of Automobile Engineering, Manav Rachna International Institute of Research and Studies
Faridabad, India

Acknowledgments

The editors are grateful to CRC Press for showing their interest to publish this book in the buzz area of manufacturing strategies and systems. The editors express their personal adulation and gratitude to Ms. Cindy Renee Carelli (Executive Editor), CRC Press, for giving consent to publish our work. She undoubtedly imparted the great and adept experience in terms of systematic and methodical staff who have helped the editors to compile and finalize the manuscript. The editors also extend their gratitude to Ms. Kaitlyn Fisher, CRC Press, for supporting during her tenure.

The editors wish to thank all the chapter authors for contributing their valuable research and experience to compile this volume. The chapter authors, corresponding authors in particular, deserve special acknowledgments for bearing with the editors, who persistently kept bothering them for deadlines and with their remarks.

Dr. Ajay also wishes to express his gratitude to his parents, Sh. Jagdish and Smt. Kamla, and his loving brother Sh. Parveen for their true and endless support. They have made him able to walk tall before the world regardless of sacrificing their happiness and living in a small village. He cannot close these prefatory remarks without expressing his deep sense of gratitude and reverence to his life partner Mrs. Sarita Rathee for her understanding, care, support, and encouragement to keep his morale high all the time. No magnitude of words can ever quantify the love and gratitude he feels in thanking his daughters, Sejal Rathee and Mahi Rathee, and son Kushal Rathee who are the world's best children.

Finally, the editors obligate this work to the divine creator and express their indebtedness to the "almighty" for gifting them power to yield their ideas and concepts into substantial manifestation. The editors believe that this book would enlighten the readers about each feature and characteristic of manufacturing strategies and systems.

Section I

Evolving Trends in Manufacturing Technologies

1 Densification Study of Fe-0.8%C Powder Metallurgy Steel with the Effect of Mo under Cold Upsetting Using Surface Plots and ANOVA

Rajeshkannan Ananthanarayanan,
A. K. Jeevanantham, Devi Seenivasagam,
Ravichandran Rajeeth Kumar, and
Rajkumar Shashank Rajkumar

1.1 INTRODUCTION

Powder metallurgy (P/M) is widely employed in the manufacturing of steel components, particularly for incorporating specific properties into the final product. This method is favored due to its straightforward steps and cost-effectiveness for mass production [1, 2], as well as its ability to produce unique and complex parts for job-type production [3]. Production of the steels at high temperatures is susceptible to decarbonization [4, 5]; however, in the P/M process, sintering is carried out in a controlled or protective environment that can overcome this challenge [6]. Hence, for smaller steel components in automobile applications, the P/M methods are widely employed [7, 8]. However, for synthesizing any new material using the P/M process, studying its characteristics is essential [9, 10]. Due to the unavoidable presence of porosity after the conventional P/M operation, the material may sometimes require secondary deformation processes, such as upsetting, to achieve the necessary mechanical strengthening [11, 12]. In the present study, Fe-0.8%C steel is chosen due to its high industry relevance as it is extensively applied in the automotive, construction and tooling industries [13]. Imparting Mo element with carbon steel induces hardenability and reduces temper embrittlement. It is reported that Mo of 1–2% in high-temperature steel increases corrosion resistance and high-temperature strength [14].

In P/M, the produced parts experience significant plastic deformation, which is further intensified during secondary deformation processes. Hence, the deformation characteristics in association with densification become imperative to study [15, 16]. It was reported in a study that in Fe-0.5%C steel preforms with the varying amount

DOI: 10.1201/9781032725086-2

of Mn from 0 to 2% while subjected to cold upsetting, the increase of Mn helped to form undissolved carbides and as a consequence, there is a retardation in deformation and densification [17]. In another study, the process parameters such as aspect ratios and lubricant conditions influence were studied with Fe-0.5%C steel subjected to cold upsetting. It was reported as density increases when friction during deformation increases and is further supported by the decreasing aspect ratios [18–20]. Similar results were observed in other articles too [13, 14]. In the present study, the densification characteristics are studied against deformation with respect to various strains with varying Mo from 0 to 2% using a 3D surface plot. The 3D surface plot is reported to be better used to visualize the variations of two factors simultaneously affecting a third parameter [21]. To study the behavior of two face-centered cubic (FCC) catalysts at the same operational attrition conditions, such as flow rate and time, the 3D surface plot was used to represent the effect of each operational parameter and their interactions [22].

It is also observed in the literature that analysis of variance (ANOVA) is extensively used to identify the significance of the main and interaction effect of the process parameters. It was reported that an ANOVA test was conducted for density, hardness and compressive strength at two different temperatures and times to predict its statistical significance [23]. Likewise, another study has reported that the compositions, aspect ratios, initial densities and lubricants' individual and their interactional effects were studied using ANOVA against strain hardening exponent and strength coefficient factor of sintered Al-Cu-TiC preforms during cold upsetting [24].

Therefore, Fe-0.8%C steel alloy with varying Mo content from 0 to 2% was produced using the conventional P/M route and then subjected to cold upsetting to enhance its density. The densification characteristics were analyzed in relation to deformation by examining various strains, including axial, hoop, mean and effective strains under triaxial conditions, utilizing 3D surface plots and 2D contour plots. Additionally, the effects of these strains and the alloy composition were investigated using ANOVA. The ANOVA results were further used to develop models for predicting densification based on the specified process parameters.

1.2 MATERIALS AND METHODS

The research methodology used for the present investigation is shown in Figure 1.1. The research began with a review of the literature on the significance of carbon steels and their synthesis using P/M processing. This was followed by studying the plastic deformation of porous materials and reviewing the use of surface and contour plots for result presentation. Additionally, the use of ANOVA was reviewed, highlighting its effectiveness in modeling and predicting the main and interaction effects of various factors. The next step in this investigation involves conducting experiments on the selected carbon steel, which includes primary P/M operations and secondary operations such as cold upsetting. The results from the cold upsetting will be used to study densification in relation to deformation.

As a first step in experiment methods, the elemental powders are gathered. The atomized iron powder of particle size of 150 μm was procured from M/s: the Sundram

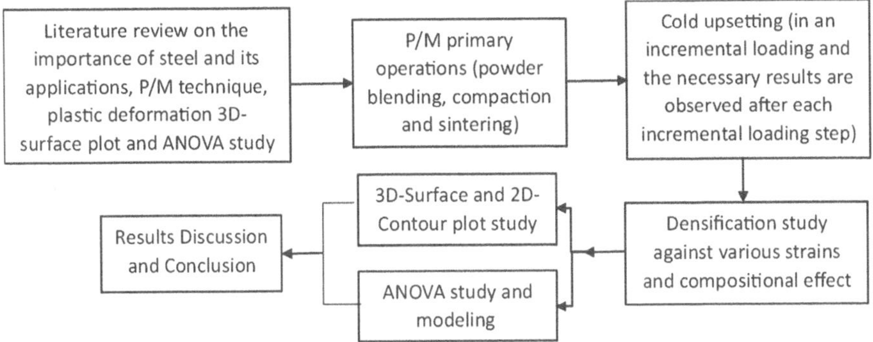

FIGURE 1.1 Research methodology.

TABLE 1.1
Sieve Analysis of Atomized Iron Powder

Sieve size (mm)	150	+106	+90	+63	+45	+37
Wt. % Ret.	32.04	9.46	22.12	23.2	6.01	7.00

Fasteners Limited, Hyderabad, India. The sieve analysis (as shown in Table 1.1) has revealed a particle size range from 150 to 37 µm. The scanning electron microscopic (SEM) characterization endorses the spherical geometry of powder (as shown in Figure 1.2), and the purity of powder was found to be 99.7%. Also, the alloying elements, such as 2–3µm carbon and molybdenum of less than 37µm, were procured from M/s: the Metal Powder Company, Madurai, India.

The elemental powders of appropriate amount were measured using an electronic weighing machine of 10^{-4} g accuracy to formulate a Fe-0.8%C, Fe-0.8%C-1%Mo,

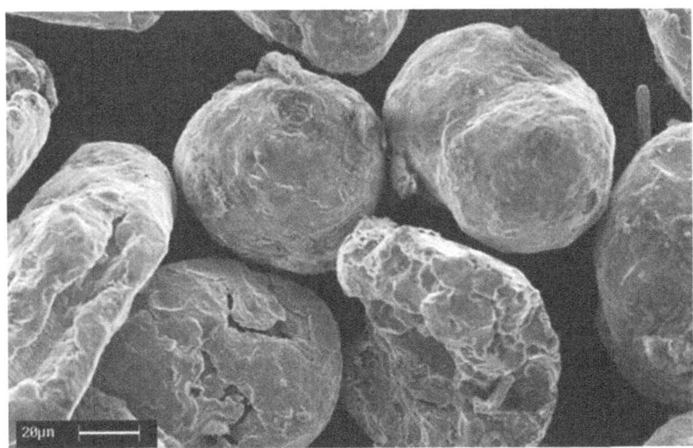

FIGURE 1.2 SEM image of atomized iron powder.

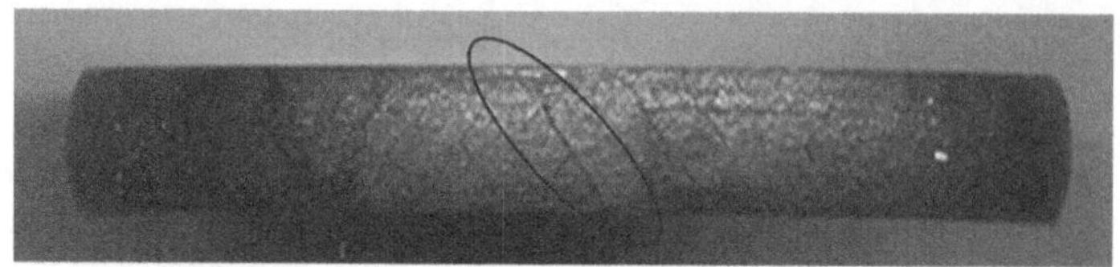

FIGURE 1.3 Visible crack on the bulge surface of a deformed specimen.

Fe-0.8%C-1.5%Mo and Fe-0.8%C-2%Mo to carry out the blending process using the Retsch ball mill. The homogeneity of the powder blend was ensured by obtaining a consistent flow rate in successive intervals of time from the ball mill [25]. The blended powder was then poured into the compaction die-set for compaction operation, carried out in a 1 MN capacity hydraulic press. The green compact diameter of 24 mm and the height of 10 mm with an approximation of 86% fractional density were obtained from compaction in the pressure range of 430±10 MPa.

The green compacts are immediately taken to apply flux coating to prevent oxidation at elevated temperatures. This coating was applied twice perpendicular to each other on the compacts in a 24-hour interval and was allowed to dry at atmospheric conditions. The sintering operation on the flux-coated specimen was carried out in two intervals, one as a drying operation at 700°C for 30 minutes and the next at 1150±10°C for 90 minutes. Further, it was left in the furnace to cool to atmospheric conditions.

These specimens are gently machined to get a 0.4 aspect ratio, then subjected to cold upsetting operations at incremental loading of 4 tons. The compression loading was applied on the specimen between two flat dies at 40 kN/s. After each incremental loading step, the deformed dimensions, both diametrical and height measurements, were made, in addition to the density measurement using Archimedes' principle. Due to the cold deformation being conducted under dry friction conditions, the preforms experienced bulging. The deformation process was concluded when a fracture appeared in the hoop region. Figure 1.3 shows the visible crack on the bulge surface of the deformed specimen for the Fe-0.8%C-1%Mo preform, while Figure 1.4 illustrates the preform both before and after deformation, along with the notations used in the theoretical analysis.

1.3 RESULTS AND DISCUSSION

Densification is studied using surface and contour plots against various strains, followed by an ANOVA study that has been carried out to elucidate the influence of individual and mixed parameters. The density of the preforms, measured from their packed density, ρ_f, relative to the theoretical density, ρ_{th}, of the chosen composition, is referred to as fractional theoretical or relative density, R, and is defined as follows:

$$R = \rho_f / \rho_{th}. \tag{1.1}$$

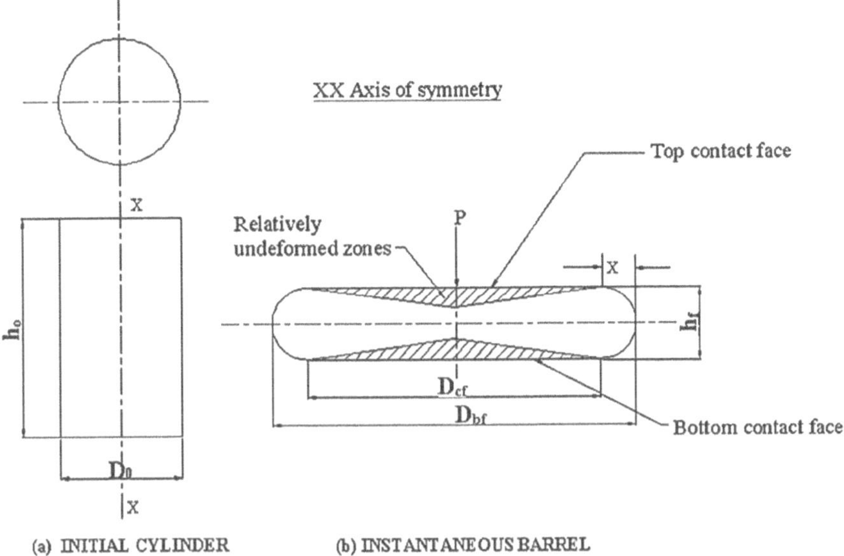

FIGURE 1.4 Schematic view of preform (a) before and (b) after deformation [26].

1.3.1 Densification against Axial Strain

The axial strain, ε_h, is the principal deformation that undergoes by the cylindrical preform in the direction of the applied load itself, and it is of compressive nature. This strain is the main cause for the pore closure in the deforming preform [27], which is determined as follows:

$$\varepsilon_h = ln\left(h_f/h_o\right), \tag{1.2}$$

where h_f is the preform height deformed at any given instant and h_o is the initial height of the preform.

The surface and the contour plot are shown in Figure 1.5a and b, respectively, against axial strain with the influence of varying Mo from 0 to 2% in Fe-0.8%C steel. The results (Figure 1.5a) indicate that continuous straining in the axial direction consistently increases the preform density for all the Mo-varied compacts. However, there is accelerated densification from 0 to 0.4ε_h, following further straining that retards densification; this behavior is observed to be common for all types of compacts. Interestingly, the inclusion of Mo in Fe-0.8%C shows some differences in the relative density; for instance, from 0 to 0.4 axial strain, 2Mo densifies at the lowest rate, while 1.5Mo densifies at the highest rate, followed by 1Mo and 0Mo. But, from 0.6 to 0.8 axial strain 2Mo shifts to the peak densification, followed by 1.5Mo and 0Mo, in the case of 1Mo, densifies at the least. In summary, the baseline compacts (Fe0.8%-0%Mo) exhibit consistent densification rates from the beginning to the end of axial strain. However, with the addition of 1Mo, although there is a slight increase in densification rate at the onset of strain, it shows lower densification at later strain

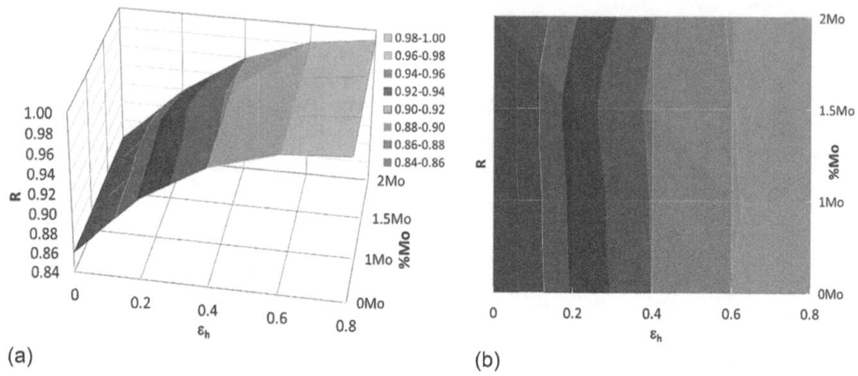

FIGURE 1.5 (a) Surface and (b) contour plot for density against axial strain and %Mo variation.

values, similar to the trend observed with 1.5Mo. Notably, 2Mo behaves uniquely: initially, it densifies at a slower rate, but later in the strain process, it accelerates, reaching peak densification by the end of axial strain. This behavior is discernible from the contour plot (Figure 1.5b) where, for any given strain, if the mode is on the left-hand side, it indicates an increase in densification, while on the right-hand side, it suggests a decrease in densification.

1.3.2 Densification against Hoop Strain

The hoop strain, ε_θ, appears on the circumferential portion of the deforming preform, whose nature is tensile, and it is the prime cause for the preform failure that ultimately demands to stop the deforming operation as soon as a visible fracture appears (refer to Figure 1.3). The deformation in the hoop region is the consequence of the deformation in contact diameter, D_{cf}, and bulge diameter, D_{bf}, at any given instant of forging as shown in Figure 1.4b. The expression for it is derived elsewhere and is given by [28]:

$$\varepsilon_\theta = \left(2D_{bf}^2 + D_{cf}^2\right)/\left(3D_o^2\right), \tag{1.3}$$

where D_o is the original contact diameter after the sintering operation. Figure 1.6a and b is the surface and contour plots drawn for relative density against hoop strain with the influence of Mo variation. It can be noted from Figure 1.6a that the general trend of densification is enhanced when the hoop strain increases. However, up to $0.3\varepsilon_\theta$, density has an accelerated trend, and thereafter, it retards. But with the addition of Mo, although it appears a minute change among them, it is noteworthy that baseline compact (0Mo) and 1Mo compacts generally show a common densification trend from the beginning to the end of the hoop strain as per the contour plot in Figure 1.6b. Nevertheless, once the Mo percentage is raised to 1.5, the initial densification is appreciable against hoop strain, but in the later strain stage

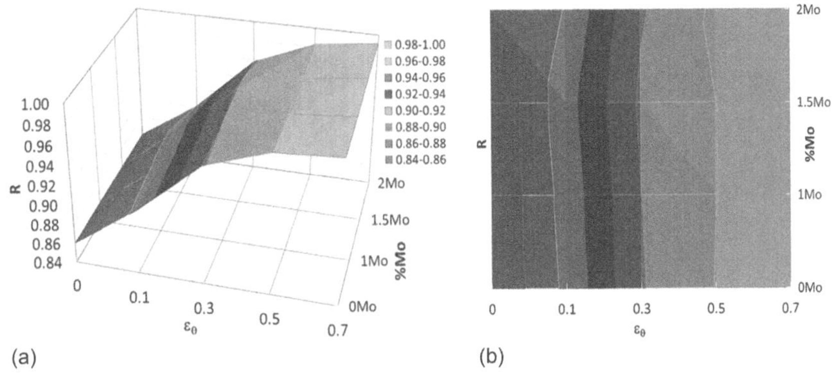

FIGURE 1.6 (a) Surface and (b) contour plot for density against hoop strain and %Mo variation.

(i.e., after $0.3\varepsilon_\theta$), it almost coincides with 1Mo. While the 2Mo shows a slow pace in densification during the initial stages of ε_θ, at a later stage, the densification is profound compared to any other Mo varied compacts. Hence, the future research interest is to test from 1.5Mo and above to check where exactly densification peaks and how much it strains before it fractures to produce fracture-free high-density components.

1.3.3 Densification against Mean Strainc

The mean or hydrostatic strain, ε_m, in a way, represents the average of the axial, hoop and radial strain as it is determined for triaxial condition [29], which can be expressed as follows:

$$\varepsilon_m = (\varepsilon_z + \varepsilon_r + \varepsilon_\theta)/3. \tag{1.4}$$

Since the preform is a cylindrical shape, and for the triaxial state, $\varepsilon_r = \varepsilon_\theta$, substituting this in Eq. (1.4) gives

$$\varepsilon_m = (\varepsilon_z + 2\varepsilon_\theta)/3. \tag{1.5}$$

The surface and contour plots for the densification attained against mean strain with the effect of Mo variation are shown in Figure 1.7a and b, respectively. The mean strain characteristics are directly influenced more by hoop than axial strain characteristics; hence, it can be observed from Figure 1.7a that there is a continuous improvement in density for all the Mo-varied sintered eutectoid steel against increasing mean strain. However, as anticipated, adding 1Mo and 1.5Mo with Fe-0.8%C enhanced densification for any given mean strain until $0.35\varepsilon_m$ (evident in Figure 1.7b), after that, the trend is reversed, that is, 2Mo peaks the densification followed by 1.5Mo, whereas the 1Mo and 0Mo coincide.

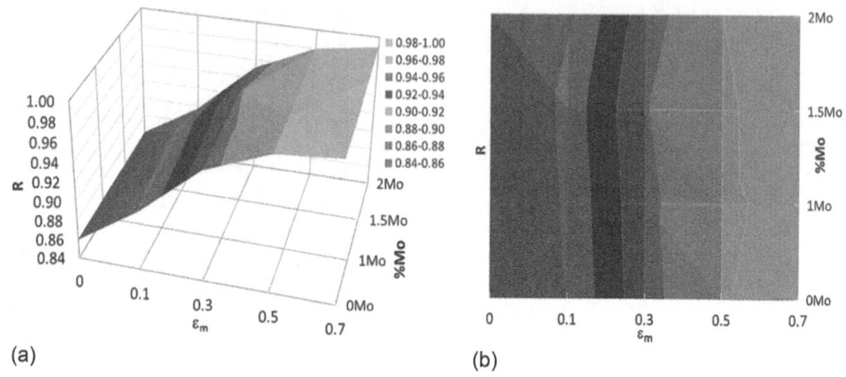

FIGURE 1.7 (a) Surface and (b) contour plot for density against mean strain and %Mo variation.

1.3.4 Densification against Effective Strain

The effective strain, ε_{eff}, is the one experienced by the whole body of the preform that encompasses the axial, radial or hoop strain and the relative density in a complex way; hence, it is derived for triaxial condition, the derivation for it is explained elsewhere [30], and it can be expressed as follows:

$$\varepsilon_{eff} = \left\{ \left[2 / \langle 3(2+R^2) \rangle \right] \left[2\varepsilon_h^2 + 2\varepsilon_\theta^2 - 4\varepsilon_h \varepsilon_\theta \right] + \left[\left(\langle 2\varepsilon_\theta - \varepsilon_h \rangle^2 / 3 \right) \left(1 - R^2 \right) \right] \right\}^{0.5}. \quad (1.6)$$

The surface and contour diagrams are plotted for relative density against effective strain with the influence of the Mo effect in Fe-0.8%C steels, shown in Figure 1.8a and b, respectively. Figure 1.8 indicates that the maximum span of effective strain experienced by the preform is to densify from 0.96 to 0.98 and thereafter. During

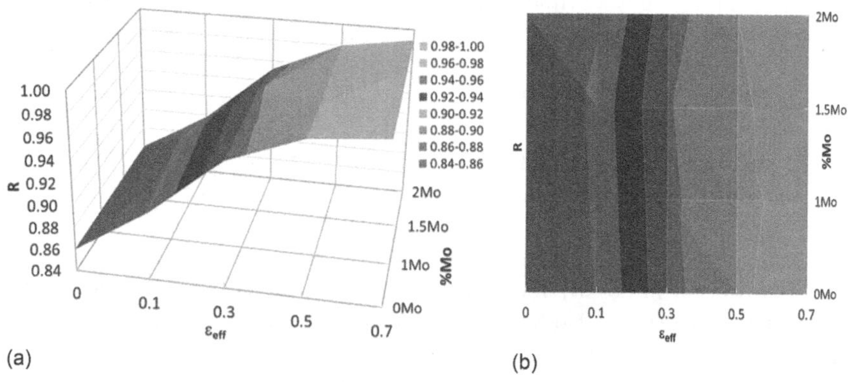

FIGURE 1.8 (a) Surface and (b) contour plot for density against effective strain and %Mo variation.

this span of strain, there is a shift in the order of densification from 1.5Mo, 1Mo, 0Mo and 2Mo to 2Mo, 1.5Mo, 0Mo and 1Mo. From this order of densification, it can be observed from Figure 1.8b that 0Mo maintains the same position throughout, irrespective of the strain induced; however, the 2Mo shows a massive shift from least to topmost in densification as strain increases. While 1Mo and 0Mo almost follow a similar path with negligible variation in densification, on the other hand, 1.5Mo shows a slight but notable drop from its peak densification next to 2Mo. This confirms that only an addition of 1.5Mo and above shows slightly significant densification as strain increases, whereas less than 1.5Mo, more or less, follows the 0Mo path. Further, it was noticed from the experiment that there is an early failure in 1Mo and 1.5Mo, which is not the case for 0Mo and 2Mo preforms; therefore, it is noteworthy to conduct further experimental research from 1.5Mo to 2.5Mo (at least) to examine the maximum densification attainment and strain retainment in the preform while cold deformation.

1.3.5 ANOVA Study for Various Strains, Compositional Effect and Its Mixture

ANOVA analysis was conducted to assess densification during cold forging across various strains, considering the influence of Mo variation from 0 to 2% in Fe-0.8%C steels. The ANOVA results for axial, hoop, mean and effective strains are presented in Tables 1.2–1.5. In all strain cases, the obtained model exhibited a high coefficient of determination R^2 value of 0.99. Additionally, Tables 1.2–1.5 show relatively high F-values with p-values well below 0.05, confirming the significance of the models. The model equations for all strains are provided in the last row of the respective tables. It is noted in the ANOVA tables that the linear mixture indicates the mixture of Fe-0.8%C as 'A' and Mo as 'B'. Though the linear mixture influence is insignificant in all the cases of strains, its influence in combination with the corresponding strains, particularly with axial (in Table 1.2) and effective strains (in Table 1.5), plays

TABLE 1.2
ANOVA for Axial Strain and Linear Mixture

Source	Sum of Squares	df	Mean Square	F-Value	p-Value
Model	3.29E-02	6	5.48E-03	946.69	<0.0001
Linear mixture	2.57E-06	1	2.57E-06	0.335	0.5726
C-axial strain	0.0291	1	0.0291	3791.52	<0.0001
AC^2	0.0017	1	0.0017	217.66	<0.0001
BC^2	0.0019	1	0.0019	253.95	<0.0001
ABC(A – B)	0.0001	1	0.0001	10.29	0.0069
ABC^3(A – B)	0.0001	1	0.0001	10.52	0.0064
Residual	0.0001	13	7.69E-06		
Cor total	3.30E-02	19			
Model equation	$0.9617A + 0.9611B + 0.0623C - 0.0398AC^2 - 0.0365BC^2 + 0.1833ABC(A-B) - 0.20196ABC^3(A-B)$				

TABLE 1.3
ANOVA for Hoop Strain and Linear Mixture

Source	Sum of Squares	df	Mean Square	F-Value	p-Value
Model	2.55E-02	3	8.50E-03	688.32	<0.0001
Linear mixture	8.17E-06	1	8.17E-06	0.3579	0.558
C-hoop strain	0.0194	1	0.0194	849.62	<0.0001
C^2	0.0061	1	0.0061	267.68	<0.0001
Residual	0.0004	16	0.000025		
Cor total	2.59E-02	19			
Model equation	$0.9745A + 0.9762B + 0.0556C - 0.0564C^2$				

TABLE 1.4
ANOVA for Mean Strain and Linear Mixture

Source	Sum of Squares	df	Mean Square	F-Value	p-Value
Model	2.56E-02	3	8.53E-03	977.47	<0.0001
Linear mixture	3.02E-06	1	3.02E-06	0.1909	0.668
C-mean strain	0.0206	1	0.0206	1301.72	<0.0001
C^2	0.005	1	0.005	317.99	<0.0001
Residual	0.0003	16	0.00001875		
Cor total	2.59E-02	19			
Model equation	$0.9705A + 0.9715B + 0.0573C - 0.0512C^2$				

TABLE 1.5
ANOVA for Effective Strain and Linear Mixture

Source	Sum of Squares	df	Mean Square	F-Value	p-Value
Model	2.57E-02	4	6.43E-03	1114.98	<0.0001
LM	7.43E-07	1	7.43E-07	0.0713	0.7931
C-effective strain	0.0222	1	0.0222	2135.2	<0.0001
AC^2	0.0015	1	0.0015	141.32	<0.0001
BC^2	0.002	1	0.002	188.66	<0.0001
Residual	0.0002	15	1.3333E-05		
Cor total	2.59E-02	19			
Model equation	$0.9664A + 0.9669B + 0.05595C - 0.0461AC^2 - 0.04601BC^2$				

a relatively significant role. Therefore, it can be concluded that the predicted model can produce the densification results within the selected levels of parameters with high accuracy. Further, it can be concluded that all the strains contribute significantly to densification; however, the variations of Mo are relatively negligible with hoop and mean strain in contributing to densification, but with axial and effective strains, its role is inevitable in densification.

1.4 FUTURE SCOPE AND DIRECTIONS

The study on the densification of Fe-0.8%C powder metallurgy steel with the effect of Mo under cold upsetting using surface plots and ANOVA offers promising avenues for future research and application. Future investigations could focus on several directions. First, exploring higher concentrations of Mo beyond 2% could provide insights into the upper limits of densification enhancement and mechanical properties. Additionally, studying the combined effects of Mo with other alloying elements could elucidate synergistic effects on densification and mechanical behavior. Moreover, considering alternative processing techniques or parameter variations in cold upsetting, such as varying strain rates or temperatures, could offer novel insights into densification mechanisms and optimize manufacturing processes further. Investigating the influence of different preform geometries or compaction methods on densification behavior could also enhance understanding and applicability in diverse industrial contexts. Furthermore, extending the study to evaluate long-term mechanical performance, including fatigue and fracture resistance, of densified components could provide comprehensive insights into their suitability for demanding applications. Additionally, exploring the scalability of the optimized densification process for large-scale production and its economic feasibility would be beneficial for industrial adoption. In summary, future research could delve into optimizing alloy compositions, exploring alternative processing parameters, assessing mechanical performance, and addressing scalability concerns to advance the utilization of densified P/M steel components in various industries. Such endeavors hold the potential to revolutionize manufacturing practices and enhance the performance of steel components in automotive, construction and tooling applications.

1.5 CONCLUSIONS

Sintered Fe-0.8%C with varying amounts of Mo from 0 to 2% is subjected to cold forging, whose densification characteristics are studied against various strains using surface plot and ANOVA; the following are its major findings:

- The densification is found to be increasing for any strains (be it an axial, hoop, mean or effective) and compositional variations. However, the axial strain is found to have undergone large compared to the other strains, irrespective of the compositions. Further, it was inferred from the study that the densification accelerated during the initial phase of strains, while at the later stage, it retards, or the densification curve flattens.

- The compositional effect in densification was found to be negligible for 1Mo as it follows an almost similar pattern of 0Mo; however, from 1.5Mo and above (in particular, the 2Mo), there is a significant shift from lowest to the highest densification against any strains as per the surface and contour plots. On the other hand, the ANOVA study reveals that only strains are making a statistically significant impact. Still, the compositions only get significant when interacting with axial and effective strain components but not with hoop and mean strain.
- The model predicted out of ANOVA is found to have excellent agreement with experimental values; hence it predicts density for a given parameter condition to the levels considered in the present investigation.
- The novelty lies in the application of surface plots and ANOVA to explore densification trends, shedding light on the interaction between Mo content and densification across various strain types. The primary objective was to assess densification behavior under cold upsetting conditions, considering Mo variation, to enhance the mechanical strength of the components. The study provides significant insights into the densification behavior of Fe-0.8%C steel under cold upsetting, considering Mo variation. The findings contribute to the optimization of powder metallurgy processes, particularly in industries such as automotive and construction, by enhancing the production of high-density steel components. The study's comprehensive methodology and analysis offer a valuable framework for future research in this field.

REFERENCES

1. Araoyinbo, A. O., F. A. Ishola, E. Y. Salawu, M. B. Biodun, and A. U. Samuel. 2022. "Overview of Powder Metallurgy Process and Its Advantages." 020153. Athens, Greece. https://doi.org/10.1063/5.0092502.
2. Srivastava, A. K., A. Kumar, P. Kumar, P. Gautam, and N. Dogra. December 2023. "Research Progress in Metal Additive Manufacturing: Challenges and Opportunities." International Journal on Interactive Design and Manufacturing (IJIDeM). https://doi.org/10.1007/s12008-023-01661-6.
3. Angelo, P. C., R. Subramanian, and B. Ravisankar. 2022. Powder Metallurgy: Science, Technology and Applications. Second. PHI Learning Pvt. Ltd.
4. Vollertsen, F. 2014. "High–Temperature Sintering of a Hypereutectoid Steel to Theoretical Density." International Journal of Materials and Product Technology 6 (4): 371–80. https://doi.org/10.1504/IJMPT.1991.036642.
5. Saxena, A., T. Kumar Gupta, R. Srivastava, A. K. Srivastava, and A. Kumar. 2024. "A Comparison of Different Experimental Approaches in the Determination of Dynamic Fracture Toughness (J1d) Behavior for RHA Steel Cost Sustainable MMAW Weldments." Journal of Adhesion Science and Technology, 1–22. https://doi.org/10.1080/01694243.2024.2334267.
6. Mao, Y., J. Yuan, Y. Heng, K. Feng, D. Cai, and Q. Wei. 2023. "Effect of Hot Isostatic Pressing Treatment on Porosity Reduction and Mechanical Properties Enhancement of 316L Stainless Steel Fabricated by Binder Jetting." Virtual and Physical Prototyping 18 (1): e2174703. https://doi.org/10.1080/17452759.2023.2174703.

7. Ramakrishnan, P. 2013. "17 - Automotive Applications of Powder Metallurgy." In Advances in Powder Metallurgy, edited by Isaac Chang and Yuyuan Zhao, 493–519. Woodhead Publishing Series in Metals and Surface Engineering. Woodhead Publishing. https://doi.org/10.1533/9780857098900.4.493.
8. Kumar, A., V. Kumar Shrivastava, P. Kumar, A. Kumar, and V. Gulati. 2024. "Predictive and Experimental Analysis of Forces in Die-Less Forming Using Artificial Intelligence Techniques." Proceedings of the Institution of Mechanical Engineers, Part E: Journal of Process Mechanical Engineering, February, 09544089241235473. https://doi.org/10.1177/09544089241235473.
9. Rajeshkannan, A., E. Guyot, and A. K. Jeevanantham. 2020. "Effect of Preform Geometry and Molybdenum Addition on Work Hardening Behavior of Fe-0.8%C Steel Preforms during Cold Upsetting." Materials Today: Proceedings, 2nd International Conference on Materials Manufacturing and Modelling, ICMMM – 2019, VIT University, Vellore, March 29–31, 2019, 22 (January): 1822–28. https://doi.org/10.1016/j.matpr.2020.03.016.
10. Kumar, A., P. Kumar, H. Singh, A. Haleem, and R. K. Mittal. 2023. "Chapter 4 - Integration of Reverse Engineering With Additive Manufacturing." In Advances in Additive Manufacturing, edited by Ajay Kumar, Ravi Kant Mittal and Abid Haleem, 43–65. Additive Manufacturing Materials and Technologies. Elsevier. https://doi.org/10.1016/B978-0-323-91834-3.00028-4.
11. Seetharam, R., S. Kanmani Subbu, and M. J. Davidson. 2017. "Hot Workability and Densification Behavior of Sintered Powder Metallurgy Al-B4C Preforms During Upsetting." Journal of Manufacturing Processes 28: 309–18. https://doi.org/10.1016/j.jmapro.2017.06.012.
12. Kumar, A., P. Kumar, R. K. Mittal, and H. Singh. 2023. "Chapter 9 - Preprocessing and Postprocessing in Additive Manufacturing." In Advances in Additive Manufacturing, edited by Ajay Kumar, Ravi Kant Mittal and Abid Haleem, 141–65. Additive Manufacturing Materials and Technologies. Elsevier. https://doi.org/10.1016/B978-0-323-91834-3.00005-3.
13. Sugimoto, K., T. Hojo, and A. Srivastava. 2019. "Low and Medium Carbon Advanced High-Strength Forging Steels for Automotive Applications." Metals 9 (12): 1263. https://doi.org/10.3390/met9121263.
14. Chen, J., S. Tang, Z. Liu, and G. Wang. 2013. "Influence of Molybdenum Content on Transformation Behavior of High Performance Bridge Steel During Continuous Cooling." Materials & Design 49: 465–70. https://doi.org/10.1016/j.matdes.2013.01.017.
15. Kandavel, T. K., R. Chandramouli, and M. Ravichandran. 2010. "Experimental Study on the Plastic Deformation and Densification Characteristics of Some Sintered and Heat Treated Low Alloy Powder Metallurgy Steels." Materials & Design 31 (1): 485–92. https://doi.org/10.1016/j.matdes.2009.06.048.
16. Kumar, A., P. Kumar, R. K. Mittal, and V. Gambhir. 2023. "Chapter 12 - Materials Processed by Additive Manufacturing Techniques." In Advances in Additive Manufacturing, edited by Ajay Kumar, Ravi Kant Mittal and Abid Haleem, 217–33. Additive Manufacturing Materials and Technologies. Elsevier. https://doi.org/10.1016/B978-0-323-91834-3.00014-4.
17. Vishnuraj, J. T., T. K. Kandavel, I. Sai Kishan, and G. Akash. 2018. "A Study on Deformation and Densification Characteristics of P/M Fe-C-Mn Alloy Steels under Cold Upset." Materials Today: Proceedings, International Conference on Advanced Materials (SCICON '16), December 19–21, 2016, 5 (8, Part 3): 16740–47. https://doi.org/10.1016/j.matpr.2018.06.039.
18. Narayan, S., and A. Rajeshkannan. "Densification Behaviour of Sintered Aluminum Composites During Hot Deformation." AIMS Materials Science 5 (5): 902–15. https://doi.org/10.3934/matersci.2018.5.902.

19. Nizam Khan, M., S. Narayan, and A. Rajeshkannan. "Influence of Process Parameters on the Workability Characteristics of Sintered Al and Al–Cu Composites During Cold Deformation." AIMS Materials Science 6 (3): 441–53. https://doi.org/10.3934/matersci.2019.3.441.
20. Khan, M. N., S. Narayan, and A. Rajeshkannan. 2019. "Workability of Sintered Aluminium Composite Preforms of Varying Cu and TiC Contents during Cold Deformation." 040002. California, USA. https://doi.org/10.1063/1.5094322.
21. Tangwe, S. L., and M. Simon. 2018. "Evaluation of Performance of Air Source Heat Pump Water Heaters Using the Surface Fitting Models: 3D Mesh Plots and 2D Multi Contour Plots Simulation." Thermal Science and Engineering Progress 5: 516–23. https://doi.org/10.1016/j.tsep.2018.01.014.
22. Afshar Ebrahimi, A., and S. Foroutan Ghazvini. 2018. "Experimental Attrition Study of FCC Catalysts Through 2D/3D Contour Plots and Response Surface Models." Powder Technology 336: 80–84. https://doi.org/10.1016/j.powtec.2018.05.046.
23. Esamael, S. K., and A. A. Fatalla. 2022. "Evaluation on Processing Parameter's Effects on Some Mechanical Properties of Pure Magnesium Bulk Prepared by Powder Metallurgy." Materials Today: Proceedings, Third International Conference on Aspects of Materials Science and Engineering 57: 622–29. https://doi.org/10.1016/j.matpr.2022.02.016.
24. Jeevanantham, A. K., D. R. Seenivasagam, and R. Ananthanarayanan. 2020. "Strain Hardening Analysis and Modelling for Sintered Al-Cu-TiC Preforms With Varying Process Parameters During Cold Upsetting." Journal of Materials Research and Technology 9 (5): 12007–18. https://doi.org/10.1016/j.jmrt.2020.08.065.
25. Narayan, S., and A. Rajeshkannan. 2011. "Densification Behaviour in Forming of Sintered Iron–0.35% Carbon Powder Metallurgy Preform During Cold Upsetting." Materials & Design 32 (2): 1006–13. https://doi.org/10.1016/j.matdes.2010.08.010.
26. Ananthanarayanan, R. 2010. "Workability Studies on Cold Upsetting of Sintered Copper Alloy Preforms." Materials Research 13 (4): 457–64. https://doi.org/10.1590/S1516-14392010000400006.
27. Rajeshkannan, A., and S. Shanmugam. 2018. "Study on Densification Behaviour Under Cold Forging of Sintered High Carbon Alloy Steels." Key Engineering Materials 777: 311–15. https://doi.org/10.4028/www.scientific.net/KEM.777.311.
28. Narayanasamy, R., V. Anandakrishnan, and K. S. Pandey. 2008b. "Effect of Geometric Work-Hardening and Matrix Work-Hardening on Workability and Densification of Aluminium–3.5% Alumina Composite During Cold Upsetting." Materials & Design 29 (8): 1582–99. https://doi.org/10.1016/j.matdes.2007.11.006
29. Rajeshkannan, A., and U. Mehta. 2014. "Deformation Study of Sintered Iron–Carbon–Silicon–Copper Steel Compacts During Cold Forging." Materials and Manufacturing Processes 29 (4): 442–47. https://doi.org/10.1080/10426914.2013.864401
30. Narayanasamy, R., V. Anandakrishnan, and K. S. Pandey. 2008a. "Comparison of Workability Strain and Stress Parameters of Powder Metallurgy Steels AISI 9840 and AISI 9845 During Cold Upsetting." Materials & Design 29 (10): 1919–25. https://doi.org/10.1016/j.matdes.2008.04.023.

2 On the Use of Sustainable Bio-Dielectric Fluid in the Spark Erosion Machining

Mohd Yunus Khan, Sudhakar Rao Patange, and Bahadur Singh Pabla

2.1 INTRODUCTION

Traditional machining, like turning, drilling, and milling, is challenging with hard and brittle materials. Non-traditional machining is utilized when conventional machining is not feasible, satisfactory, or cost-effective. Hard and brittle materials are difficult to clamp using conventional methods or when the shape of the work sample is too complex, and the workpiece needs to be more flexible or thin. Nonconventional machining can be effectively employed. These techniques are chosen to manufacture tools, dies, molds, and other precision items from modern materials (composites, super-alloys, and ceramics). These methods enable the creation of complex forms with higher surface finishes [1–7].

Spark erosion, or electrical discharge machining (EDM), is widely used. In this method, the workpiece is eroded without cutting forces. A succession of sparks causes the high-precision material expulsion. To achieve better erosion, the work sample used in the spark machining must be electrically conductive and submerged in the dielectric fluid [8]. Spark machining has several applications in the automotive, aerospace, mold and die industry, and healthcare sectors [9–11]. Hard metals like titanium, pre-hardened steel, and other materials that are challenging to work on conventional machines can be machined using spark erosion [12, 13]. The dielectric fluid serves many purposes and is a crucial requirement of the spark machining process [8]. The principle of die-sinker EDM is depicted in Figure 2.1.

A dielectric medium must exist between the two electrodes during spark machining to initiate regulated electrical sparks that remove material. Because dielectrics have varying compositions and cooling speeds, selecting the right dielectric is essential for machining. The volition of dielectric fluid rests upon the work sample, dimensions, geometry, tolerance, desired material expulsion, and surface quality. It is recommended that the dielectric fluid be kept the same and chosen based on the machine's most often-used application.

High dielectric strengths allow the dielectric fluids to continue to be electrically non-conductive. After the operation, the debris is removed from the machining area [14]. The dielectric absorbs heat during the process, and the work sample is

DOI: 10.1201/9781032725086-3

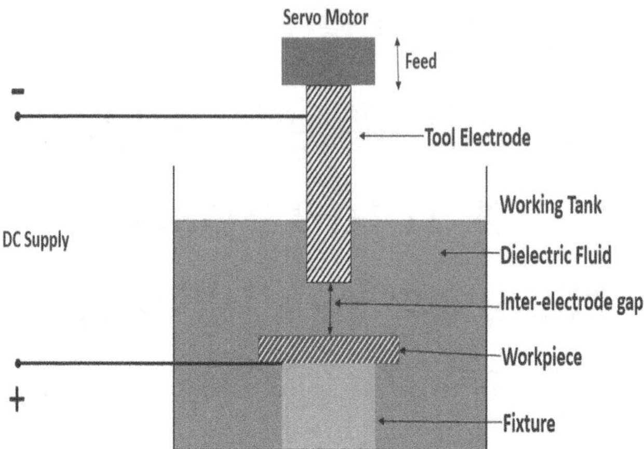

FIGURE 2.1 Schematic outline of die-sinker EDM.

cooled [15–17]. Additionally, dielectric fluid prevents oxidation by preventing the ambient environment from coming into touch with the workpiece [18]. To achieve better performance and control the over-discharge created in spark machining, the proper selection of the dielectric medium is crucial [17, 18]. Before being used again, the dielectric fluid needs to be adequately filtered after the machining process. The dielectric fluid will remain clean by having debris removed from it. The dielectric fluid should possess specific qualities to work well throughout the spark erosion. It must have adequate strength to protect a tool and workpiece until the break-down voltage is reached. After the spark is released, it should quickly de-ionize. It must have adequate viscosity to effectively cool the workpiece and clear waste from the cutting zone. Forcing debris out of the machining area should be carried out by dielectric fluid to prevent arcing [8]. It ought to be chemically inactive, non-toxic, and pollution-free. Kerosene, de-ionized water, mineral oil, synthetic oil, and air are often dielectric fluids for EDM [19–21].

The present study presents in-depth information about bio-dielectrics for spark machining. The review work has been shown in tabular form to depict a description of the research progress. This will help determine the impact of sustainable dielectric on EDM output. The information presented in this chapter can be utilized to research spark machining using bio-dielectric fluid. The spark machining process can be hybridized with other processes. Hybridization will further improve the machining performance. The combined effect of hybridization and bio-dielectric fluid on EDM performance can be conducted. Such a combination will not only improve the performance of EDM but also be promising for the environment. Studies on bio-dielectric use are reviewed, along with the effects on spark machining performance.

2.2 TYPES OF DIELECTRIC MEDIUM

The spark erosion process uses a variety of dielectric media types. Significant categories of dielectric medium are discussed below.

2.2.1 Hydrocarbon-Based Fluid

The earliest dielectric fluid used in spark machining is an oil with a hydrocarbon base. Kerosene is a typical example [14]. The re-usability and non-degradability of these oils are significant problems. They release toxic gases, dangerous fumes, and aerosols that are bad for the environment and the operator [22–26].

2.2.2 Synthetic Oil

The spark erosion process extensively uses synthetic oils [27–29]. Although costly, these fluids don't adversely affect health much [16].

2.2.3 Water-Based Additives Fluids

For spark erosion, de-ionized water is a commonly utilized dielectric [30, 31]. It costs little and produces no pollution. It is easily available. EDM also uses dielectric fluids based on emulsions [26, 32, 33]. Glycol-in-water [34], water-in-oil [35, 36], glycerin-in-water [37], and others are examples of these dielectrics.

2.2.4 Gaseous Dielectrics

For spark machining, oxygen and air are popular dielectric materials [38, 39]. No contamination is caused by these dielectric materials [14, 40]. The output of the spark erosion machining can be further enhanced by using additives such as glycerin in the air dielectric [41, 42].

2.2.5 Biodiesel-Based Dielectric Fluids

Environmentally friendly biodiesel is being tested to replace conventional dielectric fluid used in spark erosion machining [43]. Research has been done on biodiesel produced from the inedible oils jatropha [44], polanga [45, 46], and neem [47, 48].

2.2.6 Powder-Mixed Dielectric Fluid

This type of dielectric fluid has powder mixed into it in different amounts [15]. When powder-mixed dielectric fluid is employed, the performance of the spark machining process is enhanced [49–55]. Some frequently utilized powders are silicon, silicon carbide, nickel, aluminum, titanium, titanium carbide, tungsten, chromium, and others [55–58].

2.3 STUDIES ON BIO-DIELECTRIC FLUID

Sustainability is defined as the "capacity to withstand economic, environmental, and social forces while addressing the requirements of the present until the satisfaction of the needs of the future." Its main objective is to conserve the environment for the next generations. In manufacturing, spark erosion is a crucial procedure for cutting challenging materials. It is essential to look for a dielectric that is environmentally responsible and sustainable [59]. One possibility for the same is a dielectric made of vegetable oil. This topic provides an overview of numerous studies on various dielectric fluids and their impact on the effectiveness of the spark machining process.

A brief overview of the previously published studies in bio-dielectric fluid for spark machining is presented in Table 2.1. The literature review in tabular format presents a synopsis of included articles better as it gives more meaningful information about the study's objective and conclusions. Research articles were selected to cover a wide range of bio-dielectric fluids arranged chronologically.

There are several variables that influence the outcome of spark machining. The dielectric fluid used during machining is substantial and chiefly affects the machining characteristics. The selection of dielectric fluid requires special attention for the proper operation of the spark machining. Different dielectric fluids have been tested in discharge machining. Most of the investigations have been conducted using hydrocarbon and synthetic oils. There are certain demerits associated with the use of these dielectrics. Bio-dielectric is one of the better options to achieve green and sustainable manufacturing. Various bio-dielectric fluids which have been tested. It includes biodiesels derived from various vegetable oils. Using inedible and waste oils is considerably more crucial than edible ones. The bio-dielectric fluids are not harmful to the environment. The literature review showed that the performance of EDM improved with bio-dielectrics. Nearly all studies showed increased material expulsion and reduced tool wearing with the bio-dielectric operation. At the same time, the surface finish improved [80–84].

2.4 CONCLUSIONS AND RECOMMENDATIONS FOR THE FUTURE WORK

Synthetic and hydrocarbon fluids are primarily employed as EDM dielectrics. However, these fluids harm both nature and the machine operator. Plant-based oils are biodegradable; therefore, waste produced during machining can be naturally broken down. So, bio-dielectric is an excellent choice for creating a sustainable EDM process. The questions of sustainability and pollution can be resolved by using bio-dielectric technology. The design of the EDM machine doesn't need to be drastically changed to use bio-dielectric fluid. However, some adjustments are made to the EDM flushing system to account for the greater viscosity values of the plant oil-based dielectric. An exhaustive assessment of the at-hand literature on bio-dielectric fluids in EDM has been conducted. Literatures highlighted on bio-dielectric use and impacts this has on spark machining performance is discussed. Recent studies have centered on using jatropha and waste cooking oil-based bio-dielectric fluid. The findings of these investigations have been favorable. Other oils, especially inedible ones, should be evaluated to determine the efficacy of EDM. The new bio-dielectric fluid can be a primary dielectric choice for green EDM replacing conventional dielectric fluids.

The influence of powder addition in bio-dielectric fluid can be considered for future research. A hybrid spark erosion machining process can be employed along with bio-dielectric fluid. A systematic study of the hybridization process combined with sustainable dielectric fluid can be taken up for further investigation.

TABLE 2.1
Literature Review

S. No.	Author [Ref. No.]	Objective	Major Findings
1.	Priadi et al. [60]	Layered electrodes and jatropha dielectric were combined to improve spark machining performance.	They discovered that bio-dielectric can be used as the dielectric, producing a smoother surface with higher white layer hardness values.
2.	Triyono et al. [61]	The effects of various tools and jatropha dielectric on the machining output of spark machining were evaluated.	Tool wear increased along with material removal when vegetable oil-based was used.
3.	Valaki et al. [20]	The effectiveness of jatropha bio-dielectric for spark erosion operation was examined.	Lower surface roughness and more significant material removal were seen. When bio-dielectric was employed, the workpiece's surface hardness also increased.
4.	Valaki and Rathod [62]	Waste vegetable oil-based dielectric for spark machining was evaluated.	They found that waste oil performed better during spark machining and recommended it as a potential substitute for synthetic and hydrocarbon fluids.
5.	Mali and Kumar [63]	The effectiveness of used vegetable oil, pongamia pinnata, and blended used oil as dielectric was examined.	This experimental effort proved the viability of using these fluids as potential substitutes for the dielectric.
6.	Ng et al. [23]	Spark erosion machining using bio-dielectric made from canola and sunflower oils was carried out.	According to the study, bio-dielectric fluids have a strong chance of taking the role of traditional fluids in sustainable spark machining operations.
7.	Das et al. [64]	Neem oil biodiesel as dielectric fluid was used, and discussion on many sustainability difficulties with spark machining was presented.	Bio-dielectric fluid showed greater material removal and less surface roughness than kerosene. According to scanning electron micrographs, neem biodiesel can ensure fine surface textures.
8.	Reddy et al. [65]	The effectiveness of spark machining with a dielectric fluid based on sunflower oil was investigated.	It was discovered that bio-dielectric fluid performed better than ordinary fluid.

(Continued)

TABLE 2.1 ***(Continued)***
Literature Review

S. No.	Author [Ref. No.]	Objective	Major Findings
9.	Ahmad et al. [66]	Refined, bleached, and deodorized palm oil was tested as a EDM fluid.	They noticed that using bio-dielectric fluid significantly increased the rate of material removal.
10.	Khan et al. [16]	Jatropha biodiesel's characteristics were studied.	They suggested it might work well as a dielectric fluid for spark machining.
11.	Mishra and Routara [45]	The usefulness and environmental effects of polanga oil bio-dielectric in spark machining were investigated.	They noticed that using bio-oil increased production. With non-edible biodiesel, improvements in material expulsion, surface finish, and surface hardness were seen. With bio-dielectric, they discovered a substantial decrease in aerosol emissions.
12.	Radu et al. [67]	The viability of several vegetable oils as the dielectric fluid for spark machining was determined.	The results from this experimental trial were favorable.
13.	Singaravel et al. [25]	Multiple bio-dielectric fluids used in spark erosion machining were considered.	The outcome demonstrated that bio-dielectric fluids could be used for environmentally friendly production because they have characteristics similar to kerosene.
14.	MangapathiRao et al. [68]	The effectiveness of the spark machining procedure utilizing sunflower oil as a dielectric and Al_2O_3 powder mixed to it was examined.	The findings indicated that adding bio-dielectric fluid to powder produced surfaces with less roughness. Additionally, improved material removal and reduced tool wear were noted.
15.	Nagabhooshanam et al. [69]	The effectiveness of spark machining with different concentrations of nano-zirconia distributed in canola biodiesel was evaluated.	According to the findings, nano-zirconia in 0.5 weight percent was added to the bio-dielectric fluid to improve surface topography and finish. Increased metal removal was also noted.
16.	Singh et al. [70]	Inedible pongamia and jatropha bio-dielectrics were evaluated in spark erosion machining.	Kerosene's limits as a dielectric fluid was discussed. It was concluded that the bio-dielectric outperformed kerosene regarding material expulsion, tool wearing, and surface roughness.

TABLE 2.1 ***(Continued)***
Literature Review

S. No.	Author [Ref. No.]	Objective	Major Findings
17.	Yadav et al. [71]	The impact of utilizing a dielectric based on used vegetable oils in the spark erosion machining process was examined.	The study demonstrates that bio-dielectric performed better than traditional fluids and can be used in place of them with good results.
18.	Khan et al. [72]	The feasibility of used oil-bio-dielectric for spark erosion machining was established.	The results indicated better machining performance with bio-dielectric fluid compared to EDM oil.
19.	Boopathi et al. [73]	Dry machining was conducted using compressed air and sunflower oil.	Significant increment in material expulsion and surface finish was observed with air-oil mist.
20.	Chakraborty et al. [74]	Biodiesel derived from jatropha and rice bran oils was used for spark machining.	Both biodiesels showed superior performance in contrast to EDM oil. Jatropha biodiesel showed better surface quality than rice bran oil.
21.	Ishfaq et al. [75]	The future of biodegradable dielectrics at various Cu-powder concentrations for machining of Inconel 600 was evaluated.	Reduced electrode wear and overcuts were observed when using a biodegradable dielectric fluid. The powder concentration also influenced the machining characteristics.
22.	Khan et al. [76]	Machining of Inconel 625 using used oil biodiesel was conducted.	Better surface finishes with bio-dielectric fluid while machining Inconel 625 were obtained.
23.	Pillai and Kumar [77]	Micromachining of Inconel 718 with coconut biodiesel and graphene nano-powder dispersion was conducted.	Nano-powder addition in bio-dielectric showed a significant increase in material expulsion and surface quality.
24.	Sethy et al. [78]	Diverse biodegradable oils were used for machining Ti alloy.	Mixed vegetable oil was found to be the best in terms of material expulsion. Sunflower oil performed best when overall performance was compared.
25.	Singh et al. [79]	Spark machining of Al-SiC-N hybrid composite was done using various bio-dielectrics.	Inedible oil-derived biodiesels showed remarkable material expulsion, tool wearing, and surface quality performance.

REFERENCES

1. Gupta, Kapil, and Munish Kumar Gupta. "Developments in nonconventional machining for sustainable production: A state-of-the-art review." *Proceedings of the Institution of Mechanical Engineers, Part C: Journal of Mechanical Engineering Science* 233, no. 12 (2019): 4213–4232.
2. Kumar, Ajay, Vishal Gulati, and Parveen Kumar. "Effects of process parameters on surface roughness in incremental sheet forming." *Materials Today: Proceedings* 5, no. 14 (2018): 28026–28032. https://doi.org/10.1016/j.matpr.2018.10.043
3. Kumar, Pankaj, Sayed Shah Hussain, Ajay Kumar, Ashish Kumar Srivastava, Manowar Hussain, and Purushottam Kumar Singh. "10 Finite element method investigation on delamination of 3D printed hybrid composites during the drilling operation." *3D Printing Technologies: Digital Manufacturing, Artificial Intelligence, Industry* 4.0 (2024): 223.
4. Gangwar, Sukhdev, Subhas Chandra Mondal, Ajay Kumar et al. "Performance analysis and optimization of machining parameters using coated tungsten carbide cutting tool developed by novel S3P coating method." *International Journal on Interactive Design and Manufacturing* (2024). https://doi.org/10.1007/s12008-024-01852-9
5. Kumar, Ajay, and Vishal Gulati. "Experimental investigation and optimization of surface roughness in negative incremental forming." *Measurement* 131 (2019): 419–430. https://doi.org/10.1016/j.measurement.2018.08.078
6. Kumar, Ajay, Vishal Gulati, Parveen Kumar, Vinay Singh, Brijesh Kumar, and Hari Singh. "Parametric effects on formability of AA2024-O aluminum alloy sheets in single point incremental forming." *Journal of Materials Research and Technology* 8, no. 1 (2019): 1461–1469. https://doi.org/10.1016/j.jmrt.2018.11.001
7. Rani, Sangeeta, Khushboo Tripathi, and Ajay Kumar. "Machine learning aided malware detection for secure and smart manufacturing: A comprehensive analysis of the state of the art." *International Journal on Interactive Design and Manufacturing* (2023). https://doi.org/10.1007/s12008-023-01578-0
8. Khan, Mohd Yunus, and P. Sudhakar Rao. "Electrical discharge machining: Vital to manufacturing industries." *International Journal of Innovative Technology and Exploring Engineering* 8, no. 11 (2019): 1696–1701. https://doi.org/10.35940/ijitee.K1516.0981119
9. Aggoune, Samia, Farida Hamadi, Cherifa Abid et al. "Instabilities in the formation of single tracks during selective laser melting process." *International Journal on Interactive Design and Manufacturing* (2024). https://doi.org/10.1007/s12008-024-01887-y
10. Khan, Mohd Yunus, and P. Sudhakar Rao. "Optimization of process parameters of electrical discharge machining process for performance improvement." *International Journal of Innovative Technology and Exploring Engineering* 8, no. 11 (2019): 3830–3836. https://doi.org/10.35940/ijitee.K2262.0981119
11. Khan, Mohd Yunus, and P. Sudhakar Rao. "Hybridization of electrical discharge machining process." *International Journal of Engineering and Advanced Technology* 9, no. 1 (2019): 1059–1065. https://doi.org/10.35940/ijeat.A9477.109119
12. Ajay, Parveen, S. Ahmad, J. Sharma, and V. Gambhir (Eds.). (2023). *Handbook of Sustainable Materials: Modelling, Characterization, and Optimization* (1st ed.). CRC Press. https://doi.org/10.1201/9781003297772
13. Singh Hari, Kumar Ajay, Kumar Parveen, and Bandar AlMangour (Eds.). (2023). *Handbook of Smart Manufacturing: Forecasting the Future of Industry 4.0* (1st ed.). CRC Press. https://doi.org/10.1201/9781003333760
14. Singh, Ankit Kr, Rahul Mahajan, Anmol Tiwari, Dhruva Kumar, and Ranjan Kumar Ghadai. "Effect of dielectric on electrical discharge machining: A review." In *IOP*

Conference Series: Materials Science and Engineering, vol. 377, no. 1, p. 012184. IOP Publishing, 2018. https://doi.org/10.1088/1757-899X/377/1/012184

15. Singh, Sharanjit, and Arvind Bhardwaj. "Review to EDM by using water and powder-mixed dielectric fluid." *Journal of Minerals and Materials Characterization and Engineering* 10, no. 02 (2011): 199. https://doi.org/10.4236/jmmce.2011.102014
16. Khan, Mohd Yunus, Patange Sudhakar Rao, and Bahadur Singh Pabla. "Investigations on the feasibility of *Jatropha curcas* oil based biodiesel for sustainable dielectric fluid in EDM process." *Materials Today: Proceedings* 26 (2020): 335–340. https://doi.org/10.1016/j.matpr.2019.11.325
17. Jagtap, V. L. "A review of EDM process for difficult to cut materials." *International Research Journal of Multidisciplinary Studies* 2 (2016): 2454–8499.
18. Rizwee, Mumtaz, Patange Sudhakar Rao, and Mohd Yunus Khan. "Recent advancement in electric discharge machining of metal matrix composite materials." *Materials Today: Proceedings* 37 (2021): 2829–2836. https://doi.org/10.1016/j.matpr.2020.08.657
19. Khan, Mohd Yunus, Patange Sudhakar Rao, and Bahadur Singh Pabla. "Review of electrical discharge machining process with nanopowder and CNT mixed dielectric fluid." In *Proceedings of 4th International Online Multidisciplinary Research Conference (IOMRC-2020), Hyderabad, October*, vol. 23. 2020.
20. Valaki, Janak B., Pravin P. Rathod, and Bharat C. Khatri. "Environmental impact, personnel health and operational safety aspects of electric discharge machining: A review." *Proceedings of the Institution of Mechanical Engineers, Part B: Journal of Engineering Manufacture* 229, no. 9 (2015): 1481–1491. https://doi.org/10.1177/0954405414543314
21. Mittal, Ravi Kant, Abid Haleem, and Ajay Kumar (Eds.). *Advances in Additive Manufacturing: Artificial Intelligence, Nature-Inspired, and Biomanufacturing*. Elsevier, 2022. https://doi.org/10.1016/C2020-0-03877-6
22. Valaki, Janak B., Pravin P. Rathod, and C. D. Sankhavara. "Investigations on technical feasibility of *Jatropha curcas* oil based bio dielectric fluid for sustainable electric discharge machining (EDM)." *Journal of Manufacturing Processes* 22 (2016): 151–160. https://doi.org/10.1016/j.jmapro.2016.03.004
23. Ng, Pei Shan, S. A. Kong, and S. H. Yeo. "Investigation of biodiesel dielectric in sustainable electrical discharge machining." *The International Journal of Advanced Manufacturing Technology* 90 (2017): 2549–2556. https://doi.org/10.1007/s00170-016-9572-6
24. Das, Shirsendu, Swarup Paul, and Biswanath Doloi. "Feasibility assessment of some alternative dielectric mediums for sustainable electrical discharge machining: A review work." *Journal of the Brazilian Society of Mechanical Sciences and Engineering* 42 (2020): 1–21. https://doi.org/10.1007/s40430-020-2238-1
25. Singaravel, B., K. Chandra Shekar, G. Gowtham Reddy, and S. Deva Prasad. "Experimental investigation of vegetable oil as dielectric fluid in electric discharge machining of Ti-6Al-4V." *Ain Shams Engineering Journal* 11, no. 1 (2020): 143–147. https://doi.org/10.1016/j.asej.2019.07.010
26. Shaik, Mahaboob Basha, and Himanshu Patel. "A review on dielectric fluids used for sustainable electro discharge machining." *Indian Journal of Scientific Research* 17, no. 2 (2017): 40–46.
27. Kumar, Rachakonda Ranjith, Anup Kumar Jana, Sarat Chandra Mohanty, K. Mangapathi Rao, V. Guru Shanker, and D. Rohit Reddy. "Optimizing process parameters of die sinking EDM in AISI D2 steel by using TOPSIS using EDM oil as dielectric." In *IOP Conference Series: Materials Science and Engineering*, vol. 1057, no. 1, p. 012060. IOP Publishing, 2021. https://doi.org/10.1088/1757-899X/1057/1/012060
28. Surya, Vishwa R., KM Vinay Kumar, R. Keshavamurthy, G. Ugrasen, and H. V. Ravindra. "Prediction of machining characteristics using artificial neural network in

wire EDM of Al7075 based in-situ composite." *Materials Today: Proceedings* 4, no. 2 (2017): 203–212. https://doi.org/10.1016/j.matpr.2017.01.014

29. Mohamed, Abdul Rahman, Banu Asfana, and Yeakub Ali Mohammad. "Investigation of recast layer of non-conductive ceramic due to micro-EDM." *Advanced Materials Research* 845 (2014): 857–861. https://doi.org/10.4028/www.scientific.net/AMR.845.857
30. Al-Amin, Md, Ahmad Majdi Abdul Rani, Abdul Azeez Abdu Aliyu, Muhammad Al'Hapis Abdul Razak, Sri Hastuty, and Michael G. Bryant. "Powder mixed-EDM for potential biomedical applications: A critical review." *Materials and Manufacturing Processes* 35, no. 16 (2020): 1789–1811. https://doi.org/10.1080/10426914.2020.1779939
31. Zhang, Yanzhen, Yonghong Liu, Renjie Ji, and Baoping Cai. "Study of the recast layer of a surface machined by sinking electrical discharge machining using water-in-oil emulsion as dielectric." *Applied Surface Science* 257, no. 14 (2011): 5989–5997. https://doi.org/10.1016/j.apsusc.2011.01.083
32. Zhang, Yanzhen, Yonghong Liu, Yang Shen, Renjie Ji, Xiaolong Wang, and Zhen Li. "Die-sinking electrical discharge machining with oxygen-mixed water-in-oil emulsion working fluid." *Proceedings of the Institution of Mechanical Engineers, Part B: Journal of Engineering Manufacture* 227, no. 1 (2013): 109–118. https://doi.org/10.1177/0954405412464146
33. Zan, Shusong, Zhenlong Wang, Yuchao Jia, Guanxin Chi, and Yukui Wang. "Study of graphite tool wear in EDM with water-based dielectrics and EDM oil." *Procedia CIRP* 95 (2020): 414–418. https://doi.org/10.1016/j.procir.2020.02.270
34. Liu, Yonghong, Yanzhen Zhang, Renjie Ji, Baoping Cai, Fei Wang, Xiaojie Tian, and Xin Dong. "Experimental characterization of sinking electrical discharge machining using water in oil emulsion as dielectric." *Materials and Manufacturing Processes* 28, no. 4 (2013): 355–363. https://doi.org/10.1080/10426914.2012.700162
35. Zhang, Yanzhen, Yonghong Liu, Renjie Ji, Baoping Cai, and Yang Shen. "Sinking EDM in water-in-oil emulsion." *The International Journal of Advanced Manufacturing Technology* 65 (2013): 705–716. https://doi.org/10.1007/s00170-012-4210-4
36. Paswan, Kamlesh, Surya Pratap Singh, Somnath Chattopadhyaya, and Alokesh Pramanik. "Experimental investigation on the effects of aqueous solution in electric discharge machining." *Materials Today: Proceedings* 27 (2020): 2975–2980. https://doi.org/10.1016/j.matpr.2020.04.904
37. Leão, Fábio N., and Ian R. Pashby. "A review on the use of environmentally-friendly dielectric fluids in electrical discharge machining." *Journal of Materials Processing Technology* 149, no. 1–3 (2004): 341–346. https://doi.org/10.1016/j.jmatprotec.2003.10.043
38. Islam, Md Mofizul, Chang Ping Li, and Tae Jo Ko. "Dry electrical discharge machining for deburring drilled holes in CFRP composite." *International Journal of Precision Engineering and Manufacturing-Green Technology* 4 (2017): 149–154. https://doi.org/10.1007/s40684-017-0018-x
39. Dhakar, Krishnakant, and Akshay Dvivedi. "Experimental investigation on near-dry EDM using glycerin-air mixture as dielectric medium." *Materials Today: Proceedings* 4, no. 4 (2017): 5344–5350. https://doi.org/10.1016/j.matpr.2017.05.045
40. Wu, Xinlei, Yonghong Liu, Xuexin Zhang, Hang Dong, Chao Zheng, Fan Zhang, Qiang Sun, Hui Jin, and Renjie Ji. "Sustainable and high-efficiency green electrical discharge machining milling method." *Journal of Cleaner Production* 274 (2020): 123040. https://doi.org/10.1016/j.jclepro.2020.123040
41. Dhakar, Krishnakant, and Akshay Dvivedi. "Influence of glycerin-air dielectric medium on near-dry EDM of titanium alloy." *International Journal of Additive and Subtractive Materials Manufacturing* 1, no. 3–4 (2017): 328–337. https://doi.org/10.1504/IJASMM.2017.089928

42. Yadav, Vineet Kumar, Pradeep Kumar, and Akshay Dvivedi. "Effect of tool rotation in near-dry EDM process on machining characteristics of HSS." *Materials and Manufacturing Processes* 34, no. 7 (2019): 779–790. https://doi.org/10.1080/10426914.2019.1605171
43. Khan, Mohd Yunus, P. Sudhakar Rao, B. S. Pabla, and Suresh Ghotekar. "Innovative biodiesel production plant: Design, development, and framework for the usage of biodiesel as a sustainable EDM fluid." *Journal of King Saud University-Science* 34, no. 6 (2022): 102203. https://doi.org/10.1016/j.jksus.2022.102203
44. Basha, S. M., H. K. Dave, and H. V. Patel. "Experimental investigation of *Jatropha curcas* bio-oil and biodiesel in electric discharge machining of Ti-6Al-4V." *Materials Today: Proceedings* 38 (2021): 2102–2109. https://doi.org/10.1016/j.matpr.2020.04.536
45. Mishra, B. P., and B. C. Routara. "Evaluation of technical feasibility and environmental impact of *Calophyllum inophyllum* (Polanga) oil based bio-dielectric fluid for green EDM." *Measurement* 159 (2020): 107744. https://doi.org/10.1016/j.measurement.2020.107744
46. Asgar, Md Ehsan, and Ajay Kumar Singh Singholi. "Study of the effect of dielectric on performance measure in EDM." In *Advances in Industrial and Production Engineering: Select Proceedings of FLAME 2020*, pp. 843–850. Springer Singapore, 2021. https://doi.org/10.1007/978-981-33-4320-7_75
47. Das, Shirsendu, Swarup Paul, and Biswanath Doloi. "Feasibility investigation of neem oil as a dielectric for electrical discharge machining." *The International Journal of Advanced Manufacturing Technology* 106 (2020): 1179–1189. https://doi.org/10.1007/s00170-019-04736-5
48. Viswanth, V. Srinivas, R. Ramanujam, and G. Rajyalakshmi. "Improving productivity with eco-friendly dielectrics in sustainable EDM machining of AISI 2507 super duplex stainless steel." *International Journal of Precision Technology* 9, no. 2–3 (2020): 130–151. https://doi.org/10.1504/IJPTECH.2020.112061
49. Chaudhari, Rakesh, Yug Shah, Sakshum Khanna, Vivek K. Patel, Jay Vora, Danil Yurievich Pimenov, and Khaled Giasin. "Experimental investigations and effect of nano-powder-mixed EDM variables on performance measures of nitinol SMA." *Materials* 15, no. 20 (2022): 7392. https://doi.org/10.3390/ma15207392
50. Marashi, Houriyeh, Davoud M. Jafarlou, Ahmed AD Sarhan, and Mohd Hamdi. "State of the art in powder mixed dielectric for EDM applications." *Precision Engineering* 46 (2016): 11–33. https://doi.org/10.1016/j.precisioneng.2016.05.010
51. Peças, Paulo, and Elsa Henriques. "Effect of the powder concentration and dielectric flow in the surface morphology in electrical discharge machining with powder-mixed dielectric (PMD-EDM)." *The International Journal of Advanced Manufacturing Technology* 37 (2008): 1120–1132. https://doi.org/10.1007/s00170-007-1061-5
52. Pecas, P., and E. Henriques. "Electrical discharge machining using simple and powder-mixed dielectric: The effect of the electrode area in the surface roughness and topography." *Journal of Materials Processing Technology* 200, no. 1–3 (2008): 250–258. https://doi.org/10.1016/j.jmatprotec.2007.09.051
53. Kansal, H. K., Sehijpal Singh, and Pradeep Kumar. "Technology and research developments in powder mixed electric discharge machining (PMEDM)." *Journal of Materials Processing Technology* 184, no. 1–3 (2007): 32–41. https://doi.org/10.1016/j.jmatprotec.2006.10.046
54. Kansal, H. K., Sehijpal Singh, and Pradeep Kumar. "Parametric optimization of powder mixed electrical discharge machining by response surface methodology." *Journal of Materials Processing Technology* 169, no. 3 (2005): 427–436. https://doi.org/10.1016/j.jmatprotec.2005.03.028
55. Ishfaq, Kashif, Muhammad Asad Maqsood, Saqib Anwar, Muhammad Harris, Abdullah Alfaify, and Abdul Wasy Zia. "EDM of Ti6Al4V under nano-graphene mixed

dielectric: A detailed roughness analysis." *The International Journal of Advanced Manufacturing Technology* 120, no. 11 (2022): 7375–7388. https://doi.org/10.1007/s00170-022-09207-y

56. Singh, Shankar, Sachin Maheshwari, and Poorn Chandra Pandey. "Effect of SiC powder-suspended dielectric fluid on the surface finish of 6061Al/Al_2O_3P/20p composites during electric discharge machining." *International Journal of Machining and Machinability of Materials* 4, no. 2–3 (2008): 252–274. https://doi.org/10.1504/IJMMM.2008.023196
57. Agrawal, Jai Prakash, Nalin Somani, and Nitin Kumar Gupta. "A systematic review on powder-mixed electrical discharge machining (PMEDM) technique for machining of difficult-to-machine materials." *Innovation and Emerging Technologies* 11 (2024): 2440002. https://doi.org/10.1142/S2737599424400024
58. Le, Van Tao, Long Hoang, Mohd Fathullah Ghazali, Van Thao Le, Manh Tung Do, Trung Thanh Nguyen, and Truong Sơn Vu. "Optimization and comparison of machining characteristics of SKD61 steel in powder-mixed EDM process by TOPSIS and desirability approach." *The International Journal of Advanced Manufacturing Technology* 130, no. 1 (2024): 403–424.
59. Viswanth, V. Srinivas, R. Ramanujam, and G. Rajyalakshmi. "A review of research scope on sustainable and eco-friendly electrical discharge machining (E-EDM)." *Materials Today: Proceedings* 5, no. 5 (2018): 12525–12533. https://doi.org/10.1016/j.matpr.2018.02.234
60. Priadi, Dedi, Triyono Triyono, Eddy S. Siradj, and Winarto Winarto. "Surface modification of SKD 61 by electrical discharge coating (EDM/EDC) with multilayer cylindrical electrode and *Jatropha curcas* as dielectric fluid." *Applied Mechanics and Materials* 319 (2013): 96–101. https://doi.org/10.4028/www.scientific.net/AMM.319.96
61. Triyono, Triyono, Dedi Priadi, Eddy S. Siradj, and Winarto Winarto. "SKD 61 material surface treatment with electric discharge machining using Cu, CuCr & graphite electrodes and dielectric fluid *Jatropha curcas*." *Advanced Materials Research* 789 (2013): 307–312. https://doi.org/10.4028/www.scientific.net/AMR.789.307
62. Valaki, Janak B., and Pravin P. Rathod. "Assessment of operational feasibility of waste vegetable oil based bio-dielectric fluid for sustainable electric discharge machining (EDM)." *The International Journal of Advanced Manufacturing Technology* 87 (2016): 1509–1518. https://doi.org/10.1007/s00170-015-7169-0
63. Mali, Harlal Singh, and Nitesh Kumar. "Investigating feasibility of waste vegetable oil for sustainable EDM." In *Proceedings of all India Manufacturing Technology Design and Research Conference*, pp. 405–410. 2016.
64. Das, Shirsendu, Swarup Paul, and Biswanath Doloi. "Investigation of the machining performance of neem oil as a dielectric medium of EDM: A sustainable approach." In *IOP Conference Series: Materials Science and Engineering*, vol. 653, no. 1, p. 012017. IOP Publishing, 2019. https://doi.org/10.1088/1757-899X/653/1/012017
65. Reddy, G. Gowtham, Balasubramaniyan Singaravel, and K. Chandra Shekar. "Experimental investigation of sunflower oil as dielectric fluid in die sinking electric discharge machining process." In *Materials Science Forum*, vol. 969, pp. 715–719. Trans Tech Publications Ltd, 2019. https://doi.org/10.4028/www.scientific.net/MSF.969.715
66. Ahmad, Said, Richard Ngalie Chendang, Mohd Amri Lajis, Aiman Supawi, and Erween Abd Rahim. "Machinability performance of RBD palm oil as a bio degradable dielectric fluid on sustainable electrical discharge machining (EDM) of AISI D2 steel." In *Advances in Material Sciences and Engineering*, pp. 509–517. Springer Singapore, 2020.
67. Radu, Maria-Crina, Raluca Tampu, Valentin Nedeff, Oana-Irina Patriciu, Carol Schnakovszky, and Eugen Herghelegiu. "Experimental investigation of stability of

vegetable oils used as dielectric fluids for electrical discharge machining." *Processes* 8, no. 9 (2020): 1187. https://doi.org/10.3390/pr8091187
68. MangapathiRao, K., D. Vinaykumar, K. Chandra Shekar, and Rachakonda Ranjith Kumar. "Investigation and analysis of EDM process—A new approach with Al_2O_3 nano powder mixed in sunflower oil." In *IOP Conference Series: Materials Science and Engineering*, vol. 1057, no. 1, p. 012059. IOP Publishing, 2021. https://doi.org/10.1088/1757-899X/1057/1/012059
69. Nagabhooshanam, N., S. Baskar, K. Anitha, and S. Arumugam. "Sustainable machining of Hastelloy in EDM using nanoparticle-infused biodegradable dielectric fluid." *Arabian Journal for Science and Engineering* 46, no. 12 (2021): 11759–11770. https://doi.org/10.1007/s13369-021-05652-1
70. Singh, Nishant K., Lalit Yadav, and Sachin Lal. "Experimental investigation for sustainable electric discharge machining with *Pongamia* and *Jatropha* as dielectric medium." *Advances in Materials and Processing Technologies* 8, no. 2 (2022): 1447–1466. https://doi.org/10.1080/2374068X.2020.1860499
71. Yadav, Avinash, Yashvir Singh, Satyendra Singh, and Prateek Negi. "Sustainability of vegetable oil based bio-diesel as dielectric fluid during EDM process–a review." *Materials Today: Proceedings* 46 (2021): 11155–11158. https://doi.org/10.1016/j.matpr.2021.01.967
72. Khan, Mohd Yunus, Patange Sudhakar Rao, BahadurSingh Pabla, and K. Ashok Kumar. "Feasibility of used cooking oil-based biodiesel (UCOB) as a dielectric for electrical discharge machining." *Journal of King Saud University-Science* 34, no. 8 (2022): 102305. https://doi.org/10.1016/j.jksus.2022.102305
73. Boopathi, Sampath, Abdulrahman Saad Alqahtani, Azath Mubarakali, and Parthasarathy Panchatcharam. "Sustainable developments in near-dry electrical discharge machining process using sunflower oil-mist dielectric fluid." *Environmental Science and Pollution Research* (2023): 1–20. https://doi.org/10.1007/s11356-023-27494-0
74. Chakraborty, Tapas, Deepti Ranjan Sahu, Amitava Mandal, and Bappa Acherjee. "Feasibility of *Jatropha* and Rice bran vegetable oils as sustainable EDM dielectrics." *Materials and Manufacturing Processes* 38, no. 1 (2023): 50–63. https://doi.org/10.1080/10426914.2022.2089891
75. Ishfaq, Kashif, Muhammad Sana, Mudassar Rehman, Saqib Anwar, Abdullah Yahia Alfaify, and Abdul Wasy Zia. "Role of biodegradable dielectrics toward tool wear and dimensional accuracy in Cu-mixed die sinking EDM of Inconel 600 for sustainable machining." *Journal of the Brazilian Society of Mechanical Sciences and Engineering* 45, no. 4 (2023): 235. https://doi.org/10.1007/s40430-023-04126-9
76. Khan, Mohd Yunus, Patange Sudhakar Rao, and Bahadur Singh Pabla. "Investigation of surface characteristics of Inconel-625 by EDM with used cooking oil-based biodiesel as dielectric fluid." *Archives of Metallurgy and Materials* (2023): 1225–1232. https://doi.org/10.24425/amm.2023.146186
77. Pillai, K.V. Arun, and P. Saravana Kumar. "Effect of graphene dispersed coconut biodiesel on the Micro ED milling characteristics of Inconel 718 alloy." *Surface Topography: Metrology and Properties* 11, no. 2 (2023): 025021. https://doi.org/10.1088/2051-672X/acd79a
78. Sethy, Sunita, Rajesh Kumar Behera, Kamalakanta Muduli, Jayakrishna Kandasamy, J. Paulo Davim, and Jaydev Rana. "Bio-dielectrics to improve the performance of electro discharge machining–an investigation for cleaner production opportunities." *Advances in Materials and Processing Technologies* (2023): 1–20. https://doi.org/10.1080/2374068X.2023.2215607
79. Singh, Nishant Kumar, Yashvir Singh, Erween Abd Rahim, T. Senthil Siva Subramanian, and Abhishek Sharma. "Electric discharge machining of hybrid composite with bio-dielectrics for sustainable developments." *Australian Journal of Mechanical Engineering* (2023): 1–18. https://doi.org/10.1080/14484846.2023.2249577

80. Natarajan, Manikandan, Thejasree Pasupuleti, Mahmood M. S. Abdullah, Faruq Mohammad, Jayant Giri, Rajkumar Chadge, Neeraj Sunheriya, Chetan Mahatme, Pallavi Giri, and Ahmed A. Soleiman. "Assessment of machining of Hastelloy using WEDM by a multi-objective approach." *Sustainability* 15, no. 13 (2023): 10105. https://doi.org/10.3390/su151310105
81. Kamble, Prashant D., Jayant Giri, Emad Makki, Neeraj Sunheriya, Shilpa B. Sahare, Rajkumar Chadge, Chetan Mahatme, Pallavi Giri, Sathish T., and Hitesh Panchal. "An application of hybrid Taguchi-ANN to predict tool wear for turning EN24 material." *AIP Advances* 14, no. 1 (2024). https://doi.org/10.1063/5.0186432
82. Giri, Jayant, Neeraj Sunheriya, T. Sathish, Yash Kadu, Rajkumar Chadge, Pallavi Giri, A. Parthiban, and Chetan Mahatme. "Optimization of process parameters to improve mechanical properties of fused deposition method using Taguchi method." *Interactions* 245, no. 1 (2024). https://doi.org/10.1007/s10751-024-01925-x
83. Tufail, M. S., Jayant Giri, Emad Makki, T. Sathish, Rajkumar Chadge, and Neeraj Sunheriya. "Machinability of different cutting tool materials for electric discharge machining: A review and future prospects." *AIP Advances* 14, no. 4 (2024). https://doi.org/10.1063/5.0201614
84. Narasimhamu, Katta Lakshmi, Manikandan Natarajan, Pasupuleti Thejasree, Emad Makki, Jayant Giri, Neeraj Sunheriya, Rajkumar Chadge, Chetan Mahatme, Pallavi Giri, and T. Sathish. "Development of hybrid optimization model using grey-ANFIS-jaya algorithm for CNC drilling of aluminium alloy." *Journal of Engineering* 2024 (2024): 1–12. https://doi.org/10.1155/2024/1476770

3 Analysing Mechanical Behaviour of LM26/ZrB_2 Composite Using Finite Element Modelling (FEM)

Vijayan Singarajan Nagammal, Samson Jerold Samuel Chelladurai, Saiyathibrahim Abdul Pari, Jai Aultrin Kamalan Sundarabai, Makeshkumar Mani, and Infant Jegan Rakesh Antony John Bosco

3.1 INTRODUCTION

Aluminium-based metal matrix composites (AMMCs) research has expanded in recent decades to create novel materials for applications. AMMCs are widely used in the aerospace, automotive, and military industries because of their low weight, high strength-to-weight ratio, strong wear resistance, increased corrosion resistance, and high-temperature stability [1–5]. It is commonly believed that the additional reinforcements provide excellent characteristics for both aluminium and aluminium alloys. Aluminium or its alloys are the most often used matrix materials in producing AMMCs for demanding applications due to their availability, low cost, and recyclable nature [6, 7]. Research on AMMCs has grown significantly in recent decades to develop innovative materials for various applications. AMMCs are extensively utilized in the aerospace, automotive, and military sectors due to their lightweight nature, superior strength-to-weight ratio, excellent wear resistance, enhanced corrosion resistance, and remarkable temperature stability. There is a widely held belief that the inclusion of extra reinforcements greatly enhances the properties of both aluminium and aluminium alloys. Aluminium and its alloys are commonly utilized as matrix materials in the production of AMMCs for demanding applications. This is because they are readily available, inexpensive, and can be recycled. AMMCs are created using any of these approaches, including solid and liquid state procedures. Casting is a liquid state technique that is commonly used to manufacture metal matrix composites because it is a low-cost and practical way to achieve the desired form and size [8, 9]. LM26 aluminium alloy is a common material used for die-casting due to its excellent castability, machining qualities, conductivity, and minimal die-casting faults. The addition of ZrB_2 reinforcement to aluminium alloys using liquid metallurgy leads to enhanced mechanical properties, including higher hardness and ultimate tensile strength (UTS) [10, 11]. The process of creating these materials requires significant financial and time constraints. Consequently, there

DOI: 10.1201/9781032725086-4

has been a surge in interest in numerical analysis. The primary rationale for using this numerical analysis approach is its high degree of concordance with empirical findings. Furthermore, numerical analysis may be utilized to generate solutions for concerns that may arise in these AMMCs in the future [12]. The influence of microstructural characteristics of an aluminium composite during localized stress build-up and subsequent failure was studied. A numerical investigation revealed that the overall behaviour of aluminium composites is mostly related to the reinforcement that is distributed inside them [13]. ANSYS was used to investigate the performance (stress, strain distribution, and deformation) of an AA6082 alloy-based composite model used in a pressure vessel with internal pressures [14].

A static analysis was performed on a piston constructed of Al-12.5Si alloy and reinforced with a variety of particles (15% cenosphere and 3% SiC). Stress, deformation, and strain were determined and compared to models constructed using the LM26 and A4032 [15]. A composite model comprising an Al6064 matrix and SiC reinforcement particles was employed to simulate a tensile test in order to ascertain the stress-strain relationship and deformation of the composite. The finite element method is a valuable technique for forecasting the stress, strain, deformation, mass, and factor of safety of car components made of aluminium-based hybrid composites, including connecting rods [16]. A group of researchers conducted a study on the mechanical properties and fracture behaviour of the Al_3Ti/A356 composite using finite element method. They specifically examined the impact of microstructural factors, including the irregular spatial distribution of reinforcements, varying characteristics of reinforcements' shape, and the anisotropic orientation [17, 18]. While considering the previous works, it can be said that there are only a few numerical analyses available with respect to AMMCs [19–23]. Hence, it is decided to give importance to the numerical analysis of aluminium alloy-based composites. As a result, in this study, it is decided to develop an LM26/ZrB_2 composite model that properly accounts for the volume percentage of ZrB_2 particles and thoroughly analyses the effect on the hardness and tensile behaviour.

3.2 METHODOLOGY AND MODELLING

The composite models created in this research included LM26 (Al-Si10Cu3Mg1) aluminium alloy as the main material, with varying amounts of reinforcement (ZrB_2) ranging from 0 to 2.5%, 5%, 7.5%, and 10% by volume. Five compositions were thoroughly examined, and the following composite models were created: LM26 model, LM26 + 2.5% ZrB_2 model, LM26 + 5% ZrB_2 model, LM26 + 7.5% ZrB_2 model, and LM26 + 10% ZrB_2 model. The ANSYS R15.0 software was utilized to analyse the mechanical properties of all composite models, including tensile strength and hardness. Table 3.1 displays the conventional chemical composition of the LM26 aluminium alloy utilized in this investigation. The composite models used for tensile testing were carefully constructed following the guidelines outlined in the ASTM E8 standard to study the behaviour of materials under tension. Hardness measurements were obtained from models with dimensions of 25 × 25 × 10 mm.

Figure 3.1 depicts a process flow chart depicting the current numerical investigation of LM26/ZrB_2 composites. All three-dimensional composite models with

TABLE 3.1
Chemical Composition of LM26 Alloy

Elements	Cu	Mg	Si	Fe	Mn	Ni	Zn	Pb	Sn	Ti	Al
Composition %	2.0–4.0	0.5–1.5	8.5–10.5	1.2 max	0.5 max	1.0 max	1.0 max	0.2 max	0.1 max	0.2 max	Remaining

various volume fractions of ZrB_2 reinforcement were created using modelling software and exported to analysis software (ANSYS R15.0) to determine the best volume fraction of composite that results in better mechanical properties based on numerical analysis after selecting the analysis method, properties, and degrees of freedom.

3.2.1 Finite Element Modelling (FEM)

Finite element modelling (FEM) is a very effective numerical method that is frequently employed to predict the behaviour of a composite model under external stress conditions prior to its implementation in real-world applications [24–30]. Prior to this, composite models with diverse reinforcements were created. By employing the meshing option, the pre-processor stage subdivides each composite model into numerous finite elements, so ensuring precise outcomes. The model was subjected to boundary conditions to restrict any rotation or displacement. Additionally, a load of 4000 N was applied in the linear direction (y-coordinate) to test its tensile strength, while a load of 1500 N was applied to test its hardness. The model inputs for assessing von Mises stress, equivalent elastic strain, and total deformation of the composite models were Young's modulus, Poisson's ratio, and density. The highest achieved value is taken into account when discussing the stress, strain, and deformation of the composite model caused by the applied load. Figures 3.2 and 3.3 depict a composite model that combines load circumstances in order to analyse the tensile and hardness properties.

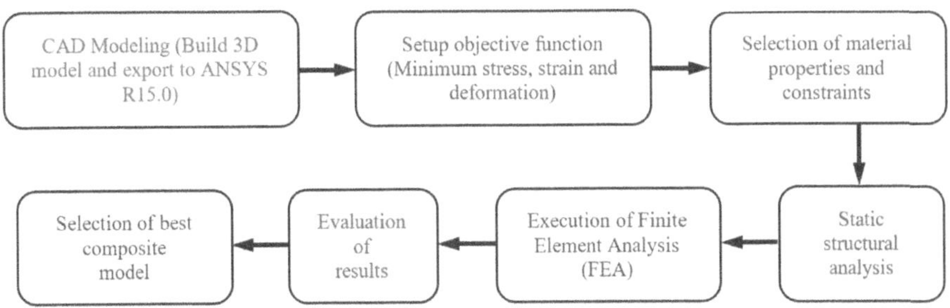

FIGURE 3.1 Process flow chart of the present numerical study of LM26/ZrB_2 composites.

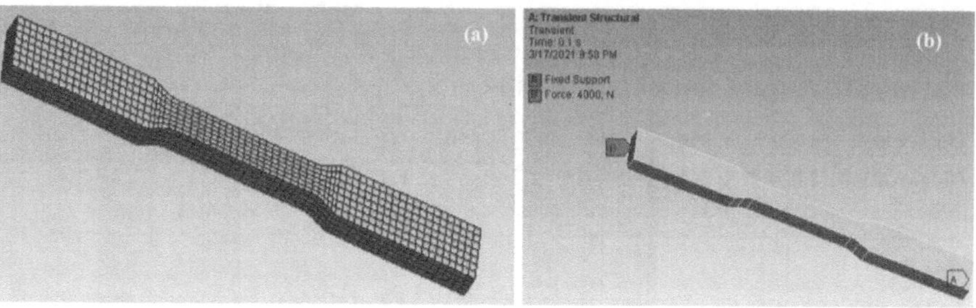

FIGURE 3.2 FE model of tensile specimen (a) meshing and (b) constraints with load.

3.3 RESULTS AND DISCUSSION

FEM approach was implemented for various developed LM26/ZrB_2 composite models, and the tensile behaviour and hardness variations were obtained through observing von Mises stress, elastic strain, and total deformation. The following sections discuss the analysis and observations of each composite model with respect to the abovementioned mechanical properties and present the better composite model that would be beneficial and meet the industrial requirements.

3.3.1 Hardness Analysis

The ability of a substance to resist indentation, abrasion, and wear is referred to as its hardness. All composite models have been examined using ANSYS R15.0 to predict their ability to withstand external stress and detect hardness changes. According to Table 3.2, indentation generated stress waves and a maximum stress of 14.077 MPa was recorded for the LM26 + 0% ZrB_2 or unreinforced LM26 models owing to the application of static load. In the same scenario, the LM26 + 10% ZrB_2 model had the lowest stress (12.31 MPa). The inclusion of additional ZrB_2 reinforcements leads to an increase in the hardness of LM26/ZrB_2 composites. The imposed load was adequately withstood by the equally distributed reinforcements without producing

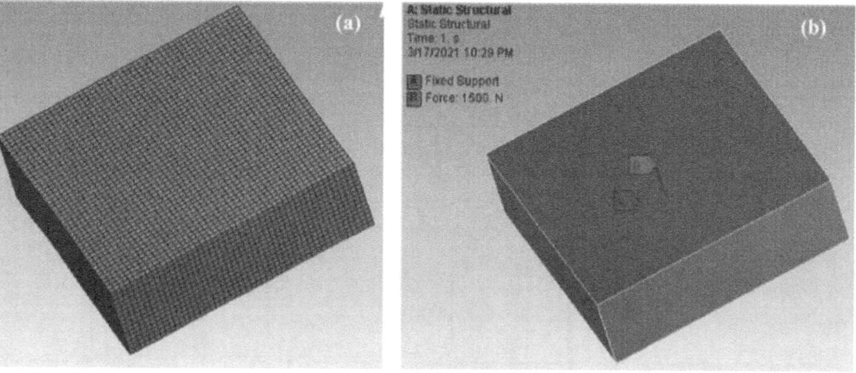

FIGURE 3.3 FE model of hardness specimen (a) meshing and (b) constraints with load.

TABLE 3.2
Hardness Analysis Results of Developed Composite Models

Model	von Mises Stress (MPa)	Equivalent Elastic Strain (×10^{-4} mm/mm)	Total Deformation (×10^{-4} mm)
LM26	14.077	1.9826	4.2942
LM26 + 2.5% ZrB_2	13.713	1.6673	3.6684
LM26 + 5.0% ZrB_2	13.003	1.3906	3.1606
LM26 + 7.5% ZrB_2	12.654	1.0612	2.4953
LM26 + 10% ZrB_2	12.31	0.12346	0.28533

excessive stress within the matrix, which is advantageous in places where the load is applied continuously. Figure 3.4 clearly shows that the stress caused by the indentation decreases consistently when incorporating reinforcement up to 10% in the composite models. It was also discovered that the amount of strain created inside the model because of the external load steadily decreased.

The equivalent elastic strain and total deformation of the unreinforced LM26 model in response to indentation were found to be 1.9826×10^{-4} mm/mm and 4.2942×10^{-4} mm, respectively, indicating that it could not withstand indentation and that plastic deformation would increase further as the indentation level increased. After incorporating ZrB_2 reinforcement ranging from 2.5% to 10% into the composite model, the corresponding elastic strain and total deformation rapidly decreased from 1.6673×10^{-4} mm/mm and 3.6684×10^{-4} mm to 0.12346×10^{-4} mm/mm and 0.28533×10^{-4} mm, respectively. Figure 3.5 shows that the addition of reinforcement particles decreases the von Mises stress, elastic strain, and overall deformation of the composite model.

The distribution of ZrB_2 reinforcement particles inside the matrix (LM26) is principally responsible for the increase in hardness by reducing the α-Al phase and resisting deformation. The ductile nature of the LM26 aluminium alloy is said to have been greatly modified by this addition, and such reinforcements provide support to sustain the applied [31].

3.3.2 Tensile Analysis

FEM was used to simulate the tensile load on all five created models, with volume fractions of ZrB_2 ranging from 0% to 10% up to the failure point. The study results summarized in Table 3.3 show that both stress and displacement did not differ considerably amongst the composite models, indicating that the results are deemed reliable.

The LM26 + 7.5% ZrB_2 composite model outperformed all other models in the findings of tensile behaviour simulations. This model demonstrated a relatively low strain value of 1.4503×10^{-3} mm/mm; nonetheless, LM26 + 10% ZrB_2 demonstrated an increasing trend of the tensile parameters investigated. When comparing composite models containing 7.5% and 10% ZrB_2, a 13.18% increase in strain and

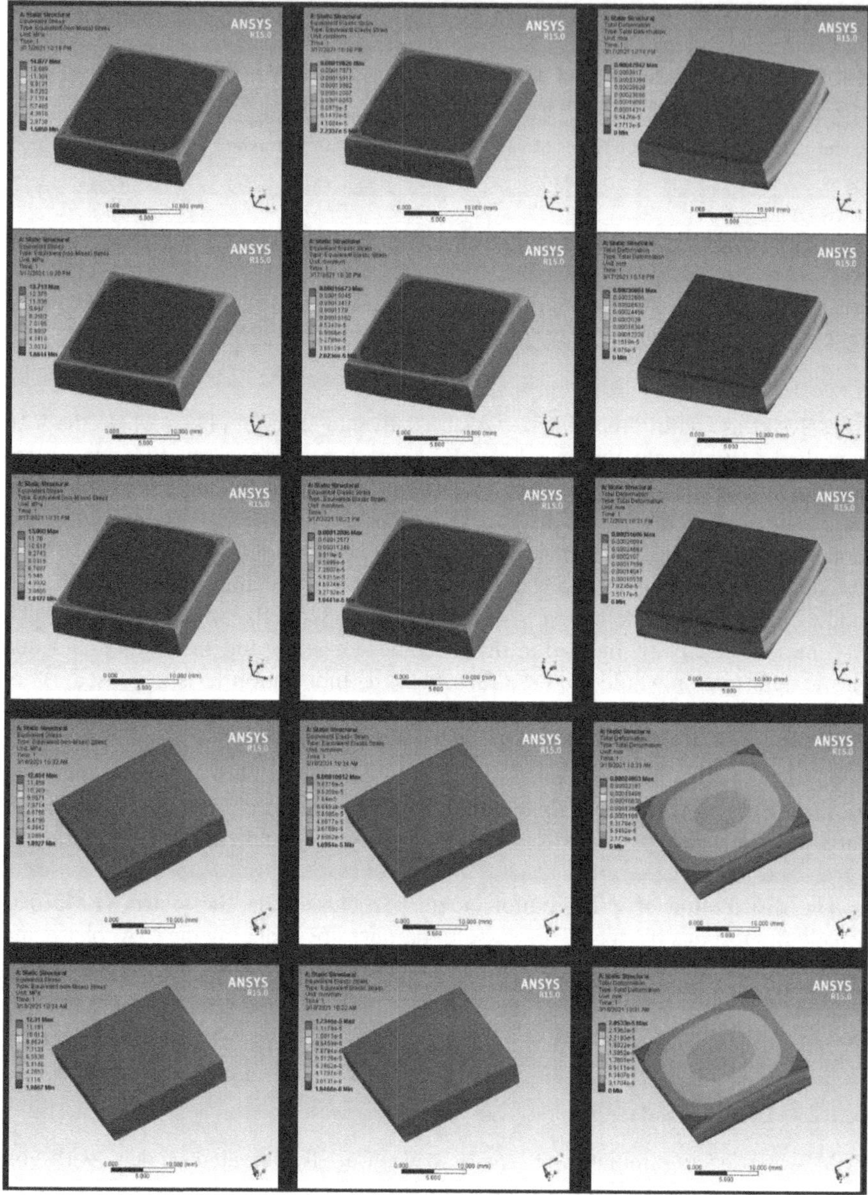

FIGURE 3.4 FEM images of hardness analysis results.

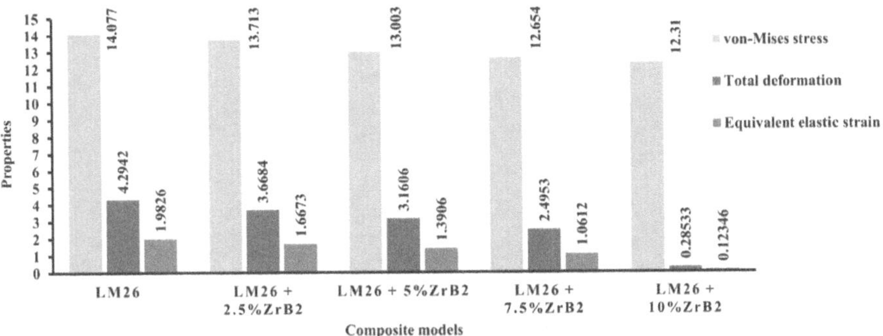

FIGURE 3.5 Graphical representation of hardness analysis results.

deformation was noticed for the LM26 + 10% ZrB_2 model. Clustering of ZrB_2 is responsible for the increase in strain and deformation under tensile load.

The deformation of composite models under stress is essential in any tensile study. Figure 3.6 depicts the deformation of the five composite models tested. The largest deformation (0.14036 mm) occurred during transient loading circumstances for the unreinforced model, which is due to the poor load-bearing capability of the matrix (ductile).

It was also discovered that the von Mises stress created during tensile loading was greatest at the neck of the composite models developed. The LM26 + 7.5% ZrB_2 composite model showed a stress value of 168.21 MPa, which is lower than all other models and helps to postpone fracture during continuous load application. Tension strain is generated in the longitudinal and transverse directions because of the constant uniaxial force applied. All contours include red legends denoting the greatest stress locations and blue legends showing the least stress experienced zones. The tensile behaviour of composite models clearly demonstrated that the applied load is the primary cause of the total deformation seen. The contour analysis of LM26/ZrB_2 composite models revealed that stress and strain were experienced over the whole span of the tensile models. The study used identical load and boundary conditions to those employed in experimental tensile strength testing conducted on a universal testing machine (UTM). While the load-bearing capability of the composite models

TABLE 3.3
Tensile Test Results of Developed Composite Models

Model	von Mises Stress (MPa)	Equivalent Elastic Strain ($\times 10^{-3}$ mm/mm)	Total Deformation (mm)
LM26	168.47	2.3732	0.14036
LM26 + 2.5% ZrB_2	168.40	2.0477	0.12115
LM26 + 5.0% ZrB_2	168.27	1.8	0.10658
LM26 + 7.5% ZrB_2	168.21	1.4503	0.085895
LM26 + 10% ZrB_2	168.22	1.6414	0.097221

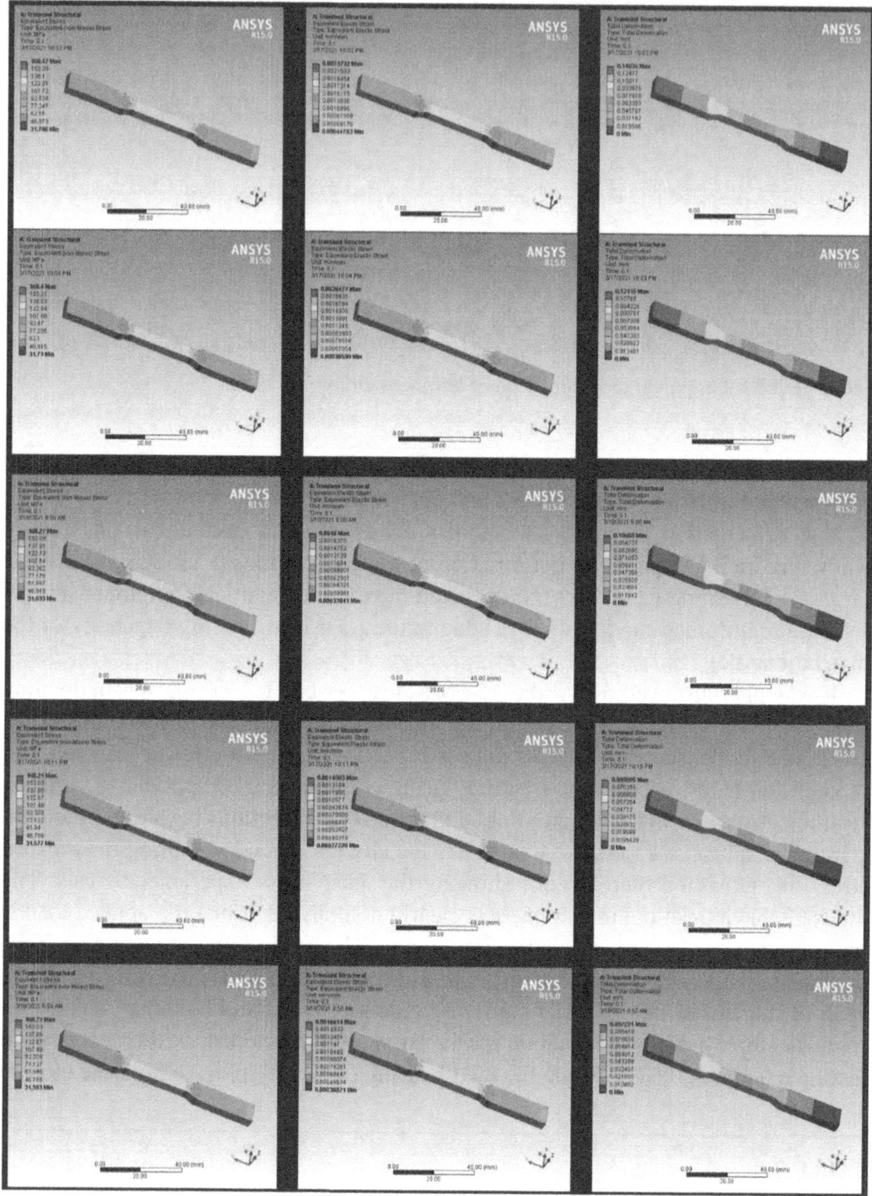

FIGURE 3.6 FEM images of tensile analysis results.

was significantly enhanced, the introduction of a greater volume fraction of ZrB_2 (exceeding 7.5%) led to an escalation in overall deformation. This leads to the understanding that clustering reinforcement inside the matrix produces higher deformation in LM26/ZrB_2 composite models. The graphical representation of tensile results is depicted in Figures 3.7–3.9.

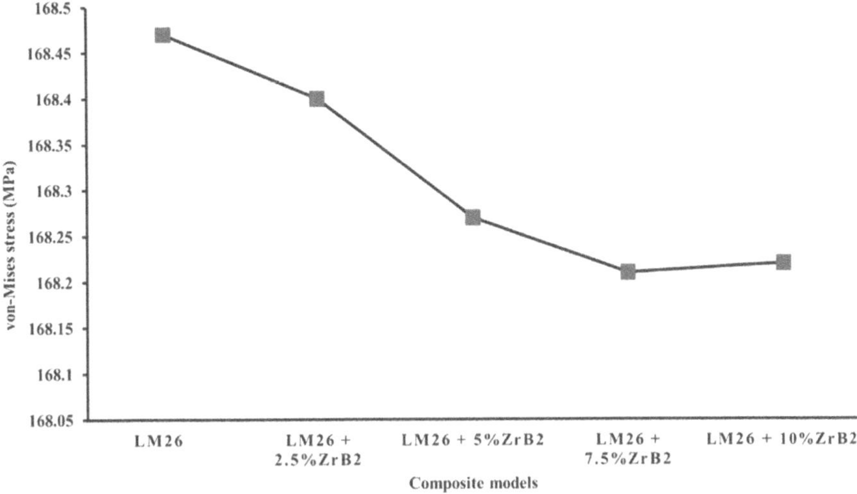

FIGURE 3.7 Graphical representation of tensile results (von Mises stress).

3.4 CONCLUSION

The objective of this research is to examine the characteristics of von Mises stress, equivalent elastic strain, and total deformation through the utilization of FEM in order to anticipate the properties of composites prior to their implementation in real-time applications. Given the limited amount of previous research, this study seeks to address that deficiency. The study examines the influence of ZrB_2 volume percentage on mechanical properties, specifically hardness and tensile strength, in LM26/ZrB_2

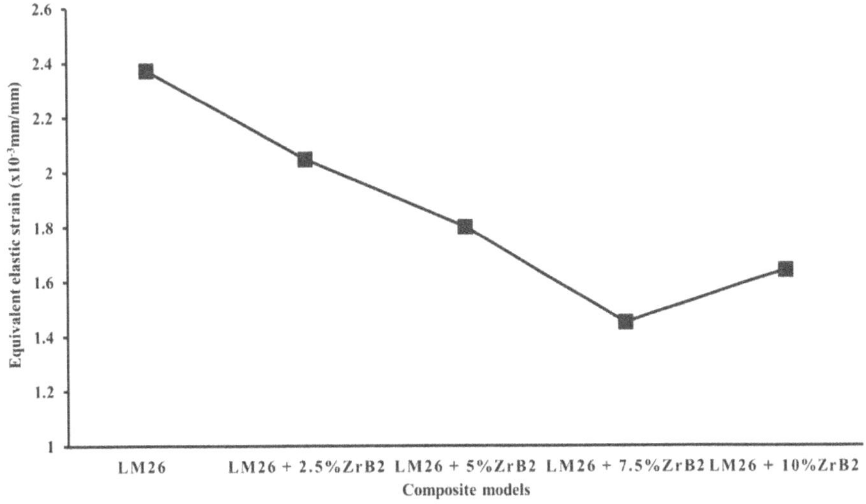

FIGURE 3.8 Graphical representation of FEM results of tensile analysis (equivalent elastic strain).

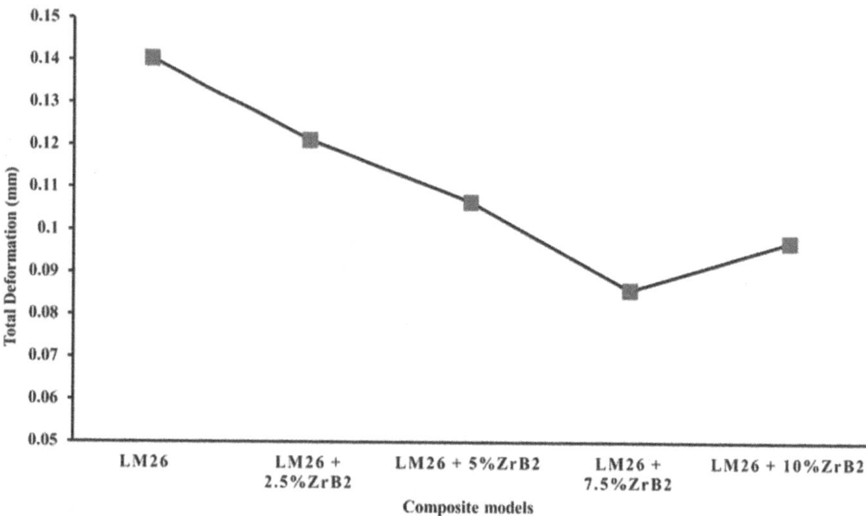

FIGURE 3.9 Graphical representation of FEM results of tensile analysis (total deformation).

composite models. Five composite models with different levels of reinforcement were examined to analyse metrics such as von Mises stress, equivalent elastic strain, and total deformation. The observations are mentioned as follows:

1. Finite element modelling (FEM) is an effective tool for analysing the behaviour of LM26/ZrB_2 composite models, resulting in time and cost savings compared to traditional production and testing methods.
2. ZrB_2 appeared to be an excellent reinforcement for improving the mechanical characteristics of composites, as its presence significantly increased both hardness and tensile strength.
3. The maximum hardness value may be achieved by incorporating 10% ZrB_2, since the corresponding LM26 + 10% ZrB_2 composite model displayed minimal values for the variables analysed, which were 12.31 MPa von Mises stress, 0.12346×10^{-4} mm/mm equivalent elastic strain, and 0.28533×10^{-4} mm total deformation.
4. Incorporating 7.5% ZrB_2 resulted in higher tensile characteristics by exhibiting the least deformation among the composite models studied. The addition of extra reinforcement resulted in agglomeration, excessive stress, and strains, all of which reduced the tensile capabilities of the composite.
5. Overall, LM26/ZrB_2 composite can be experimentally fabricated, and because of its superior characteristics, it finds application in automotive sectors.

3.5 FUTURE SCOPE OF THE WORK

The aerospace, automotive, and electronics industries use AMMCs with enhanced thermal, mechanical, and electrical characteristics. The design and analysis of composites with the aid of FEM can predict whether the specific requirements of composites are satisfied or not. FEM will be essential for forecasting how novel engineered materials will behave under various loading scenarios. The future scope of current research work is to examine the thorough comprehension of composite behaviour by integrating atomistic, microscale, and macroscale models. This method aids in capturing the impact of microstructural characteristics on the composite macroscopic characteristics. Also, combine FEM and machine learning techniques to improve composite behaviour prediction and save computing expenses. Real-time FEM simulations will also be carried out with computational resources, where immediate feedback is required during real-time applications such as manufacturing processes and structural health monitoring. To improve the accuracy of FEM results, better experimental techniques will be developed to validate the FEM results.

ABBREVIATIONS

AMMCs Aluminium-based metal matrix composites
ZrB_2 Zirconium diboride
SiC Silicon carbide
FEM Finite element modelling
UTM Universal testing machine
UTS Ultimate tensile strength

REFERENCES

1. Kumar, V. M., and C. V. Venkatesh. "A comprehensive review on material selection, processing, characterization and applications of aluminium metal matrix composites." *Materials Research Express* 6, no. 7 (2019): 072001. https://doi.org/10.1088/2053-1591/ab0ee3
2. Kumar Murmu, S., S. Chattopadhayaya, R. Cep, A. Kumar, A. Kumar, S. Kumar Mahato, A. Kumar, P. Ranjan Sethy, and K. Logesh. "Exploring tribological properties in the design and manufacturing of metal matrix composites: An investigation into the AL6061-SiC-fly ASH alloy fabricated via stir casting process." *Frontiers in Materials* 11 (2024): 1415907.
3. Saxena, A., T. K. Gupta, R. Srivastava, A. K. Srivastava, and A. Kumar. (2024). "A comparison of different experimental approaches in the determination of dynamic fracture toughness (J1d) behavior for RHA steel cost sustainable MMAW weldments." *Journal of Adhesion Science and Technology*, 1–22. https://doi.org/10.1080/01694243.2024.2334267
4. Wagih, A., H. Junaedi, H. A. Mahmoud, G. Lubineau, A. Kumar, and T. A. Sebaey. "Enhanced damage tolerance and fracture toughness of lightweight carbon-Kevlar fiber hybrid laminate." *Journal of Composite Materials* 58, no. 9 (2024): 1109–1121.
5. Ahmad, S., J. Sharma, and V. Gambhir, eds. *Handbook of Sustainable Materials: Modelling, Characterization, and Optimization*. CRC Press, 2023. https://doi.org/10.1201/9781003297772
6. Pastuszak, P. D., and A. Muc. "Application of composite materials in modern constructions." *Key Engineering Materials* 542 (2013): 119–129. 10.4028/www.scientific.net/KEM.542.119
7. Tripathi, A., N. K. Jha, R. N. Hota, A. Kumar, and R. Tyagi. "Green sound-absorbing material prepared by using natural fiber for building acoustics." *Proceedings of*

the Institution of Mechanical Engineers, Part E: Journal of Process Mechanical Engineering (2024). https://doi.org/10.1177/09544089241253973

8. Garg, P., A. Jamwal, D. Kumar, K. Kumar Sadasivuni, C. Mustansar Hussain, and P. Gupta. "Advance research progresses in aluminium matrix composites: Manufacturing & applications." *Journal of Materials Research and Technology* 8, no. 5 (2019): 4924–4939. https://doi.org/10.1016/j.jmrt.2019.06.028
9. Srivastava, A. K., S. Tiwari, P. Pachauri, N. Gupta, B. Sunil, and A. Kumar. "Bonding strength and microstructural features of Al5083-AZ31B alloys laminated sheet through friction stir additive manufacturing." *Journal of Adhesion Science and Technology* 38, no. 4 (2023): 583–596. https://doi.org/10.1080/01694243.2023.2240637
10. Kumar, A., P. Kumar, R. K. Mittal, and V. Gambhir. "Materials processed by additive manufacturing techniques." *Advanced Additive Manufacturing* 231 (2023): 217–233.
11. Vijayan, S. N., S. J. S. Chelladurai, and A. Saiyathibrahim. "Investigation on mechanical behavior of LM26 aluminum alloy—ZrB2 and copper-coated short steel fiber-reinforced composites using stir casting process." *International Journal of Metalcasting* (2024): 1–14. https://doi.org/10.1007/s40962-024-01303-x
12. Deghoul, N., H. Errouane, Z. Sereir, A. Chateauneuf, and S. Amziane. "Effect of temperature on the probability and cost analysis of mixed-mode fatigue crack propagation in patched aluminium plate." *International Journal of Adhesion and Adhesives* 94 (2019): 53–63. https://doi.org/10.1016/j.ijadhadh.2019.05.004
13. Peng, Y., H. Zhao, J. Ye, M. Yuan, L. Tian, Z. Li, and Z. Wang. "Multiscale 3D finite element analysis of aluminum matrix composites with nanoµ hybrid inclusions." *Composite Structures* 288 (2022): 115425. https://doi.org/10.1016/j.compstruct.2022.115425
14. Mohanavel, V., S. Prasath, M. Arunkumar, G. M. Pradeep, and S. Surendra Babu. "Modeling and stress analysis of aluminium alloy based composite pressure vessel through ANSYS software." *Materials Today: Proceedings* 37 (2021): 1911–1916. https://doi.org/10.1016/j.matpr.2020.07.472
15. Vinoth, M. A., L. R. Arun, and V. K. Uppinal. "Development and assessment of piston by using Al-Si hybrid metal matrix composites reinforced with SiC and cenosphere particulates." *International Journal of Engineering Sciences & Research Technology (IJERT)* 3, no. 7 (2014): 1234–1238.
16. Yuan, M. N., Y. Q. Yang, C. Li, P. Y. Heng, and L. Z. Li. "Numerical AIS of the stress–strain distributions in the particle reinforced metal matrix composite SiC/6064Al." *Materials & Design* 38 (2012): 1–6. https://doi.org/10.1016/j.matdes.2011.12.043
17. Singh, A., A. K. Srivastava, A. Kumar et al. "Design and development of plentiful fly ash-based glass powder-reinforced plastic composite bricks for low water absorption and high compressive and flexural strength." *International Journal on Interactive Design and Manufacturing* (2023). https://doi.org/10.1007/s12008-023-01580-6
18. Ma, S., X. Zhuang, and X. Wang. "3D micromechanical simulation of the mechanical behavior of an in-situ Al_3Ti/A356 composite." *Composites Part B: Engineering* 176 (2019): 107115. https://doi.org/10.1016/j.compositesb.2019.107115
19. Kamble, P. D., J. Giri, E. Makki, N. Sunheriya, S. B. Sahare, R. Chadge, C. Mahatme, P. Giri, S. T., and H. Panchal. "An application of hybrid Taguchi-ANN to predict tool wear for turning EN24 material." *AIP Advances*, 14, no. 1 (2024). https://doi.org/10.1063/5.0186432
20. Giri, J., N. Sunheriya, T. Sathish, Y. Kadu, R. Chadge, P. Giri, A. Parthiban, and C. Mahatme. "Optimization of process parameters to improve mechanical properties of fused deposition method using Taguchi method." *Interactions*, 245, no. 1 (2024). https://doi.org/10.1007/s10751-024-01925-x

21. Tufail, M. S., J. Giri, E. Makki, T. Sathish, R. Chadge, and N. Sunheriya. "Machinability of different cutting tool materials for electric discharge machining: A review and future prospects." *AIP Advances*, 14, no. 4 (2024). https://doi.org/10.1063/5.0201614
22. Narasimhamu, K. L., M. Natarajan, P. Thejasree, E. Makki, J. Giri, N. Sunheriya, R. Chadge, C. Mahatme, P. Giri, and T. Sathish. "Development of hybrid optimization model using grey-ANFIS-Jaya algorithm for CNC drilling of aluminium alloy." *Journal of Engineering*, 2024 (2024): 1–12. https://doi.org/10.1155/2024/1476770
23. Natarajan, M., T. Pasupuleti, J. Giri, H. A. Al-Lohedan, L. N. Katta, F. Mohammad, N. Sunheriya, R. Chadge, C. Mahatme, P. Giri, S. Mallik, and T. Sathish. "Optimization of wire spark erosion machining of Grade 9 titanium alloy (Grade 9) using a hybrid learning algorithm." *AIP Advances*, 14, no. 1 (2024). https://doi.org/10.1063/5.0177658
24. Vijayan, S. N., S. J. S. Chelladurai, S. Karthik, R. M. Saliek, S. M. Mansoor, and L. Neelakandan. "Numerical analysis of impact behaviour of LM26/ZrB2 composite using finite element method." In *AIP Conference Proceedings*, vol. 2527, no. 1. AIP Publishing, 2022. https://doi.org/10.1063/5.0108205
25. Vijayan, S. N., and M. Makeshkumar. "Material specific product design analysis for conditional failures–A case study." *International Journal of Engineering Science and Technology* 4, no. 03 (2012): 976–984.
26. Sönmez, Atilla Uygar, and S. N. Vijayan. "Numerical analysis of frictional drag reduction of watercraft using water lubrication technique." *Engineering and Technology Quarterly*: 18.
27. Choudhary, T., M. Kumar Sahu, V. Shende, and A. Kumar. "Computational analysis of a heat transfer characteristic of a wavy and corrugated channel." *Materials Today: Proceedings* 56 (2022): 263–273, https://doi.org/10.1016/j.matpr.2022.01.121
28. Goyat, V., T. A. Enab, G. Ghangas, S. Kadiyan, and A. Kumar. "On stress concentration analysis of inverse distance weighted function based finite FGM panel with circular hole under biaxial loading." *Multidiscipline Modeling in Materials and Structures* 18, no. 4 (2022): 708–733. https://doi.org/10.1108/MMMS-04-2022-0070
29. Kumar, P., S. Shah Hussain, A. Kumar, A. Kumar Srivastava, M. Hussain, and P. K. Singh. "10 Finite element method investigation on delamination of 3D printed hybrid composites during the drilling operation." In *3D Printing Technologies: Digital Manufacturing, Artificial Intelligence, Industry 4.0* 223, 2024. https://doi.org/10.1515/9783111215112-010
30. Singh, H., and B. AlMangour, eds. *Handbook of Smart Manufacturing: Forecasting the Future of Industry 4.0*. CRC Press, 2023. https://doi.org/10.1201/9781003333760
31. Kanta, D. D., M. P. Chandra, S. Saranjit, and T. R. Kumar. "Properties of ceramic-reinforced aluminium matrix composites—A review." *International Journal of Mechanical and Materials Engineering* 9 (2014): 1–12. https://doi.org/10.1186/s40712-014-0012-9

4 Various Challenges in Metal and Ceramic Particle-Filled Composite Filament-Based 3D Printing

A Review

Siva Prasad Tadi, Shanmuka Srinivas Maddula, Venkata Krishna Devara, Bhargav Chandan Palivela, and Ravi Sankar Mamilla

4.1 INTRODUCTION

The manufacturing industry faces challenges of cost, time, aesthetics, precision, and mechanical properties [1]. Additive manufacturing (AM) offers solutions, but it requires extensive preprocessing and post-processing for desired quality. AM process fabricates 3D components by depositing material based on a 3D CAD (computer-aided design) model, involving file format conversion and slicing [2]. Moreover, this process enhances manufacturing flexibility by producing parts from diverse materials, reducing costs and lead times, including biomaterials and smart materials, with an emphasis on patient-specific biocomponents and responsive materials [3, 4]. As metal 3D printing is growing at faster rates, researchers are looking at cost-effective versions of metal parts fabrication. One such economical alternative form is to print the 'green parts' containing metal particles encapsulated in a polymer matrix [5]. This green part undergoes de-binding and sintering, to obtain the final metallic feature, as shown in Figure 4.1. This process involves using a composite filament containing metal particles dispersed within a polymer matrix. Metal particle-filled composite filaments have gained significant attention [6] in recent years due to their ability to enhance the properties of 3D-printed objects. It is less expensive than beam-based AM due to less heat input [7]. Additionally, it prevents beam-induced melting and solidification from creating high-temperature gradients and anisotropy [8]. However, the successful implementation of metal and ceramic particle-filled composite filaments in 3D printing requires overcoming various challenges [9]. Regardless of the advantages of this version, there are a few drawbacks that call for further study

DOI: 10.1201/9781032725086-5

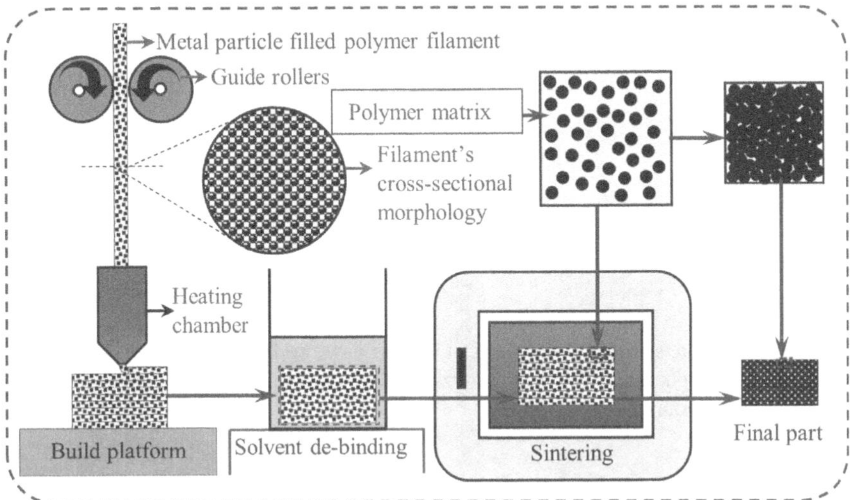

FIGURE 4.1 Overview of the particle-filled polymer composite 3D printing.

and scientific advancement [10]. It is very important to identify the process drawbacks affiliated with this 3D printing technology. This knowledge helps improve the printing quality measures. Obstacles in material processing help researchers to think about novel processing technologies, which reduce the printing issues. This study helps scientific advancement in the field of composite filament-based 3D printing and drives innovation, expanding the applications and improving the superiority of the prints. The sections of this review are divided according to the challenges that occurred in sequence. Section 4.2 highlights the problems that occurred during filament fabrication, Section 4.3 describes the role of printing parameters, Section 4.4 describes the issues after green state printing, and Section 4.5 elaborates on the miscellaneous challenges.

4.2 CHALLENGES IN FILAMENT FABRICATION

The important feedstock for this 3D printing is particle-filled polymer filament. These filaments are usually fabricated by extruders. The polymer granules and metal particles are compounded in a dual screw extruder and then cut into small pellets. These small pellets are extruded as a filament in a single screw extruder. The selection of powder particles influences the end properties of the filament. The porosity and sphericity of the particles depend on the method adopted for atomizing them. Chen et al. [11] experimented with three methods of atomization and reported the effect of the atomization process on particle sphericity and porosity, as shown in Figure 4.2. Kukla et al. [12] prepared 316L particle-filled polymer filaments with two different particle size distributions. It was found that the apparent shear viscosity of feedstock, the secant modulus, and the elongation at break of filaments reduce when the average particle size increases from 5.5 to 8.6 μm. The recommendation is to use powders with a narrower size distribution [13].

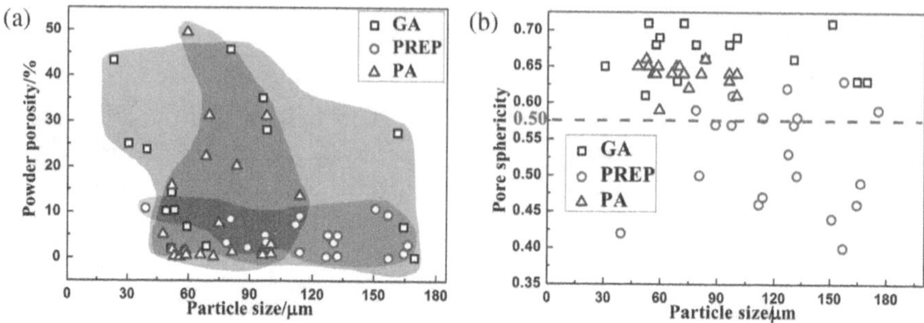

FIGURE 4.2 Pore structural properties of various gas atomized powders (a) powder porosity and (b) pore sphericity. GA: gas atomized, PREP: plasma rotating electrode processed, and PA: plasma atomized powders [11].

High metal powder content leads to excessive rigidity, brittleness, and low ductility, which causes filaments to break during winding. When the infill content in composite filament is nearer to 60 vol.%, continuous winding becomes problematic. Whereas with infill of 50–55 vol.%, continuous spooling is found to be possible [14]. Roshchupkin et al. [15] proposed optimum infill powder content, which ranges from 50 to 65 vol.% for successful printing. Higher solid loading helps prevent distortion during de-binding operation, but excessive solid loading hampers the filament because of its brittleness [16]. The concern with increasing solid loading is that it enhances viscosity and degrades flexibility. So, selecting a suitable binder system is challenging [17]. Proper selection of the metal type is essential to avoid certain problems. For example, one may avoid silver alloys for printing at high temperatures because of their tendency for agglomeration, which may clog the extruder nozzle and exhibit discontinuity in printing [18].

A consistent filament diameter is necessary for smooth feeding. If the filament is improperly or incompletely solidified, problems like inconsistent diameter, wrinkles, and stretched filaments may result [19]. If pores are present in the filament, their density is reduced, causing low filament strength. When the filament is passed through the pinch rollers, it may buckle, and cracks might propagate. Such filament, when prints the green part, gaps are present between the layers, and the resulting green density of the part dwindles [16]. Because of the internal stresses present in part, cracks may proliferate when de-bind the polymer.

The buckling of the filament must be prevented to facilitate the harmony of the feeding and extrusion process. Buckling frequently occurs when the filament is turned flat and is used to happen at excessive pinch forces [20]. The filament buckling could also happen when feeding the highly flexible filaments [21]. If oversized particles cannot disperse within the polymer matrix, they might form aggregation and lead to buckling at the nozzle [22]. Buckling is minimized by maintaining a small gap between the rollers and replacing toothed rollers with belt drives, enhancing the contact area between the filament and the belt [23]. High viscosity, stiffness, and insufficient roundness make the filament challenging to feed.

4.2.1 Co-processing Requirements for Multi-Material Filaments

Another challenge in metal cum ceramic fused filament fabrication is to select a proper pair (metal and ceramic) of powders that permit co-processing in the sequence of thermal treatment. For this to happen, powders of comparable thermal expansion coefficients and similar shrinkage behavior during sintering must be carefully chosen. Abel et al. [17] proposed a specialized milling practice, namely attrition milling followed by planetary ball milling, to match the shrinkage behavior of stainless steel with zirconia. The optimum milling time was chosen to match the thermal behavior of stainless steel powder. This milling practice would benefit multi-component materials with a significant difference in their coefficient of thermal expansion (CTE) values.

4.3 SELECTION OF PRINTING PARAMETERS

The printing parameters, such as layer thickness, printing speed, extrusion multiplier, and infill density, are meticulously chosen to prevent unnecessary defects. The printing process starts with the replica CAD file. Surface defects are characterized by the staircase effect and chordal effect. Staircase defects can be reduced by decreasing layer thickness. While approximating the actual surfaces of the object, it approximates with first-order piecewise linear approximation, which causes the loss of topological and geometrical data, and introduces an error called chordal error. The chordal error can be minimized by assuming the smallest chord length before converting the CAD file into an STL file.

4.3.1 Effect of Printing Feed Rate

The printing feed rate defines the speed at which the print head moves over the build platform per unit time. Singh et al. studied the influence of the feed rate on the printing process by adopting two types of powder filaments, namely fine powders and coarse powders [24]. For the experimentation, four samples with 10 × 10 × 5 mm (length × width × height) were printed with different feed rates with the aim to print the parts with 100% density and a mass of 1.5 g.

The results are observed for both the fine and coarse powder filament printing. It can be observed from Figure 4.3 that the fine powder filaments had good printability up to a feed rate of 2 mm/s. However, increasing the feed rate resulted in nonuniform material deposition and eventually no printing at the feed rate of 10 mm/s. The coarse powder filaments exhibited good printability up to a feed rate of approximately 8 mm/s, beyond which nonuniform material deposition and printing stoppage occurred. The dimensional deviation from the CAD file increased with higher feed rates.

4.3.2 Effect of Nozzle Temperature and Fan Speed

Nozzle temperatures over the melting point are required for the filament to become sufficiently less viscous to be extruded. Thompson et al. used 316L particle-filled polymer filaments to print blade features [25]. In their study, a box geometry with

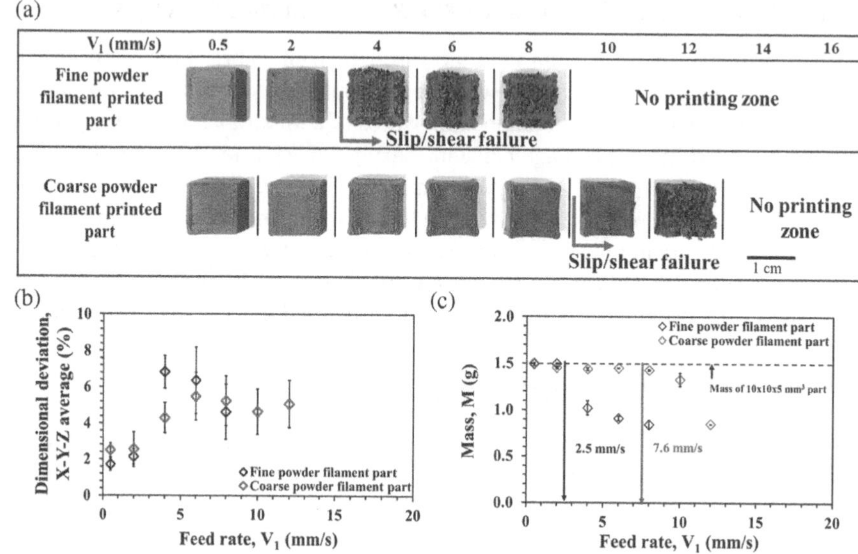

FIGURE 4.3 Effect of powder size on print quality [24].

a wall thickness of 5 mm and an exterior length of 25 mm was selected for the purpose of optimizing printing parameters. Initial printing tests revealed that, for consistently reproducible printing of the material employed, a nozzle temperature of 290°C was optimal. The viscosity of the densely loaded filament was too high for continuous extrusion at lower nozzle temperatures (less than 270°C). Higher nozzle temperatures (from 295°C) inhibited successful printing by prohibiting newly deposited strains from adhering firmly to lower print layers. The ideal temperature range of 270–290°C was found to have the best nozzle temperature.

Printing tests, as depicted in Figure 4.4, showed that an excessive fan speed produced rough surfaces and even clogged the nozzle (Figure 4.4a). Curly structures, as seen in Figure 4.4c, were produced by using a fan speed that was too low, which prevented the printed layers from cooling. The optimum fan speed was selected to minimize the defects.

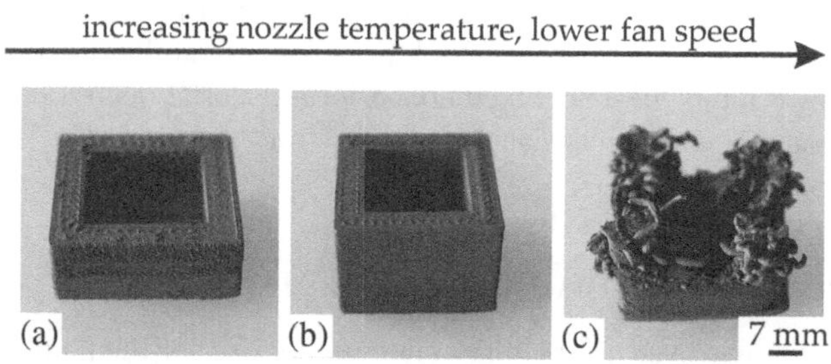

FIGURE 4.4 Printing inspections with (a) full fan speed, (b) optimized fan speed, and (c) no fan [25].

4.3.3 Effect of Infill Density and Infill Pattern

Infill density refers to the percentage of solid structure present inside the print, and infill pattern refers to the shape of the internal structures that connect the walls of the printed feature. The selection of infill pattern and infill density decides the printing time, print quality, and mechanical properties of the printed feature. Various infill patterns are illustrated in Figure 4.5. Mishra et al. investigated the effect of infill pattern and infill density on the impact energy absorbing capabilities of the composite parts and concluded that line or rectilinear infill pattern has high impact resistance [26]. The line/rectilinear infill pattern takes the shortest possible time. Further, 85% of infill density case absorbed maximum impact energy compared to other infill densities. Mamatha et al. stated that rectilinear infill with 90° filling angles gave the optimal results [27]. Though the triangular infill pattern provides high strength, the honeycomb structure gives a high strength-to-weight ratio. Compared to 50% infill density, 99% infill density showed better surface integrity, less porosity, and fewer cracks [28].

Li et al. identified that defects were much diminished and minimized when the print spacing was between 0.6 and 0.65 mm, although they could not be completely eradicated. When the raster angle was 45°/135°, there were comparatively fewer printing flaws than in other samples [5].

4.3.4 Effect of Layer Thickness

Layer thickness is defined as the height of each layer deposited one above the other. Lower layer thickness promotes the green part's superior surface finish and low porosity [29]. A histogram of layer thickness is plotted after collecting the data from various researchers [14, 25, 30–64], as shown in Figure 4.6. It is evident that most of the researchers have taken 0.4 mm as nozzle diameter and layer thickness between

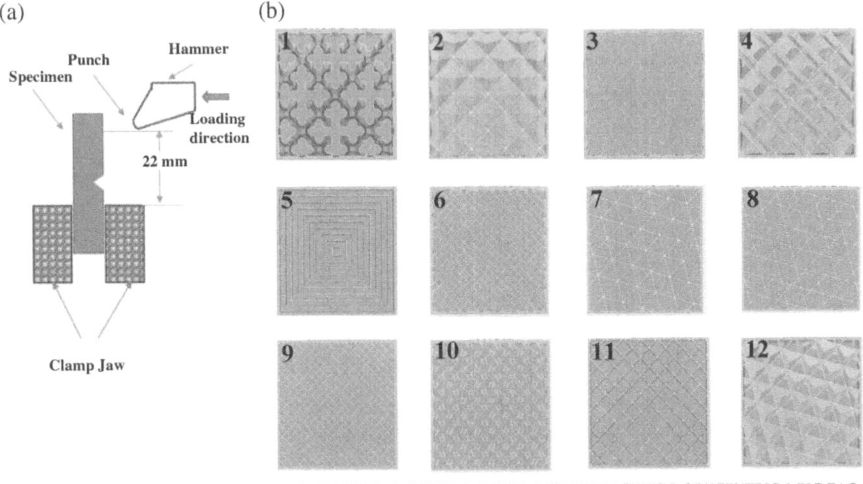

FIGURE 4.5 Various infill patterns used in 3D printing [26].

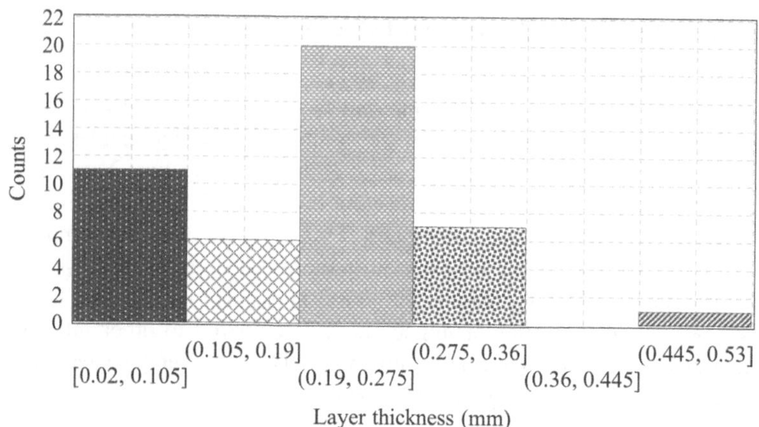

FIGURE 4.6 Ranges of layer thickness used by earlier researchers.

0.19 and 0.275 mm. It is also observed that the parts produced are superior in terms of better surface finish with 0.2 mm layer thickness. Almost 78% of the data collected is less than 0.275 mm only.

4.3.5 Effect of Extrusion Multiplier

The extrusion multiplier, or flow multiplier, indicates the amount of material coming from the nozzle. It is usually expressed as % of the theoretical discharge. Figure 4.7 indicates the various extrusion multiplier values reported by various researchers [9, 15–23, 25]. Most of them used greater than or equal to 100% multiplier. Extrusion

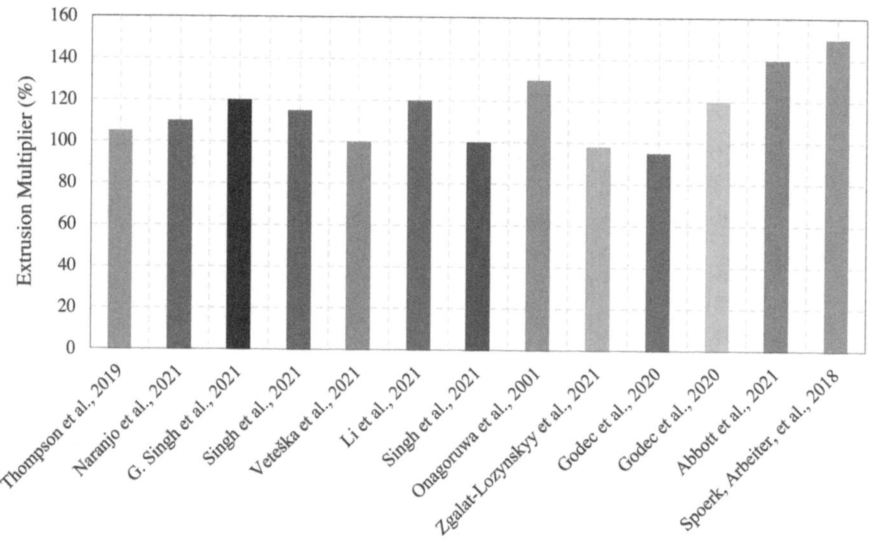

FIGURE 4.7 Extrusion multiplier reported by earlier researchers.

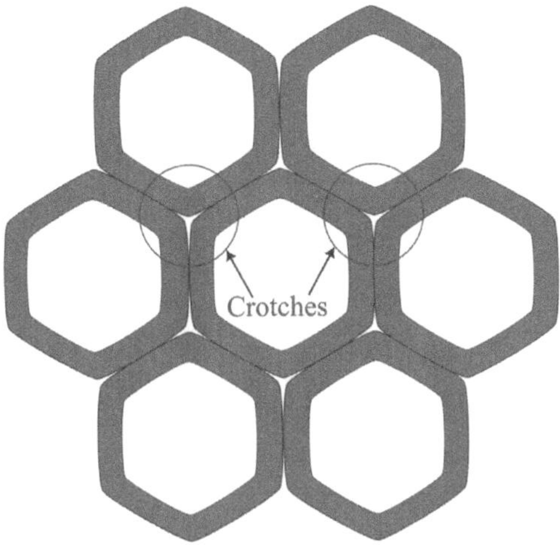

FIGURE 4.8 Schematic explaining crotch defect during infill printing.

multiplier greater than 100% means over extrusion, which means extruding the material more than the required volume per unit time.

When nozzles are clogged, the printed objects have so many gaps because of the deficiency in extruded material volume. To avoid this, an extrusion multiplier greater than 100% is recommended. When nozzles are worn out, the material coming out is more than the required volume [20]; in such cases, an extrusion multiplier of less than 100% is recommended. Higher extrusion multiplier values reduce the delamination problems. It is very important to select this parameter to avoid discontinuity in the print. Crotches are the gaps [17] formed between the adjacent layers of the given plane, as shown in Figure 4.8. These crotches can be minimized by adjusting the extrusion multiplier and adopting optimized and suitable tool path methodologies.

4.4 CHALLENGES ASSOCIATED WITH POST GREEN PART PRINTING

As mentioned in the process flow (Figure 4.1), the first printed green part would subsequently undergo de-binding and sintering. Depending on the binder used de-binding involves either solvent de-binding or thermal de-binding and/or both of the processes. The 3D printing model's shape diversity makes it challenging to rapidly predict the solvent de-binding behavior [5], and shape configuration influences solvent de-binding [65]. Solvent de-binding often consumes tens of hours to a week of processing time. There is also a lack of expertise on the de-binding behavior of the 3D-printed samples [25]. A new multi-component binder solution for manufacturing zirconia was created by Cano et al. [66]. The findings demonstrated that using the wrong binder could result in flaws, such as surface opening and cracking during solvent de-binding. According to Thompson et al. [17], for the specimen with a wall thickness of 2 mm, it took 24 hours to remove 99% of the soluble binder, while it took much longer than 57 hours for a wall thickness of 6 mm.

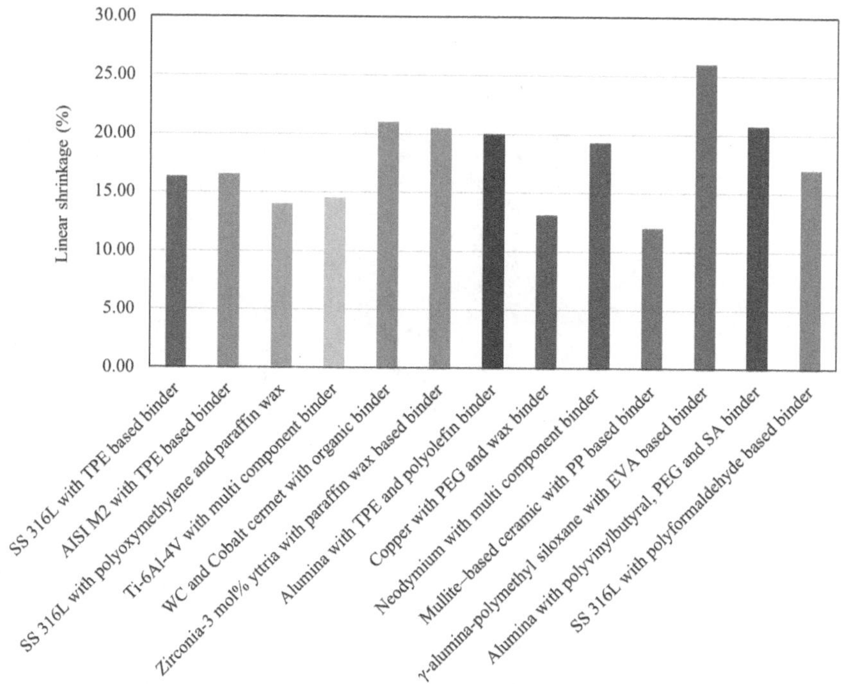

FIGURE 4.9 Different post-sintering linear shrinkage values for various materials used in metal fused deposition modeling.

After de-binding the binder, the part is sintered to attain a complete metal or ceramic part. It is evident that for every post-treatment (de-binding and sintering), the dimensions of the part shrink. Figure 4.9 explains various linear shrinkage values reported by earlier researchers for different materials. These values are significant before designing any component, otherwise resulting in dimensional inaccuracy. A bar graph of relative densities is plotted after collecting the data from various researchers [14, 25, 36, 51, 55, 67–72], as shown in Figure 4.10. Achieving greater density depends on several factors, such as proper control of input parameters, de-binding, and sintering temperatures.

4.5 MISCELLANEOUS CHALLENGES

Miscellaneous challenges refer to various additional or secondary obstacles and difficulties that may arise in a particular context or situation. These challenges are often diverse and may not fit into a specific category or have a clear classification. Such problems are classified in the following sections.

4.5.1 Multi-material Printing

Multi-material printing facilitates printing of different materials with variable printing heights, with at least two print heads. The print heads operate on and off as per the

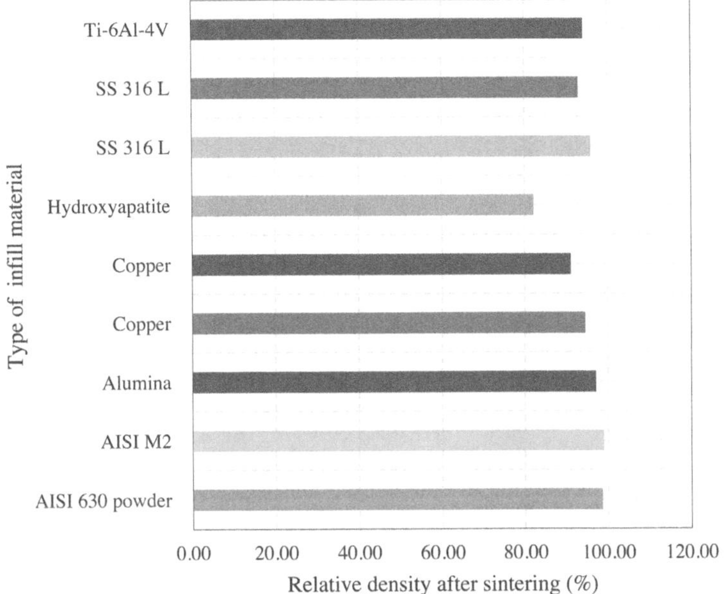

FIGURE 4.10 Relative densities of various 3D-printed composite parts after sintering.

requirements of the material transition. Whenever there is a need to change the print head to print a different material on the existing printed interface, there exists a time gap before a second print head comes into operation. It is because the second material took some time to fill in and get extruded. So, as the reason, there is no accurate and intense material transition, which results in a hideous appearance. To avoid this, small tower-like structures called 'prime pillars' are printed simultaneously with the actual components, as shown in Figure 4.11. Before printing the secondary material on the original component, a layer is first printed on the prime pillar. Prime pillar ensures the priming of the nozzle, which helps perfect continuance of material changeover [17].

Whenever the print head changes for material transition, it first prints a few layers on the 'prime pillar.' During its shift from 'prime pillar' to the actual component, it

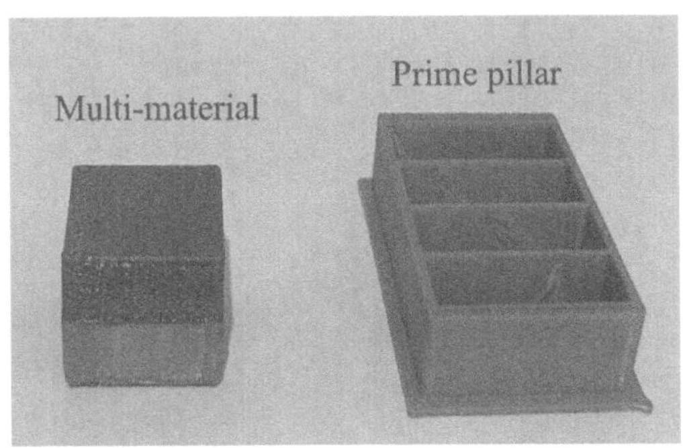

FIGURE 4.11 Multi-material printing with prime pillar.

may carry some adhering material, which hampers the precision of the component. This problem can be addressed by 'ooze-shield.' The 'ooze-shield' is a thin printed wall around the component, which can help strip off the adhered material whenever the nozzle crosses from the prime pillar to the component [17].

4.5.2 Warpage

Warpage is one of the adverse challenges in 3D printing [73]. Susceptible to warpage is dependent on the linear expansion ratio of materials. Weng et al. discussed the chances of warpage based on the linear expansion ratios. If the material has a lower linear expansion ratio, there wouldn't be much change in the volume of the component when heated and cooled to room temperature [74]. The shrinkage can also cause loss of dimensions when cooled. Better materials are those which exhibit lower thermal expansion ratios with temperatures. Figure 4.12 illustrates the warpage problem with CTE values. The surface quality of the FDM parts can also be deteriorated by shrinkage and residual stresses due to solidification and cooling [75]. Increasing the bed temperature above the glass transition temperature of the filament improves adhesion to the print bed [76].

The build platform material can influence the ease of detachability of the printed part. The bed material should be chosen carefully to prevent the curling or warpage. Spoerk et al. [76] advised that the heated bed can minimize the warpage problem. The first layer printing parameters are important for attachment and detachment problems. It is critically important that the first layer should be flawless to avoid further propagation of error to the subsequent layers. The parameter called 'print

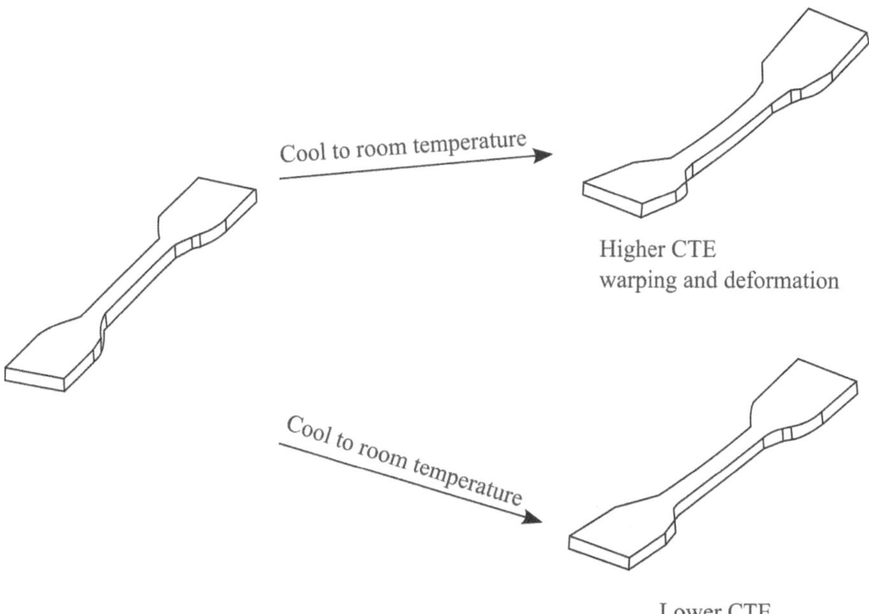

FIGURE 4.12 Effect of CTE on warpage of the printed component [74].

speed for the first layer' must be selected optimally for better adhesion. The lower values cause strong adhesion because of the availability of excess time for contact and higher side causes weaker adhesion. Nabipour et al. [77] suggested that a 6 mm thick polypropylene hatched sheet can effectively inhibit warpage by means of mechanical interlocking offered by hatching.

In some cases, the mixing of additives like plasticizers and stabilizers leads to the emission of volatile organic compounds (VOCs) during the degradation of the polymer. Potter et al. investigated samples with 90 wt.% Cu-PLA, 80 wt.% stainless steel-filled PLA, and carbon nanotubes with PLA to estimate the emissions. Stainless steel-filled PLA filament exhibited higher emissions due to the higher polymer content compared to the other combinations. Lactide acetaldehyde was gas-phase emissions from metal-containing PLA filaments during printing [78]. Polymer alone was not only responsible for emissions but also due to the catalytic activity of the metal portion. Acetaldehyde is a probable carcinogenic emission whose inhalation limit is minimal. The key challenges identified in this review are summarized in Table 4.1.

TABLE 4.1
List of Challenges Identified in Composite Filament-Based 3D Printing

S. No	Section	Key Challenges Identified	Reference
1.	Filament fabrication	Continuous winding becomes problematic as the reinforcement content in composite filament is approaching 60 vol.%.	[14]
2.	Filament fabrication	A higher solid loading prevents distortion during the de-binding process, but higher loading causes the filament surface to be brittle.	[16]
3.	Filament fabrication	The addition of metal particles to the polymer matrix may result in agglomeration unless the proper surfactants are used.	This study
4.	Printing parameters	The nozzle can be clogged due to an increased fan speed, producing rough surfaces while printing. Too low speeds result in curly structures.	[25]
5.	Printing parameters	Improper planning on extrusion multiplier can cause crotch defects.	[17]
6.	Post-printing	Choosing the wrong binder could result in flaws such as surface opening and cracking during solvent de-binding.	[25]
7.	Post-printing	CAD model should include appropriate allowances as de-binding and sintering shrink the part.	This study
8.	Miscellaneous	Multi-material printing should consider the 'prime pillar' concept to avoid poor interfacial transition.	[17]
9.	Miscellaneous	Inhaling gaseous emissions during polymer processing causes cancer.	[78]

4.6 CONCLUSION

Metal and ceramic particle-fused polymer filaments offer a very economical platform and less heat-intensive process to print metal and ceramic parts. In conclusion, metal particle-filled polymer composite 3D printing presents a promising avenue for developing advanced AM technologies. However, it also poses several challenges that must be addressed to exploit its potential fully. The challenges in metal particle-filled polymer composite 3D printing encompass various aspects, including material compatibility, particle dispersion, settling and agglomeration, printability and process optimization, post-processing, and cost scalability. These challenges require careful consideration and innovative solutions to ensure the production of high-quality and functional composite parts. Post-processing steps, such as de-binding, sintering, or surface treatment, may be required to enhance the printed objects' final properties or surface finish. Developing suitable post-processing techniques that are compatible with the composite material is crucial for achieving the desired properties [79–83]. This review presents a large database on various printing parameters, linear shrinkage values, and relative densities attained after sintering for the crucial understanding of the process overview. The key findings are observed as follows:

- Understanding the obstacles, failures, and challenges in composite 3D printing improves the quality, energy, and economy.
- Higher solid loading is required to realize fully metallic 3D prints; however, higher solid loading constrains the winding ability and flexibility of the resultant filament.
- Optimal solid loading should be chosen with suitable binder constituents without compromising flexibility and strength.
- Selecting proper printing parameters is very important to print dimensionally stable green parts.
- De-binding and sintering cycles are selected carefully with the help of proper thermal characterizations to ensure dimensional stability after sintering.

REFERENCES

1. A. Kumar, P. Kumar, R.K. Mittal, H. Singh, Preprocessing and postprocessing in additive manufacturing, in: Advances in Additive Manufacturing Artificial Intelligence, Nature-Inspired, and Biomanufacturing, Elsevier, 2023: pp. 141–165.
2. A. Kumar, P. Kumar, R.K. Mittal, H. Singh, Printing file formats for additive manufacturing technologies, in: Advances in Additive Manufacturing Artificial Intelligence, Nature-Inspired, and Biomanufacturing, Elsevier, 2023: pp. 87–102.
3. A. Kumar, P. Kumar, R.K. Mittal, V. Gambhir, Materials processed by additive manufacturing techniques, in: Advances in Additive Manufacturing Artificial Intelligence, Nature-Inspired, and Biomanufacturing, Elsevier, 2023: pp. 217–233.
4. A. Kumar, P. Kumar, H. Singh, A. Haleem, R.K. Mittal, Integration of reverse engineering with additive manufacturing, in: Advances in Additive Manufacturing Artificial Intelligence, Nature-Inspired, and Biomanufacturing, Elsevier, 2023: pp. 43–65.

5. A.K. Srivastava, S. Tiwari, P. Pachauri, N. Gupta, B. Sunil, A. Kumar, Bonding strength and microstructural features of Al5083-AZ31B alloys laminated sheet through friction stir additive manufacturing. J. Adhes. Sci. Technol. 38(4) (2024) 583–596.
6. J. Gonzalez-Gutierrez, S. Cano, S. Schuschnigg, C. Kukla, J. Sapkota, C. Holzer, Additive manufacturing of metallic and ceramic components by the material extrusion of highly-filled polymers: A review and future perspectives. Materials (Basel). 11 (2018). https://doi.org/10.3390/ma11050840
7. S.P. Tadi, S.S. Maddula, R.S. Mamilla, Sustainability aspects of composite filament fabrication for 3D printing applications. Renew. Sustain. Energy Rev. 189 (2024) 113961. https://doi.org/10.1016/j.rser.2023.113961
8. S. Aggoune, F. Hamadi, C. Abid et al. Instabilities in the formation of single tracks during selective laser melting process. Int. J. Interact Des. Manuf. (2024). https://doi.org/10.1007/s12008-024-01887-y
9. P. Ajay, S. Ahmad, J. Sharma, V. Gambhir (Eds.), Handbook of Sustainable Materials: Modelling, Characterization, and Optimization (1st ed.), CRC Press, 2023. https://doi.org/10.1201/9781003297772
10. S. Ajay, H. Parveen, B. AlMangour (Eds.), Handbook of Smart Manufacturing: Forecasting the Future of Industry 4.0 (1st ed.), CRC Press, 2023. https://doi.org/10.1201/9781003333760
11. G. Chen, S.Y. Zhao, P. Tan, J. Wang, C.S. Xiang, H.P. Tang, A comparative study of Ti-6Al-4V powders for additive manufacturing by gas atomization, plasma rotating electrode process and plasma atomization. Powder Technol. 333 (2018) 38–46. https://doi.org/10.1016/j.powtec.2018.04.013
12. C. Kukla, J. Gonzalez-Gutierrez, I. Duretek, S. Schuschnigg, C. Holzer, Effect of particle size on the properties of highly-filled polymers for fused filament fabrication, AIP Conf. Proc. 1914 (2017). https://doi.org/10.1063/1.5016795
13. J.-M. Ting, R.Y. Lin, Effect of particle size distribution on sintering. J. Mater. Sci. 30 (1995) 2382–2389. https://doi.org/10.1007/BF01184590
14. J.A. Naranjo, C. Berges, A. Gallego, G. Herranz, A novel printable high-speed steel filament: Towards the solution for wear-resistant customized tools by AM alternative. J. Mater. Res. Technol. 11 (2021) 1534–1547. https://doi.org/10.1016/j.jmrt.2021.02.001
15. S. Roshchupkin, A. Kolesov, A. Tarakhovskiy, I. Tishchenko, A brief review of main ideas of metal fused filament fabrication. Mater. Today Proc. 38 (2020) 2063–2067. https://doi.org/10.1016/j.matpr.2020.10.142
16. P. Singh, Q. Shaikh, V.K. Balla, S. V. Atre, K.H. Kate, Estimating powder-polymer material properties used in design for metal fused filament fabrication (DfMF3). JOM. 72 (2020) 485–495. https://doi.org/10.1007/s11837-019-03920-y
17. J. Abel, U. Scheithauer, T. Janics, S. Hampel, S. Cano, A. Müller-Köhn, A. Günther, C. Kukla, T. Moritz, Fused filament fabrication (FFF) of metal-ceramic components. J. Vis. Exp. 2019 (2019). https://doi.org/10.3791/57693
18. A. Kumar, P. Kumar, N. Sharma, A.K. Srivastava, 3D Printing Technologies: Digital Manufacturing, Artificial Intelligence, Industry 4.0, De Gruyter, 2024. https://doi.org/10.1515/9783111215112
19. A. Kumar, R.K. Mittal, A. Haleem, Advances in additive manufacturing artificial intelligence. Nature-Insp. Biomanufact. (2023). https://doi.org/10.1016/c2020-0-03877-6
20. G. Goyal, A. Kumar, A Gupta, 16 Recent developments in 3D printing: A critical analysis and deep dive into innovative real-world applications, in: 3D Printing Technologies: Digital Manufacturing, Artificial Intelligence, Industry 4.0, De Gruyter, 2024: p. 335.
21. G. Goyal, A. Kumar, D Sharma, 12 recent applications of rapid prototyping with 3D printing: A review, in: 3D Printing Technologies: Digital Manufacturing, Artificial Intelligence, Industry 4.0, De Gruyter, 2024: p. 245.

22. T. Ladipo, Solid additives and their lubrication effects on polyetheretherketone polymers - A review. Int. J. Eng. Res. Technol. 13 (2021) 4262–4268.
23. S. Cano, J. Gonzalez-Gutierrez, C. Kukla, S. Schuschnigg, C. Holzer, Fused filament fabrication of metals and ceramics for special applications, In: Business Digitalization. New Business Models, Smart Production and the Human side of Digitalization, Leykam, 2019: pp. 181–196.
24. P. Singh, V.K. Balla, A. Tofangchi, S.V. Atre, K.H. Kate, Printability studies of Ti-6Al-4V by metal fused filament fabrication (MF3). Int. J. Refract. Met. Hard Mater. 91 (2020) 105249. https://doi.org/10.1016/j.ijrmhm.2020.105249
25. Y. Thompson, J. Gonzalez-Gutierrez, C. Kukla, P. Felfer, Fused filament fabrication, debinding and sintering as a low cost additive manufacturing method of 316L stainless steel. Addit. Manuf. 30 (2019) 100861. https://doi.org/10.1016/j.addma.2019.100861
26. P.K. Mishra, P. Senthil, S. Adarsh, M.S. Anoop, An investigation to study the combined effect of different infill pattern and infill density on the impact strength of 3D printed polylactic acid parts. Compos. Commun. 24 (2021) 100605. https://doi.org/10.1016/j.coco.2020.100605
27. S. Mamatha, P. Biswas, P. Ramavath, D. Das, R. Johnson, Effect of parameters on 3D printing of alumina ceramics and evaluation of properties of sintered parts. J. Asian Ceram. Soc. (2021) 1–7. https://doi.org/10.1080/21870764.2021.1920159
28. S.R. Arunkumar N, N. Sathishkumar, S.S. Sanmugapriya, Study on PLA and PA thermoplastic polymers reinforced with carbon additives by 3D printing process. Mater. Today Proc. (2021). https://doi.org/10.1016/j.matpr.2021.05.041
29. I. Buj-Corral, A. Bagheri, M. Sivatte-Adroer, Effect of printing parameters on dimensional error, surface roughness and porosity of FFF printed parts with grid structure. Polymers (Basel). 13 (2021). https://doi.org/10.3390/polym13081213
30. A. Abbott, T. Gibson, G.P. Tandon, L. Hu, R. Avakian, J. Baur, H. Koerner, Melt extrusion and additive manufacturing of a thermosetting polyimide. Addit. Manuf. 37 (2021). https://doi.org/10.1016/j.addma.2020.101636
31. Y. Abe, T. Kurose, M.V.A. Santos, Y. Kanaya, A. Ishigami, S. Tanaka, H. Ito, Effect of layer directions on internal structures and tensile properties of 17-4ph stainless steel parts fabricated by fused deposition of metals. Materials (Basel). 14 (2021) 1–12. https://doi.org/10.3390/ma14020243
32. B. Liu, Y. Wang, Z. Lin, T. Zhang, Creating metal parts by fused deposition modeling and sintering. Mater. Lett. 263 (2020) 127252. https://doi.org/10.1016/j.matlet.2019.127252
33. X. Cao, S. Xuan, S. Sun, Z. Xu, J. Li, X. Gong, 3D printing magnetic actuators for biomimetic application. ACS Appl. Mater. Interfaces. 30136 (2021). https://doi.org/10.1021/acsami.1c08252
34. P. Carreira, F. Cerejo, N. Alves, M.T. Vieira, In search of the optimal conditions to process shape memory alloys (NiTi) using fused filament fabrication (FFF). Materials (Basel). 13 (2020) 1–13. https://doi.org/10.3390/ma13214718
35. M.W.M. Cunico, Investigation of ceramic dental prostheses based on ZrSiO4-glass composites fabricated by indirect additive manufacturing. Int. J. Bioprinting. 7 (2021) 90–99. https://doi.org/10.18063/ijb.v7i1.315.
36. C. Esposito Corcione, F. Scalera, F. Gervaso, F. Montagna, A. Sannino, A. Maffezzoli, One-step solvent-free process for the fabrication of high loaded PLA/HA composite filament for 3D printing. J. Therm. Anal. Calorim. 134 (2018) 575–582. https://doi.org/10.1007/s10973-018-7155-5
37. S. Esslinger, A. Grebhardt, J. Jaeger, F. Kern, A. Killinger, C. Bonten, R. Gadow, Additive manufacturing of 0973-018-7155-5rication of high loaded PLA/HA composite filament for 3D. Materials (Basel). 14 (2021) 1–14. https://doi.org/10.3390/ma14010156
38. S. Fafenrot, N. Grimmelsmann, M. Wortmann, A. Ehrmann, Three-dimensional (3D) printing of polymer-metal hybrid materials by fused deposition modeling. Materials (Basel). 10 (2017). https://doi.org/10.3390/ma10101199

39. D. Godec, S. Cano, C. Holzer, J. Gonzalez-Gutierrez, Optimization of the 3D printing parameters for tensile properties of specimens produced by fused filament fabrication of 17-4PH stainless steel. Materials (Basel). 13 (2020). https://doi.org/10.3390/ma13030774
40. J. Gonzalez-Gutierrez, D. Godec, C. Kukla, T. Schlauf, C. Burkhardt, C. Holzer, Shaping, Debinding and sintering of steel components via fused filament fabrication, 16th Int. Sci. Conf. Prod. Eng. - CIM2017. (2017) 99–104.
41. L. Gorjan, R. Tonello, T. Sebastian, P. Colombo, F. Clemens, Fused deposition modeling of mullite structures from a preceramic polymer and produced. J. Eur. Ceram. Soc. 39 (2019) 2463–2471. https://doi.org/10.1016/j.jeurceramsoc.2019.02.032
42. Q. He, J. Jiang, X. Yang, L. Zhang, Z. Zhou, Y. Zhong, Z. Shen, Additive manufacturing of dense zirconia ceramics by fused deposition modeling via screw extrusion. J. Eur. Ceram. Soc. 41 (2021) 1033–1040. https://doi.org/10.1016/j.jeurceramsoc.2020.09.018
43. D. Jiang, F. Ning, Fused filament fabrication of biodegradable PLA/316L composite scaffolds: Effects of metal particle content. Procedia Manuf. 48 (2020) 755–762. https://doi.org/10.1016/j.promfg.2020.05.110
44. D. Jiang, F. Ning, Y. Wang, Additive manufacturing of biodegradable iron-based particle reinforced polylactic acid composite scaffolds for tissue engineering. J. Mater. Process. Technol. 289 (2021) 116952. https://doi.org/10.1016/j.jmatprotec.2020.116952
45. C. Kukla, J. Gonzalez-gutierrez, C. Burkhardt, The Production of Magnets by FFF - Fused Filament Fabrication, in: Eur. 2017, Milan, Italy, 2017.
46. T. Kurose, Y. Abe, M.V.A. Santos, Y. Kanaya, A. Ishigami, S. Tanaka, H. Ito, Influence of the layer directions on the properties of 316l stainless steel parts fabricated through fused deposition of metals. Materials (Basel). 13 (2020). https://doi.org/10.3390/ma13112493
47. W. Lengauer, I. Duretek, M. Fürst, V. Schwarz, J. Gonzalez-Gutierrez, S. Schuschnigg, C. Kukla, M. Kitzmantel, E. Neubauer, C. Lieberwirth, V. Morrison, Fabrication and properties of extrusion-based 3D-printed hardmetal and cermet components. Int. J. Refract. Met. Hard Mater. 82 (2019) 141–149. https://doi.org/10.1016/j.ijrmhm.2019.04.011
48. T. Li, J. Gonzalez-Gutierrez, I. Raguž, C. Holzer, M. Li, P. Cheng, M. Kitzmantel, L. Shi, L. Huang, Material extrusion additively manufactured alumina monolithic structures to improve the efficiency of plasma-catalytic oxidation of toluene. Addit. Manuf. 37 (2021). https://doi.org/10.1016/j.addma.2020.101700
49. M. Mohammadizadeh, H. Lu, I. Fidan, K. Tantawi, A. Gupta, S. Hasanov, Z. Zhang, F. Alifui-Segbaya, A. Rennie, Mechanical and thermal analyses of metal-PLA components fabricated by metal material extrusion. Invent. 5 (2020). https://doi.org/10.3390/inventions5030044
50. F. Ning, W. Cong, J. Qiu, J. Wei, S. Wang, Additive manufacturing of carbon fiber reinforced thermoplastic composites using fused deposition modeling. Compos. Part B. 80 (2015) 369–378. https://doi.org/10.1016/j.compositesb.2015.06.013
51. D. Nötzel, T. Hanemann, New feedstock system for fused filament fabrication of sintered alumina parts. Materials (Basel). 13 (2020) 1–12. https://doi.org/10.3390/ma13194461
52. D. Nötzel, R. Eickhoff, T. Hanemann, Fused filament fabrication of small ceramic components. Materials (Basel). 11 (2018) 1–10. https://doi.org/10.3390/ma11081463
53. H.S. Patanwala, D. Hong, S.R. Vora, B. Bognet, A.W.K. Ma, The microstructure and mechanical properties of 3D printed carbon nanotube-polylactic acid composites. (2018). https://doi.org/10.1002/pc.24494
54. M. Quarto, M. Carminati, G. D'Urso, C. Giardini, G. Maccarini, Processability of metal-filament through polymer FDM machine. Esaform 2021. 13 (2021) 1–11. https://doi.org/10.25518/esaform21.2114

55. M. Sadaf, M. Bragaglia, F. Nanni, A simple route for additive manufacturing of 316L stainless steel via fused filament fabrication. J. Manuf. Process. 67 (2021) 141–150. https://doi.org/10.1016/j.jmapro.2021.04.055
56. F.S. Senatov, K.V. Niaza, M.Y. Zadorozhnyy, A.V. Maksimkin, S.D. Kaloshkin, Y.Z. Estrin, Mechanical properties and shape memory effect of 3D-printed PLA-based porous scaffolds. J. Mech. Behav. Biomed. Mater. 57 (2016) 139–148. https://doi.org/10.1016/j.jmbbm.2015.11.036
57. G. Sodeifian, S. Ghaseminejad, A.A. Yousefi, Preparation of polypropylene/short glass fiber composite as fused deposition modeling (FDM) filament. Results Phys. 12 (2019) 205–222. https://doi.org/10.1016/j.rinp.2018.11.065
58. M. Spoerk, F. Arbeiter, I. Raguž, G. Weingrill, T. Fischinger, G. Traxler, S. Schuschnigg, L. Cardon, C. Holzer, Polypropylene filled with glass spheres in extrusion-based additive manufacturing: Effect of filler size and printing chamber temperature. Macromol. Mater. Eng. 303 (2018). https://doi.org/10.1002/mame.201800179
59. K. Sukthavorn, N. Phengphon, N. Nootsuwan, P. Jantaratana, C. Veranitisagul, A. Laobuthee, Effect of silane coupling on the properties of polylactic acid/barium ferrite magnetic composite filament for the 3D printing process. J. Appl. Polym. Sci. (2021) 1–9. https://doi.org/10.1002/app.50965
60. C.M. Shemelya, A. Rivera, A.T. Perez, C. Rocha, M. Liang, X. Yu, C. Kief, D. Alexander, J. Stegeman, H. Xin, R.B. Wicker, E. MacDonald, D.A. Roberson, Mechanical, electromagnetic, and X-ray shielding characterization of a 3D printable Tungsten–Polycarbonate polymer matrix composite for space-based applications. J. Electron. Mater. 44 (2015) 2598–2607. https://doi.org/10.1007/s11664-015-3687-7
61. A.R. Torrado Perez, D.A. Roberson, R.B. Wicker, Fracture surface analysis of 3D-printed tensile specimens of novel ABS-based materials. J. Fail. Anal. Prev. 14 (2014) 343–353. https://doi.org/10.1007/s11668-014-9803-9
62. C. Tosto, J. Tirillò, F. Sarasini, G. Cicala, Hybrid metal/Polymer filaments for fused filament fabrication (FFF) to print metal parts. Appl. Sci. 11 (2021) 1444. https://doi.org/10.3390/app11041444
63. B.E. Yamamoto, A.Z. Trimble, B. Minei, M.N. Ghasemi Nejhad, Development of multifunctional nanocomposites with 3-D printing additive manufacturing and low graphene loading. J. Thermoplast. Compos. Mater. 32 (2019) 383–408. https://doi.org/10.1177/0892705718759390
64. O.B. Zgalat-Lozynskyy, O.O. Matviichuk, O.I. Tolochyn, O.V. Ievdokymova, N.O. Zgalat-Lozynska, V.I. Zakiev, Polymer materials reinforced with silicon nitride particles for 3D printing. Powder Metall. Met. Ceram. 59 (2021) 515–527. https://doi.org/10.1007/s11106-021-00189-2
65. M. Coffigniez, L. Gremillard, M. Perez, S. Simon, C. Rigollet, E. Bonjour, P. Jame, X. Boulnat, Modeling of interstitials diffusion during debinding/sintering of 3D printed metallic filaments: Application to titanium alloy and its embrittlement. Acta Mater. 219 (2021) 117224. https://doi.org/10.1016/j.actamat.2021.117224
66. S. Cano, J. Gonzalez-Gutierrez, J. Sapkota, M. Spoerk, F. Arbeiter, S. Schuschnigg, C. Holzer, C. Kukla, Additive manufacturing of zirconia parts by fused filament fabrication and solvent debinding: Selection of binder formulation. Addit. Manuf. 26 (2019) 117–128. https://doi.org/10.1016/j.addma.2019.01.001
67. P. Veteška, Z. Hajdúchová, J. Feranc, K. Tomanová, J. Milde, M. Kritikos, Ľ. Bača, M. Janek, Novel composite filament usable in low-cost 3D printers for fabrication of complex ceramic shapes. Appl. Mater. Today. 22 (2021) 100949. https://doi.org/10.1016/j.apmt.2021.100949
68. S. Onagoruwa, S. Bose, A. Bandyopadhyay, Fused deposition of ceramics (FDC) and composites. Int. Solid Free. Fabr. Symp. (2001) 224–231. http://dx.doi.org/10.26153/tsw/3267

69. G. Singh, J.M. Missiaen, D. Bouvard, J.M. Chaix, Copper additive manufacturing using metal injection moulding feedstock: Adjustment of printing, debinding, and sintering parameters for processing dense and defectless parts. Int. J. Adv. Manuf. Technol. (2021). https://doi.org/10.1007/s00170-021-07188-y
70. G. Singh, J.M. Missiaen, D. Bouvard, J.M. Chaix, Copper extrusion 3D printing using metal injection moulding feedstock: Analysis of process parameters for green density and surface roughness optimization. Addit. Manuf. 38 (2021) 101778. https://doi.org/10.1016/j.addma.2020.101778
71. P. Singh, V.K. Balla, A. Gokce, S.V. Atre, K.H. Kate, Additive manufacturing of Ti-6Al-4V alloy by metal fused filament fabrication (MF3): Producing parts comparable to that of metal injection molding. Prog. Addit. Manuf. (2021). https://doi.org/10.1007/s40964-021-00167-5
72. C. Suwanpreecha, P. Seensattayawong, V. Vadhanakovint, A. Manonukul, Influence of specimen layout on 17-4PH (AISI 630) alloys fabricated by low-cost additive manufacturing. Metall. Mater. Trans. A Phys. Metall. Mater. Sci. 52 (2021) 1999–2009. https://doi.org/10.1007/s11661-021-06211-x
73. P. Mencik, V. Melcova, S. Kontarova, R. Prikryl, D. Perdochova, M. Repiska, Biodegradable composite materials based on poly(3-hydroxybutyrate) for 3D printing applications. Mater. Sci. Forum. 955 (2019) 56–61. https://doi.org/10.4028/www.scientific.net/MSF.955.56
74. Z. Weng, J. Wang, T. Senthil, L. Wu, Mechanical and thermal properties of ABS/montmorillonite nanocomposites for fused deposition modeling 3D printing. Mater. Des. 102 (2016) 276–283. https://doi.org/10.1016/j.matdes.2016.04.045
75. A. Armillotta, Assessment of surface quality on textured FDM prototypes. Rapid Prototyp. J. 12 (2006) 35–41. https://doi.org/10.1108/13552540610637255
76. M. Spoerk, J. Gonzalez-Gutierrez, J. Sapkota, S. Schuschnigg, C. Holzer, Effect of the printing bed temperature on the adhesion of parts produced by fused filament fabrication. Plast. Rubber Compos. 47 (2018) 17–24. https://doi.org/10.1080/14658011.2017.1399531
77. M. Nabipour, B. Akhoundi, A. Bagheri Saed, Manufacturing of polymer/metal composites by fused deposition modeling process with polyethylene. J. Appl. Polym. Sci. 48717 (2019) 1–9. https://doi.org/10.1002/app.48717
78. P.M. Potter, S.R. Al-Abed, F. Hasan, S.M. Lomnicki, Influence of polymer additives on gas-phase emissions from 3D printer filaments. Chemosphere. 279 (2021). https://doi.org/10.1016/j.chemosphere.2021.130543
79. C. Mahatme, J. Giri, F. Mohammad, M.S. Ali, T. Sathish, N. Sunheriya, R. Chadge, Experimental and numerical investigation of PLA based different lattice topologies and unit cell configurations for additive manufacturing. Int. J. Adv. Manufact. Technol. (2024). https://doi.org/10.1007/s00170-024-13882-4
80. J. Giri, T. Sathish, T. Sheikh, N. Sunehriya, P. Giri, R. Chadge, C. Mahatme, A Parthiban, Automatic liver segmentation using U-net deep learning architecture for additive manufacturing. Interactions. 245(1) (2024). https://doi.org/10.1007/s10751-024-01927-9
81. J. Giri, N. Sunheriya, T. Sathish, Y. Kadu, R. Chadge, P. Giri, A. Parthiban, C. Mahatme, Optimization of process parameters to improve mechanical properties of fused deposition method using Taguchi method. Interactions. 245(1) (2024). https://doi.org/10.1007/s10751-024-01925-x
82. M.S. Tufail, J. Giri, E. Makki, T. Sathish, R. Chadge, N. Sunheriya, Machinability of different cutting tool materials for electric discharge machining: A review and future prospects. AIP Advances. 14(4) (2024). https://doi.org/10.1063/5.0201614
83. N. Praveen, S.K. NG, C.D. Prasad, J. Giri, I. Albaijan, U.S. Mallik, T. Sathish, Effect of pulse time (ton), pause time (Toff), peak current (Ip) on MRR and surface roughness of Cu–Al–Mn ternary shape memory alloy using wire EDM. J. Mater. Res. Technol. (2024). https://doi.org/10.1016/j.jmrt.2024.03.122

5 Effects of WAAM Process Parameters on Mechanical and Surface Integrity

A Comprehensive Review

Pardeep Kumar, Gyander Ghangas, and Sunil Dhull

5.1 INTRODUCTION

The manufacturing technology known as additive manufacturing, or 3D printing, has revolutionized how products are created, prototyped, and developed [1]. With additive manufacturing, objects are constructed layer by layer from digital design data as opposed to traditional subtractive manufacturing techniques, which involve cutting, drilling, or machining raw materials to achieve the desired shape. A digital 3D model produced with the help of design software such as CATIA and Solid Works or acquired through 3D scanning serves as the starting point of additive manufacturing [2]. This virtual model is alienated into thin layers, termed as slicing, each of which is printed separately or stacked one above the other. During the printing process, material is deposited or solidified using a variety of methods, most frequently employing liquid resin, powdered metal or polymer, or even biological tissue [3]. The ability to generate problematic interior edifices and multifaceted geometries that are challenging or impossible to manufactured using conservative manufacturing techniques is one of the major remunerations of additive manufacturing [4]. The creation of highly customized and optimized parts is made possible by this design freedom, which enhances usefulness and performance. Additive manufacturing has a number of significant advantages [5]. Rapid prototyping is made possible by it, which cuts down on the time and money needed to develop new goods. Additionally, it facilitates just-in-time production and on-demand manufacturing, which both do away with the need for substantial inventory storage. In addition, additive manufacturing might enable decentralized production by enabling local or even on-site printing of things, which lowers the cost of logistics and shipping.

There are numerous and diverse industries where additive manufacturing is used. It is extensively employed in a variety of industries, as well as the automobile, aerospace, healthcare, consumer goods, and architecture [6]. By democratizing

 DOI: 10.1201/9781032725086-6

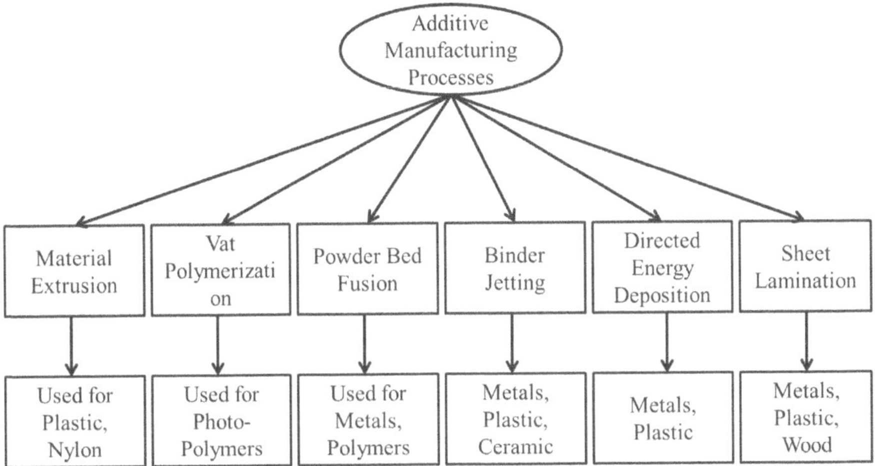

FIGURE 5.1 Numerous types of process used in additive manufacturing.

production capabilities, additive manufacturing has the ability to transform supply chains, enable mass customization, and promote innovation.

Based on the method or procedure used to manufacture objects layer by layer, additive manufacturing can be generally divided into a number of types [7, 8]. Here are a few standard categories of additive manufacturing as shown in Figure 5.1.

Material extrusion: Material is deposited using an extruder or nozzle in this technique called material extrusion. It is comparable to how a glue gun is used. Examples of the material extrusion are direct ink writing and fused deposition modeling.

Vat polymerization: Using light sources like lasers or digital light processing (DLP) projectors, a liquid photopolymer resin is selectively cured or hardened in this process. Vat polymerization developments take account of Stereolithography (SLA) and DLP, for instance.

Powder bed fusion: This method employs a bed of powdered material, such as metal or polymer, and uses laser as heat source for melting of selected powder particles together. Powder bed fusion techniques that are frequently used include selective laser sintering (SLS) and selective laser melting (SLM).

Binder jetting: To bind together thin layers of powdered material, a liquid binder is selectively placed onto the layers. Applications involving metal or sand casting frequently use this approach.

Material jetting: The process of "material jetting," which uses print heads to deposit liquid material in droplets that are later cured or solidified, is similar to inkjet printing. It is compatible with many different types of materials and enables high-resolution printing.

Sheet lamination: Using adhesive, ultrasonic welding, or other techniques, sheet lamination involves joining layers of material together. It can be carried out using supplies like paper, plastic, or metal foils.

Directed energy deposition: In this technique, material is melted or fused as it is being dropped using a focused energy source, such as an electron or laser beam. For the additive manufacturing of large, multifarious items or for repairs, it is frequently utilized [9].

5.1.1 Wire Arc Additive Manufacturing (WAAM) Process

Wire arc additive manufacturing (WAAM) is a pioneering method manufacturing that uses an electric arc welding process to construct three-dimensional things layer by layer. In terms of speed, material accessibility, and scalability, it is a kind of directed energy deposition (DED) technique that has special benefits [10]. A wire electrode is continually fed into an electric arc welding torch as shown in Figure 5.2 during the WAAM process, where the arc creates an electric current that melts the wire electrode between the wire and the work piece. Layers of the molten metal are accurately deposited onto the work piece, gradually accumulating to produce the desired object. Robotic systems are generally used to automate the procedure and ensure precise and reproducible deposition. The concept of using arc welding for additive manufacturing purposes dates back to the 1920s. However, the modern development of WAAM can be traced back to the 1990s when researchers began exploring the feasibility of using arc welding techniques for building up metal parts layer by layer. In the early 2000s, several research institutes and universities began conducting experiments and developing WAAM technology. These early studies focused on various aspects such as process optimization, control systems, and material selection. Around the mid-2000s, WAAM started gaining attention from the industry as a potential cost-effective method for producing large-scale metal components. Industries such as aerospace, automotive, and maritime recognized the potential benefits of WAAM in terms of reduced material waste, squatter lead times, and subordinate costs compared to outmoded manufacturing methods. Over

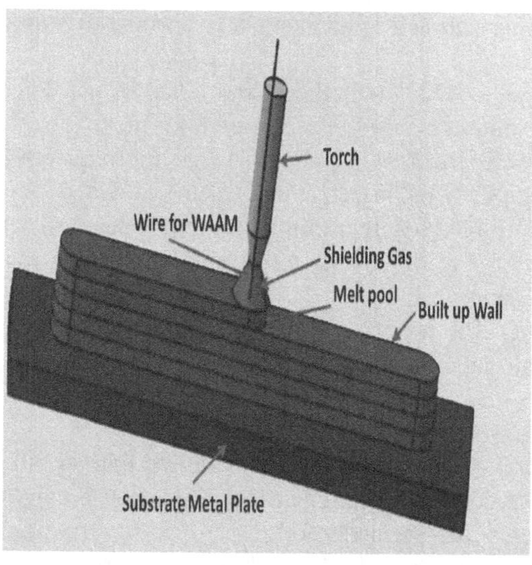

FIGURE 5.2 Schematic setup of working of WAAM process.

the years, advancements in technology and equipment have improved the precision, control, and reliability of WAAM. This has expanded its applications to various industries, including aerospace, defense, energy, and tooling. WAAM has been used to fabricate components such as aircraft parts, molds, dies, and repair or modify existing metal structures. Research and development in WAAM continue to evolve, focusing on improving the quality and performance of printed parts, expanding the range of printable materials, and enhancing process efficiency. The future of WAAM is expected to involve the integration of other manufacturing processes, such as machining and surface finishing, further enhancing the capabilities and application potential of this additive manufacturing technique [11].

The primary step in any additive manufacturing process is designing the object you want to create. This typically encompasses using CAD software to create 3D prototypes. Once the design is ready, the next step is to prepare the manufacturing setup. This involves setting up the WAAM machine, which typically consists of a robotic arm or gantry system, a power source, and a wire feedstock. In the WAAM technique, a metal is used as the feedstock in the form of wire. The specific metal chosen depends on the preferred properties of the final object. Common metals preferred in WAAM process are steel, aluminum, titanium, and nickel-based alloys. In WAAM process, material is deposited layer by layer by locating the robotic arm or gantry system over the build platform. An arc is generated between the wire feedstock and substrate plate with the help of electricity for melting the wire. The robotic arm or gantry system moves the melting wire in a controlled manner. The motion control ensures precise deposition and adherence to the design specifications. As individually layer is dropped, it rapidly cools and solidifies, bonding with the previous layers. This layer-by-layer approach allows complex geometries to be built up gradually. After the object is completely produced, the post-processing operation is required for final parts. This can include removing of support material of the structures, surface finishing, and the heat treatment process, or supplementary machining processes to accomplish the desired final properties and surface quality.

5.1.2 The Classification of WAAM

Different processing categories of additive manufacturing can be made based on the heat source. Common heat sources include electron beams, electric arcs, and lasers. Depending on the type of heat source employed, there are three main groups of metal additive manufacturing that can be commonly used: first, laser additive manufacturing (LAM); second, WAAM; and third, electron beam additive manufacturing (EBAM) (Figure 5.3).

5.1.2.1 Gas Metal Arc Welding (GMAW)

Gas metal arc welding (GMAW) is a typical additive manufacturing technique in which an electric arc is generated between electrode and the work piece. A consumable wire of metal is used as filler material in the shielding gas as shown in Figure 5.4a. This method is frequently used to connect metals in a variety of industries. Additionally, GMAW can be utilized in additive manufacturing, often known as GMAW-based additive manufacturing or GMAW-AM. In this application,

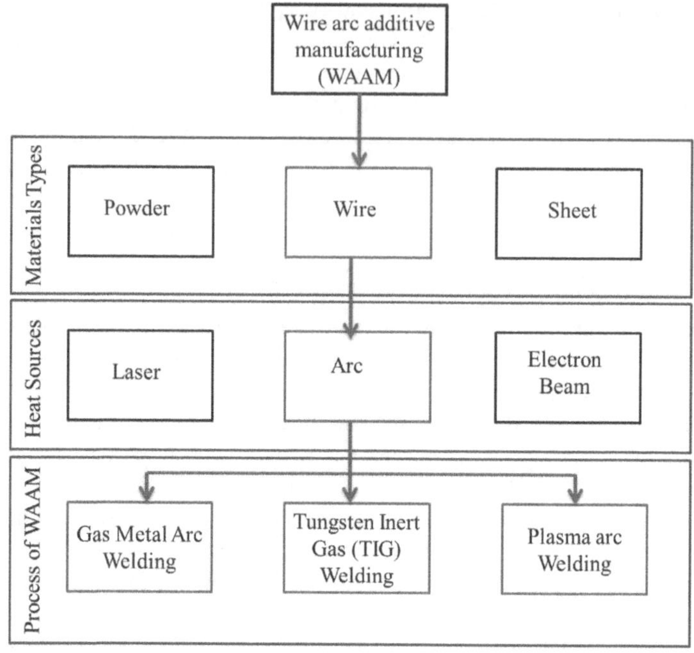

FIGURE 5.3 The cataloging of process used in WAAM.

GMAW is used to construct three-dimensional objects layer by layer rather than just for joining. To construct intricate geometries, the procedure includes depositing molten metal onto a substrate or previously applied layers.

5.1.2.2 Gas Tungsten Arc Welding (GTAW)

Tungsten inert gas (TIG) welding technique, as shown in Figure 5.4b, is a type of welding in which an electric arc is produced between the electrode and the work piece by means of a non-consumable tungsten electrode and an inactive shielding gas such as argon with 99.99% of purity. GTAW is widely used for material of non-ferrous in nature, such as Al, stainless steel, and copper alloys. Similar to GMAW-based additive manufacturing, in this application, GTAW is used to build three-dimensional things layer by layer. But there are some clear distinctions between the two procedures.

5.1.2.3 Plasma Arc Additive Manufacturing (PAAM)

In plasma arc additive manufacturing (PAAM) process, plasma arc is used as the heat source in this manufacturing process; that's why it is known as PAAM, as shown in Figure 5.4(c), which melts and deposits metal powders one layer at a time. To produce complex metal structures, PAAM incorporates features of both additive manufacturing and welding. PPAAM provides many benefits over laser and EBAM together with high welding speed, full density, and truncated cost. A process approach that associates the advantages of melting and non-melting electrodes is

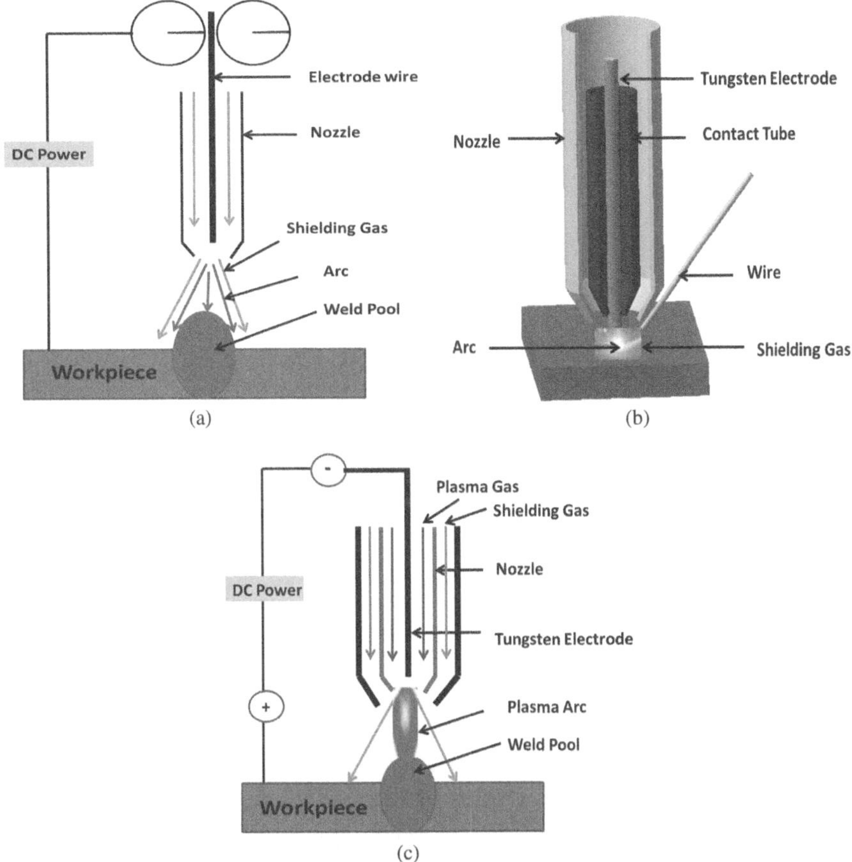

FIGURE 5.4 Schematic illustration of (a) GMAW, (b) GTAW, and (c) PAW processes.

plasma arc additive manufacturing. It is significant and beneficial for several kinds of scientific research.

5.2 WAAM MATERIAL

As a generalization, all the material that is accessible in the form of wire can be used for parts manufacturing by WAAM process. There are diverse filler metals that can be used for WAAM, such as alloy steel; aluminum; Inconel alloys such as 625, 718, and X-750; chrome-nickel; steel; or bronze [8].

Steel: Different alloys of steel, such as carbon steel, stainless steel, and tool steel, are widely used in WAAM. Steel offers good strength, durability, and versatility, making it appropriate for extensive range of industrial solicitations.

Titanium: Titanium and its numerous alloys are used in WAAM due to their outstanding mechanical characteristics such as corrosion resistance, strength-to-weight ratio, and biocompatibility. These properties of titanium mark its ideal for aerospace, medical, and automotive applications.

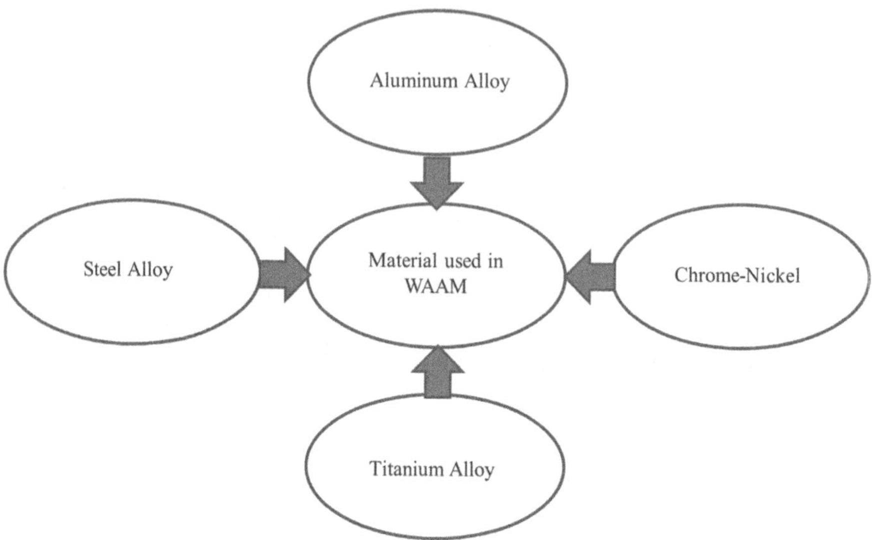

FIGURE 5.5 Different materials for WAAM process.

Aluminum: Aluminum and its alloys are commonly used in WAAM due to their lightweight nature with extraordinary thermal and electrical conductivity and good strength. Aluminum components find applications in aerospace, automotive, and other industries.

Nickel-based alloys: Alloys like Inconel and Hastelloy, which contain a significant amount of nickel, are often used in WAAM. These alloys offer excellent resistance to high temperatures, corrosion, and oxidation, making them appropriate for applications in aerospace, power generation, and chemical processing.

Copper: Copper and its alloys are used in WAAM when high thermal and electrical conductivities are required. Copper components find applications in electrical and electronic industries, heat exchangers, and cooling systems (Figure 5.5).

5.3 LITERATURE REVIEW

WAAM's ability to create large-scale, reasonably priced, and high-quality metal parts has drawn a lot of attention. Understanding the fundamental ideas, fields of study, and most current advancements in the meadow of WAAM should be made easier with the help of these carefully chosen references. WAAM has the potential to completely transform production processes across a range of industries, and research is still being done to address its limitations and broaden its uses. I'll give a quick overview of some important ideas and current advancements in WAAM through the literature below.

The results show that PAP-WAAM formed greater geometric features and microstructures than GT-WAAM and that it also had greater tensile strength. In order to examine the molten pool's size, the temperature differences between layers, and the cooling rates during deposition, a three dimensional transient finite element model was developed [12]. Researchers developed a direct energy deposition process for super alloy Inconel 718 by using plasma arc process. The deposited material in the form of wall structure was tested for both tensile strength and microstructure in both deposited conditions and normal heat treatment for the alloys. From the results it was understood that deposited structure materials showed significant cracking [13]. The researcher developed a functional graded material with the help of stainless steel of grade 316l with Inconel alloy of 625 by using diverse deposition techniques like Twin-WAAM (twin-WAAM). Microscopic examination revealed that functionally graded material (FGM) material transitions could be both direct and smooth. According to all of the data, FGM with direct interaction has superior qualities in terms of greater strength and elongation following failure [14]. Researchers focused on the microstructural representation of the urbanized sample by GWAM-based WAAM process of different materials like bimetallic, austenitic stainless steel and mild steel. Charpy impact (ASTM D6110) test was also conducted for the toughness of the deposited samples. For the residual stresses, a deep hole drilling (DHD) technique was also abandoned. According to the findings, austenitic stainless steel material experiences tensile strains while mild steel is subject to compressive stresses [15]. The researcher focused on the effect of print orientation on the mechanical properties and surface integrity of sample manufactured a thin wall structure of Inconel 625 material. The results show that component finished by GTAW had good quality [16]. Researcher focused on the microstructure and properties of samples deposited of Inconel 718 by WAAM process. Slot milling machining operations were also studied on the manufactured parts with Inconel 718 in this research. The WAAM alloys have greater properties to those that are manufactured by casting alloy and comparable to those of other WAAM techniques that are using the CMT process [17]. Researchers have concentrated on the strategy for arguing that additive manufacturing is a better option than traditional industrial machining for medium-sized aeronautical parts [18]. The microstructure and corrosion resistance of WAAM-fabricated parts of bimetallic materials were the main research areas. By using a single feeding wire, samples were created. The findings demonstrate that the bimetallic material Q345/308 via additive manufacturing sample had a higher level of toughness than the Q345 or 308 single wire feeding manufacture. The Q345 sample suffered severe electrochemical corrosion [19]. The mechanical characteristics, shape, and micro-porosity flaws of the sample prepared by WAAM are studied by the researcher. The WAAM sample was also exposed to tensile tests and hardness tests. They also have identified the mechanism of micro-porosity closure characterization of the samples by X-ray computed tomography (XCT) and finite element method (FEM) [20]. Researchers investigate the mechanical properties that comprise the hardness and tensile properties of sample manufactured by Inconel 625 alloy at different temperatures ranging from 750 to 1200°C. Different layers of the microstructure of the specimen were also studied. The experiments were carried out with diverse heights among each layer and diverse travel speeds until the optimum process parameters were obtained [21]. The consequence of progression of variables on the deposited wall of 308L steel, including voltage, welding travel speed, and welding current, was studied by the researchers. The constructed wall of the 308L steel was also examined for its microstructure and mechanical properties [22]. The cold metal transfer WAAM technique was studied by researchers to determine how the magnetic field affects the mechanical characteristics and the microstructure of Inconel 625 [23]. Nikon Epiphot 300 optical microscope was used for different forms of phase transformations. Higher bypass current's impact on the ER50-6 steel's behavior during rectification was researched [24]. Researchers investigate the mechanical properties and surface integrity of structured wall made by Inconel 718 and Inconel 625 material. WAAM process for creep resistant was used in this research. From the results it was found that Inconel 718 confirmed greater strength with a smaller amount of elongation than Inconel 625. IN718 material parts manufactured by WAAM had a higher ultimate tensile strength and yield strength performance as compare to Inconel 625 [25]. Researchers focused on the two primary consequences of oxidation of aluminum alloy by WAAM. Aluminum part surfaces that had undergone oxidation as well as oxidation abnormalities were noticed and examined. The findings demonstrate that throughout manufacturing, the aluminum's surface oxidation changed its color from translucent to white. With a decreasing gas flow rate, researchers used the WAAM technique to create four walls. Throughout WAAM, the melt pool was seen and examined with a Cavitar C200 [26]. On the basis of heat input or cooling rate, the residual stresses intensified [27]. Researchers investigate the consequence of deposition resolution on the mechanical properties and microstructure of parts printed of Inconel 625 and 316 steels deposited by CMT technique. The findings demonstrated that the quality of laves stage in Inconel 625 improved as the number of layers deposited to produce parts. The contact was flawless when 316L and Inconel 625 were deposited first and second, respectively [28]. The microstructure and mechanical characteristics such as hardness and strength of the Inconel 625 alloy-fabricated parts were studied. Results show that the ultimate tensile strength of Inconel alloy escalated from 647 MPa to 687 MPa and yield strength greater than before from 376 MPa to 400 MPa. Results indicated that the CMT-WAAM technique produced mechanical properties more effectively than casting Inconel 625 alloy, which was not the case [29]. The results obtained demonstrate the fabrication of a solitary-pass multilayer component with two different types of columnar grains with unique origins and textures. The components' cross-sectional micro-hardness is not constant from the bottom to the top. The material characteristics of steel items made by the WAAM technique are deliberate in terms of monotonic and cyclic behavior. For cyclic material properties, a monotonic tensile test was carried out. Measurements and comparisons were also made of the hardness and the residual strains created during manufacturing [30]. The specimens manufactured from Inconel 718 using WAAM were studied for their mechanical characteristics. To study the microstructure of manufactured parts, SEM and X-ray diffraction were utilized. The obtained results demonstrate that the wrought alloy IN718 has lower hardness and yield strength than the WAAM IN718 alloy [31]. The influence of an auxiliary gas-metal-arc-based WAAM approach was studied for the process parameter arc current variety from 100 to 180 A. Rendering to research findings, the material deposition of an auxiliary GMA-WAAM approach was approx. 1.7 times higher than that of ordinary process at the same deposition current. The area of the bed's heat-affected zone shrank by at least 28.6% [32]. The WAAM technology with CMT was studied by researchers to check the effect of process parameter and heat treatment process on the deposited structure in the shape of a wall. To analyze the process stability on the Inconel 718 wall structure made of a Ni-based alloy, electrical transients and melt pool imaging process were utilized. It can be concluded from the results that WAAM with CMT of IN718 has a large melt pool size and a low as-deposited hardness [33]. The methodical investigation of the effects of the three different arc modes under consideration on the aluminum alloy AlMg5Mn material deposited by GMAW process was the main emphasis of the research. The difficulties in manufacturing, such as porosity and subpar mechanical qualities, were also examined by the researchers. Additionally, the researches display the material's mechanical characteristics, such as its hardness and tensile strength [34]. Researchers focused on the characteristics and surface roughness of the post heat-treated parts by wire arc manufacturing of P91 steel. Results obtained indicate that the steel with grade P91 manufactured by WAAM process has a significantly greater strength than conventional processed wrought material. Very fine martensite and rational previous austenite grain size were used to optimize the microstructure. Excellent mechanical qualities are displayed by WAAM parts of P91 steel, which has UTS of 774 MPa, YS of 686 MPa, and ductility of 19.4% [35]. Researchers reconnoiter the effect of the cooling rate in the surface integrity and behavior of the manufactured parts by WAAM of Ti-6Al-4V alloy. Researchers regulate the cooling rate of deposited material by inter-pass temperature control. The micro- and macrostructure of manufactured Ti-6Al-4V were characterized by using SEM and EBSD [36]. Researchers concentrated on the mechanism of microstructural progression and phase drizzle of Inconel 718 thin wall manufactured by plasma arc additive manufacturing technology. Comparing the fabricated samples to castings of Inconel 718 alloy, the test results showed that the manufactured samples had much greater tensile strength and exceptional ductility. Some common heat treatment procedures were employed to improve the tensile strength and reduction the ductility [37]. The impact of heat treatment process on the mechanical characteristics and microstructure of Inconel 718 produced parts was the main research topic. In inter-dendritic zones, WAAM Inconel 718 was prone to micro-segregation, which brought about the production of undesired laves phases. The results show that the improved heat treatment also reduced grain structural anisotropy. Specialized post-deposition heat treatments that take into consideration WAAM IN718's unique microstructure were necessary to achieve the desired mechanical properties [38]. Researchers investigate the influence of WAAM parameters on the hardness and the tensile strength of parts prepared by CMT-WAAM. According to the results, castings manufactured of the Inconel 625 alloy did not match the mechanical performance of parts created by CMT-WAAM [39]. The microstructural properties of Inconel 625 fabricated parts manufactured using the WAAM process are being studied by researchers to regulate the influence of heat-treatment interval. The results of this study make it easier to choose the best heat treatment settings for WAAM Inconel 625 material. Inconel 625's microstructure was primarily made up of columnar dendrites [40]. From Table 5.1, we can take an overview of the literature analysis and investigation of how progression constraints demonstrate a momentous role to distress the mechanical characteristics and surface quality of the parts manufactured by cold metal transfer process and WAAM with different materials.

TABLE 5.1

Overview of the Literature Analysis and Investigation of How Process Variables Affect Mechanical Characteristics and Surface Quality

Sl No.	Material Used	Substrate Material	Technique Used/Heat Source	Finding/Outcomes	Process Parameters	Ref.
1	Ti6Al4V wire (diameter = 1.2 mm)	Ti6Al4V	Pulsed arc plasma WAAM (PAP-WAAM)	Parts produced by pulsed plasma arc have greater tensile strength as compare to other technique.	Current, voltage, and wire feed speed	[12]
2	Inconel 718 (diameter = 1.2 mm)		Pulsed arc plasma WAAM (PAP-WAAM)	Inconel 718's microstructure and tensile properties.	Current, wire feed speed, and torch travel speed (m/min)	[13]
3	316L and Inconel 625	Mild steel	GT-WAAM	Comparison of the properties of functional graded materials	Wire feed speed, current, and torch travel speed	[14]
4	Mild steel (G3Si1), austenitic stainless steel (SS304)	Mild steel plate (S235JR)	GMAW-based WAAM	The residual stresses and mechanical properties of bimetallic materials.	Travel speed, voltage, current, and shielding gas	[15]
5	Inconel 625	Q235 steel	GTAW with hot wire	Bottom to the top microstructure and bottom to the top hardness	Wire feed speed, travel speed, current, and shielding gas	[41]
6	Inconel 718 super alloy 1.2 mm	Wrought plate	PAW-WAAM	Fine dendrite microstructure and better mechanical properties	Deposition rate, voltage, current, and wire feed rate	[42]
7	Stainless steel 316, Inconel 718, titanium 6Al4V, and aluminum 5356		PAGM	Optimized process parameters for different materials.	Wire feed rate, torch travel speed, and shielding gas	[43]
8	Q345 and 308	Q345 steel	Hot wire GTAW	Macroscopic and microstructure observation Hardness analysis Corrosion resistance	Current (A) Wire feed speed (mm/min) Wire feeding mode	[19]

9	AA2196 Al-Li alloy	Aluminum alloy	TIG	Mechanism of micro-porosity closure	Current Deposited layer thickness Wire feed speed Wire feed angle	[20]
10	Inconel 625 alloy	Q235 steel	Multi-axis robot	Temperature effect on hardness and on the microstructure	Wire feed rate Welding speed Current Voltage Time of pass Total number of layers Height of each Layer	[21]
11	308L stainless steel	SS400 steel	GMAWAM process	The tensile strength of GMAWAM 308L	Current, voltage, and travel speed	[22]
12	Inconel 625 super alloy	Q345 steel plates	CMT-WAAM	The effect of magnetic field on microstructure and mechanical properties were investigated	Current (I) Voltage (V) Torch travel speed Wire feed rate (m/min) Contact angle	[23]
13	ER50-6 Steel	Q235	A-W GTAW	Macroscopic morphology observation Hardness and nano-indentation characterization	Current (A) Welding speed Shielding gas rate	[24]
14	Inconel 718 and Inconel 625	Mild steel plate	Plasma-transferred arc process, six-axis robot	Comparisons of the macrostructures of both the materials	Wire diameter (mm) Current Travel speed and wire feed speed	[25]

(Continued)

TABLE 5.1 *(Continued)*

Overview of the Literature Analysis and Investigation of How Process Variables Affect Mechanical Characteristics and Surface Quality

Sl No.	Material Used	Substrate Material	Technique Used/Heat Source	Finding/Outcomes	Process Parameters	Ref.
15	AW4043/AlSi5 (wt%), wire diameter 1.2 mm	AlSi1MgMn	Six-axis robot with CMT	Surface oxidation and oxidation anomalies were investigated Parameters influencing oxidation anomalies	Wire feed speed Travel speed Mode of welding CMT+P mode CMT	[26]
16	10NiMnMoCr8-7-6	Steel S690QL	Six-axis robot with WAAM	Residual stress analysis Micro-section of the weld metal Layer and component dimension	Welding speed Wire feed rate	[27]
17	316L and Inconel 625	304 Stainless steel	CMT	Microstructure of bimetallic structures Hardness and tensile tests	Voltage Wire feeding speed	[28]
18	SAL 7055-Al		350 GTAW	Evaluation of the micromechanical properties and micro-hardness.	Current Wire feed rate Deposition rate Arc Length	[29]
19	A-G-50-7-M21-4Mo	—	Metal inert gas welding robot	Local distribution of hardness Residual stresses Detailed examination of local deformation behavior	Welding current voltage Wire feed rate Inert gas quantity Feed path velocity	[30]
20	Nickel-based alloy (IN718)	Wrought IN718	Plasma arc welding	Microstructure analysis SEM and XRD analysis and fractographic analysis of the specimens	Wire feed rate m/min Current Voltage Torch travel speed mm/min	[31]

21	H08Mn2Si steel	Q235 mild steel	A-GMA-AM	Auxiliary wire optimal feeding orientation	Current Voltage and travel speed	[32]
22	Inconel 718	Structural mild steel S255	CMT-WAAM process	Analysis of the effect of process parameters.	Voltage, current, wire feed speed, and torch travel speed	[33]
23	Aluminum alloy AlMg5Mn	Aluminum alloy EN AW-5754A H111	Six-axis KUKA robot with GMAW	Analysis of geometrical properties of aluminum parts.	Welding arc CMT, CMT-ADV, CMT-PADV	[34]
24	Grade 91 steel ER90S-B91	Mild steel substrate	Six-axis robot coordinating a plasma arc welding (PAW)	Microstructure and hardness of built parts with P91 steel.	Average voltage Average current Wire feed speed Travel speed	[35]
25	Ti-6Al-4V alloy	Mild steel substrate	WAAM process with controlled cooling rates	Tensile properties and Vickers hardness of Ti-6Al-4V	Voltage Travel speed Wire feed rate and current	[36]
26	718 Super alloy	Q235A steel	Pulsed plasma arc additive manufacturing (PPAAM)	Microstructure and micro-hardness measurements	Voltage Current and Wire feed speed	[37]
27	Inconel 625	Q235 steel	CMT	Micro-hardness and mechanical properties analysis	Voltage, Wire feed speed Travel speed and current	[44]
28	Aluminum-magnesium ER5356	6061-T6 aluminum alloy	Cold metal transfer (CMT)	Aluminum parts' microstructure and mechanical characteristics.	Voltage Travel velocity Rates of current, wire feeding, and gas flow	[45]
29	Inconel 718	Mild steel	Three-axis linear CNC system with plasma arc welding	Grain morphology texture and hardness of built parts.	Travel speed Current and wire feed rate	[38]

(Continued)

TABLE 5.1 *(Continued)*

Overview of the Literature Analysis and Investigation of How Process Variables Affect Mechanical Characteristics and Surface Quality

Sl No.	Material Used	Substrate Material	Technique Used/Heat Source	Finding/Outcomes	Process Parameters	Ref.
30	ERNiCrMo-3 AMS 5837 (Inconel 625)	Q235 steel	Cold metal transfer (CMT)	Micro-hardness and mechanical properties analysis	Current Travel speed Wire feed speed and Voltage	[39]
31	Inconel 625	Low-carbon alloy steel	CMT	Microstructure of heat-treated sample of Inconel 625	Voltage Current Wire feed speed Torch travel speed Torch angle	[46]
32	Mn4Ni2CrMo Steel	Steel (S235JR)	PAW and GMAW	Metallographic and Charpy testing of samples	Voltage Wire speed Feed rate Intensity	[47]
33	Duplex stainless steel type-2209	Duplex stainless steel	MIG welding machine with an ABB 1400 robot	Micro-structure of the duplex stainless steel.	Current Travel speed	[48]
34	Inconel 718 super alloy		Three-axis CNC with plasma module	Mechanical testing and microstructural analysis Effective rolling for recrystallization		[49]

5.4 CONCLUSION AND FUTURE SCOPE

It is significant that the effects of arc voltage on mechanical characteristics and surface roughness might differ based on the kind of welding (stick, MIG, TIG, etc.) and the materials used (aluminum, steel, etc.). Arc voltage interacts with other welding factors, such as current, travel speed, and electrode/filler material, to affect the outcome. It's critical to properly choose and control welding parameters, including arc voltage, depending on the particular needs of the welding application in order to maximize mechanical qualities and surface polish. A flexible and promising method for creating complicated metal parts with significant time and cost reductions is called WAAM. Through this study, we have looked into a number of WAAM-related topics, such as process variables, material concerns, and applications.

5.4.1 Process Parameter Optimization

For WAAM to achieve the desired mechanical qualities and surface finishes, it is essential to choose and standardize procedure parameters for WAAM process in the best optimized way. From Table 5.2 research reviews of various researchers, we find the optimized process parameters and optimization techniques such as the designs of experiments (DOE), and sensitivity analyses have proven to be effective strategies for parameter optimization.

Effect of wire feed speed: From the above said literature, we find the consequence of wire feed speed such as lower heat input tends to be the result of higher wire feed speed. This may result in a more specific fusion zone and less penetration into the underlying material. A smaller HAZ is produced by faster wire feed rates since they produce less heat. Reduced speed increases heat intake, which raises the HAZ. Hardness and toughness are two mechanical qualities that can be impacted by the microstructure of the HAZ. The aggregation and dissemination of stresses in residual nature in the welded parts may change as a result of fluctuations in feed rate.

In the instruction to regulate the mechanical characteristics and surface roughness of welded connections, wire feed speed is essential. The best settings depend on a number of variables, and selecting parameters methodically is crucial to getting the results you want.

Torch travel speed: From the above said literature, we find the influence of welding speed known as torch travel speed such as the quantity of heat input into the welded joint is directly affected by torch travel speed. Lower heat input is produced by faster travel speed, while more heat input is produced by slower speed. Torch travel speed affects the heat-affected zone (HAZ) sizes, which in turn distresses the mechanical characteristics of materials. Alteration in the speed at which the torch travels can affect the amount and distribution of residual stresses in the welded joint. While slower travel rates can result in greater residual strains, faster travel speeds may lead to lower residual stresses.

Arc voltage and current effect: From the above said literature, we find the effect of arc current and arc voltage such as heat input through the welded

TABLE 5.2
Summary of Process Parameters Used in Various Research Works by Researchers. The Following Are the Main Findings from Our Study

Sl no.	Current	Voltage	Wire Feed Speed (mm/s)	Travel Speed (mm/s)	Wire Diameter	Deposited Material	Findings	Ref.
1	203 A	15.8 V	8.5 m/min	0.552	1.2 mm	Inconel 718	Consequence of WAAM parameters such as torch travel speed and WFS at microstructure and mechanical characteristics of material.	[17]
2	180 A	20 V	42 mm/s	3.8 mm/s	1.2 mm	Inconel 718	Stimulus of parameters on the microstructure of Inconel material.	[50]
3	180 A		30 and 34 mm/s	5 and 6 mm/s	1.2 mm	Inconel 718 and Inconel 625	Influence of parameters on the UTS and yield strength.	[51]
4	180 A		30 mm/s	5 mm/s	1.2 mm	Inconel 718	Tensile strength performance and microstructure.	[52]
5	120 A	19.6–14.3 V	34 mm/s	2.5 mm/s	1.2 mm	Inconel 718	Analysis was done on the micro-structural development and phase precipitation.	[53]
6	180 A		66 mm/s	12.5 mm/s	1.2 mm	Inconel 718	Inconel 718's microstructure and mechanical characteristics were examined.	[54]
7	130–160 A	26 V	148 mm/s	8.3 mm/s	0.9 mm	Inconel 718	Impact of process parameters on mechanical properties and geometric characterizations.	[55]

TABLE 5.2 ***(Continued)***
Summary of Process Parameters Used in Various Research Works by Researchers. The Following Are the Main Findings from Our Study

Sl no.	Current	Voltage	Wire Feed Speed (mm/s)	Travel Speed (mm/s)	Wire Diameter	Deposited Material	Findings	Ref.
8			6, 8, 10 m/min	0.9–2.1 m/min	0.89 mm	Inconel 718	Impact of process parameter on the microstructure of Inconel 718	[56]
9			4.5 m/min	0.3 m/min	1.2 mm	Inconel 718	The capabilities of the CMT technique have been premeditated on Inconel 718.	[57]
				Inconel 625				
10	115 A	20.2 V	84 mm/s	15 mm/s	1.2 mm	Inconel 625	Effect of process parameters at microstructure at different layers of built parts.	(Mansoor et al., 2021) [58]
11	150 A		100 mm/s	8 mm/s	1.2 mm	Inconel 625	Bottom to top microstructure of parts manufactured.	(Manufactured et al., 2022) [16]
12	148 A	14.6 V	108 mm/s	8, 9, 10 mm/s	1.2 mm	Inconel 625	Impact of travel speed on the mechanical characteristics and micro-structure.	(Ola et al., 2024) [59]
13	140 A		1.5 m/mm	0.23 m/min	1.2 mm	Inconel 625	The surface quality of the PPAD Inconel 625 sample.	(Xu et al., 2013) [60]
14	130 A	12.5 V	6.5 m/min	80 cm/min	1.2 mm	Inconel 625	Each sample's microstructure after being heated Inconel 625 sample's crystal structure.	(Newaz et al., 2019) [40]

joint is directly correlated with arc voltage. The size of the heat-effected zone (HAZ) and fusion zone can be affected by greater arc voltages, which often cause an increase in heat input. The rate at which the filler material is deposited is directly correlated with arc current. Increased currents typically lead to increased rates of deposition, which in turn affect the amount of melted metal and the overall dimensions of the weld bead.

Material compatibility: Selecting the right materials for WAAM is a very difficult issue. From this review work, we have found that several materials and their alloys, such as steel, aluminum, titanium, Inconel, and alloys based on nickel, can be successfully employed in WAAM. Exploring new material possibilities and their unique processing needs will require more investigation.

Quality and post-processing: Obtaining uniform quality in WAAM parts remains difficult, and problems like porosity and residual stresses call for more research. To guarantee the dependability and safety of the finished goods, it is crucial to develop efficient post-processing processes and non-destructive evaluation methodologies.

Applications in industries: WAAM's adaptability is demonstrated by the several areas in which it has been applied, including energy, maritime, automotive, and aerospace. It is a practical alternative for applications where standard manufacturing techniques may be too expensive or inefficient due to its capacity to generate huge and complicated components.

Future research directions: Although WAAM has achieved great progress, there are a number of exciting areas for additional study and advancement.

Advanced materials: Research and create novel WAAM-specific materials, such as high-temperature alloys, composites, and hybrid materials. Additionally, look into techniques for combining and stacking several materials in a single structure.

Enhance in-situ monitoring and control systems to find flaws and change process settings instantly, raising the caliber and consistency of WAAM components.

Standardization and certification: To ensure quality and safety in crucial applications, such as aerospace and healthcare, develop industry-wide standards and certification processes for WAAM.

Innovations in post-processing: To meet the unique problems posed by WAAM parts, develop novel post-processing procedures, such as machining, heat treatment, and surface finishing.

Sustainability assessment: Perform thorough life cycle analyses to calculate the environmental advantages and disadvantages of WAAM in comparison to conventional manufacturing techniques.

REFERENCES

1. Srivastava, A. K., Kumar, A., Kumar, P., Gautam, P., & Dogra, N. (2023). Research progress in metal additive manufacturing: Challenges and opportunities International Journal on Interactive Design and Manufacturing (IJIDeM) 1–17.

2. Kumar, A., Kumar, P., Mittal, R. K., & Gambhir, V. (2023). Integration of reverse engineering with additive manufacturing. In: Ajay Kumar, Ravi Kant Mittal and Abid Haleem (eds.), Advances in Additive Manufacturing: Artificial Intelligence, Nature-Inspired, and Biomanufacturing. Elsevier.
3. Kumar, A., Kumar, P., Mittal, R. K., & Gambhir, V. (2023). Printing file formats for additive manufacturing technologies. In: Ajay Kumar, Ravi Kant Mittal and Abid Haleem (eds.), Advances in Additive Manufacturing: Artificial Intelligence, Nature-Inspired, and Biomanufacturing. pp. 87–102. Elsevier.
4. Batista, R. C., Agarwal, A., Gurung, A., Kumar, A., Altarazi, F., Dogra, N., H. M., V., Chiniwar, D. S., & Agrawal, A. (2024). Topological and lattice-based AM optimization for improving the structural efficiency of robotic arms Frontiers in Mechanical Engineering 10 1422539.
5. Singh, A., Parveen, H., & AlMangour, B. (Eds.). (2023). Handbook of Smart Manufacturing: Forecasting the Future of Industry 4.0 (1st ed.). CRC Press. https://doi.org/10.1201/9781003333760
6. Kumar, A., Mittal, R. K., & Haleem, A. (Eds.). (2023). Advances in Additive Manufacturing: Artificial Intelligence, Nature-Inspired, and Biomanufacturing. Elsevier. https://doi.org/10.1016/C2020-0-03877-6
7. Aggoune, S. et al. (2024). Instabilities in the formation of single tracks during selective laser melting process International Journal on Interactive Design and Manufacturing. https://doi.org/10.1007/s12008-024-01887-y
8. Kumar, A., Kumar, P., Mittal, R. K., & Gambhir, V. (2023). Materials processed by additive manufacturing techniques. In: Ajay Kumar, Ravi Kant Mittal and Abid Haleem (eds.), Advances in Additive Manufacturing: Artificial Intelligence, Nature-Inspired, and Biomanufacturing. pp. 217–233. Elsevier.
9. Srivastava, A. K., Tiwari, S., Pachauri, P., Gupta, N., Sunil, B., & Kumar, A. (2024). Bonding strength and microstructural features of Al5083-AZ31B alloys laminated sheet through friction stir additive manufacturing Journal of Adhesion Science and Technology 38 (4) 583–596.
10. Kumar, A., Kumar, P., Sharma, N., & Srivastava, A. K. (Eds.). (2024). 3D Printing Technologies: Digital Manufacturing, Artificial Intelligence, Industry 4.0. https://doi.org/10.1515/9783111215112
11. Chaturvedi, M., Scutelnicu, E., Rusu, C. C., Mistodie, L. R., Mihailescu, D., & Subbiah, A. V. (2021). Wire arc additive manufacturing: Review on recent findings and challenges in industrial applications and materials characterization Metals 11(6) 939.
12. Duan, X. et al. (2023) Wire arc metal additive manufacturing using pulsed arc plasma (PAP-WAAM) for effective heat management Journal of Materials Processing Technology 311 (2023) 117806. https://doi.org/10.1016/j.jmatprotec.2022.117806
13. James, W. S. et al. (2023). High temperature performance of wire-arc additive manufactured Inconel 718. www.nature.com/scientificreportshttps://doi.org/10.1038/s41598-023-29026-9
14. Rodrigues, T. A. et al. (2022). Wire and Arc Additive Manufacturing of 316L Stainless steel/Inconel 625 Functionally Graded Material: Development and Characterization Journal of Materials Research and Technology 21 237e251. https://doi.org/10.1016/j.jmrt.2022.08.169
15. Reddy, K. et al. (2022). Wire arc additive manufactured mild steel and austenitic stainless steel components: Microstructure, Mechanical Properties and Residual Stresses Materials 15 7094. https://doi.org/10.3390/ma15207094; https://www.mdpi.com/journal/materials
16. Wang, X., Hu, Q., Li, T., Liu, W., Tang, D., Hu, Z., & Liu, K. (2022). Microstructure and fracture performance of wire arc additively manufactured Inconel 625 alloy by hot-wire GTAW Metals 12(3) 510.

17. Haldar, N., Anand, S., & Datta, S. (2023). On fabrication of Inconel 718 slab by wire arc additive manufacturing : Study of built microstructure and mechanical properties Arabian Journal for Science and Engineering. https://doi.org/10.1007/s13369-023-08095-y
18. Natarajan, M., Pasupuleti, T., Giri, J., Al-Lohedan, H. A., Katta, L. N., Mohammad, F., Sunheriya, N., Chadge, R., Mahatme, C., Giri, P., Mallik, S., & Sathish, T. (2024). Optimization of wire spark erosion machining of Grade 9 titanium alloy (Grade 9) using a hybrid learning algorithm AIP Advances 14 (1). https://doi.org/10.1063/5.0177658
19. Wang, X. et al. (2021). Microstructure and corrosion resistance in bimetal materials of Q345 and 308 steel wire-arc additive manufacturing Crystals 11 1401. https://doi.org/10.3390/cryst11111401; https://www.mdpi.com/journal/crystals
20. Xue, C. et al. (2021). Improving mechanical properties of wire arc additively manufactured AA2196 Al–Li alloy by controlling solidification defects Additive Manufacturing 43 (2021) 102019 https://doi.org/10.1016/j.addma.2021.102019
21. Mansoor, O. et al. (2021). Wire arc additive manufacturing (WAAM) of Inconel 625 alloy and its microstructure and mechanical properties International Research Journal of Engineering and Technology (IRJET) e-ISSN: 2395–0056.
22. Le, V. T. et al. (2021). Wire and arc additive manufacturing of 308L stainless steel components: Optimization of processing parameters and material properties Engineering Science and Technology, an International Journal 24 (4) 1015–1026. https://doi.org/10.1016/j.jestch.2021.01.009
23. Wanget, Y. (2021). Effect of magnetic field on the microstructure and mechanical properties of Inconel 625 superalloy fabricated by wire arc additive manufacturing Journal of Manufacturing Processes 64 (2021) 10–19. https://doi.org/10.1016/j.jmapro.2021.01.008
24. Hu, Q. et al. (2021). Microstructure and Properties of ER50-6 Steel Fabricated by Wire Arc Additive Manufacturing. Scanning 2021 7846116. https://doi.org/10.1155/2021/7846116
25. James, W. (2022). Microstructure and mechanical properties of Inconel 718 and Inconel 625 produced through the wire + arc additive manufacturing process Cranfield Online Research Data (CORD). https://doi.org/10.17862/cranfield.rd.18268613.v1
26. Hauser, T. et al. (2021). Oxidation in wire arc additive manufacturing of aluminum alloys Additive Manufacturing 41 (2021) 101958 https://doi.org/10.1016/j.addma.2021.101958
27. Schroepferet, D. et al. (2021). Process-related influences and correlations in wire arc additive manufacturing of high-strength steels 2021 IOP Conf. Series: Materials Science and Engineering 1147 (2021) 012002. https://doi.org/10.1088/1757-899X/1147/1/012002
28. Zhang, W. et al. (2021). Effect of deposition sequence on microstructure and properties of 316L and Inconel 625 bimetallic structure by wire arc additive manufacturing Journal of Materials Engineering and Performance. https://doi.org/10.1007/s11665-021-06137-w
29. Dong, B. et al. (2020). Wire arc additive manufacturing of Al-Zn-Mg-Cu alloy: Microstructures and mechanical properties Additive Manufacturing. https://doi.org/10.1016/j.addma.2020.101447
30. Wachteret, M. et al. (2020). Monotonic and fatigue properties of steel material manufactured by wire arc additive manufacturing Applied Sciences 10 5238. https://doi.org:10.3390/app10155238
31. Raoet, B. et al. (2020). High-temperature mechanical properties of IN718 alloy: Comparison of additive manufactured and wrought samples Crystals 10 689. https://doi.org:10.3390/cryst10080689
32. Han, Q. et al. (2020). Experimental investigation on improving the deposition rate of gas metal arc-based additive manufacturing by auxiliary wire feeding method Welding in the World. https://doi.org/10.1007/s40194-020-00994-0

33. Kindermann, R. M. et al. (2020). Process response of Inconel 718 to wire + arc additive manufacturing with cold metal transfer Materials & Design. https://doi.org/10.1016/j.matdes.2020.109031
34. Gierth, M. et al. (2020). Wire arc additive manufacturing (WAAM) of aluminum alloy AlMg5Mn with energy-reduced gas metal arc welding (GMAW) Materials 13 2671. https://doi.org:10.3390/ma13122671
35. Li, K. et al. (2020). Wire-arc additive manufacturing and post-heat treatment optimization on microstructure and mechanical properties of Grade 91 steel. https://doi.org/10.1016/j.addma.2020.101734
36. Yi, H.-J. et al. (2020). Effects of cooling rate on the microstructure and tensile properties of wire- arc additive manufactured Ti–6Al–4V alloy Metals and Materials International. https://doi.org/10.1007/s12540-019-00563-1
37. Wang, K. et al. (2019). Microstructural evolution and mechanical properties of Inconel 718 superalloy thin wall fabricated by pulsed plasma arc additive manufacturing Journal of Alloys and Compounds. https://doi.org/10.1016/j.jallcom.2019.152936
38. Seow, C. E. et al. (2018). Wire+Arc additively manufactured Inconel 718: Effect of post-deposition heat treatments on microstructure and tensile properties Materials & Design. https://doi.org/10.1016/j.matdes.2019.108157
39. Yangfan, W. et al. (2019). Microstructure and mechanical properties of Inconel 625 fabricated by wire-arc additive manufacturing Surface & Coatings Technology. https://doi.org/10.1016/j.surfcoat.2019.05.079
40. Newaz, A., Tanvir, M., Global, S., Ji, C., & Bates, B. (2019). Heat treatment effects on Inconel 625 components fabricated by wire + arc additive manufacturing (WAAM)—Part 1: Microstructural characterization Heat Treatment Effects on Inconel 625 Components Fabricated by Wire + Arc Additive. https://doi.org/10.1007/s00170-019-03828-6
41. Wang, X. et al. (2022). Microstructure and fracture performance of wire arc additively manufactured Inconel 625 alloy by hot-wire GTAW Metals 12 510. https://doi.org/10.3390/met12030510
42. Alonso, U. et al. (2021). Characterization of Inconel 718® super alloy fabricated by wire arc additive manufacturing: Effect on mechanical properties and machinability Journal of Materials Research and Technology 14 2665–2676. https://doi.org/10.1016/j.jmrt.2021.07.132
43. Suarez, A. et al. (2021). Wire arc additive manufacturing of an aeronautic fitting with different metal alloys: From the design to the part Journal of Manufacturing Processes 64 (2021) 188–197 https://doi.org/10.1016/j.jmapro.2021.01.012
44. Yangfanet, W. et al. (2019). Microstructure and mechanical properties of Inconel 625 fabricated by wire arc additive manufacturing Surface & Coatings Technology 374 116–123. https://doi.org/10.1016/j.surfcoat.2019.05.079
45. Chuanchu, S. et. al. (2019). Effect of heat input on microstructure and mechanical properties of Al-Mg alloys fabricated by WAAM Applied Surface Science 486 431–440. https://doi.org/10.1016/j.apsusc.2019.04.255
46. Tanviret, A. N. M. et al. (August 2019). Heat treatment effects on Inconel 625 components fabricated by wire + arc additive manufacturing (WAAM)—Part 1: Microstructural characterization The International Journal of Advanced Manufacturing Technology. https://doi.org 10.1007/s00170-019-03828-6
47. Artazaet, T. et al. (2019). Wire arc additive manufacturing of Mn4Ni2CrMo steel: Comparison of mechanical and metallographic properties of PAW and GMAW 8th Manufacturing Engineering Society International Conference https://doi.10.1016/j.promfg.2019.10.035
48. Hosseini, V. A. et al. (2019). Wire-arc additive manufacturing of a duplex stainless steel: Thermal cycle analysis and microstructure characterization Welding in the World 63 975–987. https://doi.org/10.1007/s40194-019-00735-y

49. Xu, X. et al. (2018). Enhancing mechanical properties of wire + arc additively manufactured INCONEL 718 super alloy through in-process thermo mechanical processing Materials & Design 160 1042–1051. https://doi.org/10.1016/j.matdes.2018.10.038
50. Gil, A., & Val, D. (2021). Characterization of Inconel 718 ® superalloy fabricated by wire arc additive manufacturing: Effect on mechanical properties and machinability. https://doi.org/10.1016/j.jmrt.2021.07.132
51. James, W. S., Ganguly, S., & Pardal, G. (n.d.). Microstructure and mechanical properties of Inconel 718 and Inconel 625 produced through the wire + arc additive manufacturing process. NATO. 1–16.
52. James, W. S., Ganguly, S., & Pardal, G. (2023). High temperature performance of wire – arc additive manufactured Inconel 718 Scientific Reports 1–8. https://doi.org/10.1038/s41598-023-29026-9
53. Wang, K., Liu, Y., Sun, Z., Lin, J., Lv, Y., & Xu, B. (2019). Microstructural evolution and mechanical properties of Inconel 718 superalloy thin wall fabricated by pulsed plasma arc additive manufacturing Journal of Alloys and Compounds 152936. https://doi.org/10.1016/j.jallcom.2019.152936
54. Williams, S. (2018). Investigation of process factors affecting mechanical properties of Inconel 718 superalloy in wire + arc additive manufacture process Journal of Materials Processing Technology. https://doi.org/10.1016/j.jmatprotec.2018.10.023
55. Hejripour, F., & Aidun, D. (2021). Effects of processing parameters on wire arc additive manufactured Inconel ® 718 Welding Journal 100 93–105.
56. Kindermann, R. M., Roy, M. J., Morana, R., & Prangnell, P. B. (2020). Process response of Inconel 718 to wire + arc additive manufacturing with cold metal transfer Materials & Design 195 109031. https://doi.org/10.1016/j.matdes.2020.109031
57. Benoit, A., Jobez, S., Paillard, P., Klosek, V., & Baudin, T. (2011). Study of Inconel 718 weldability using MIG CMT process Science and Technology of Welding & Joining 16 (6) 477–483. https://doi.org/10.1179/1362171811Y.0000000031
58. Mansoor, O., Wahdatullah, N., & Harshavardhana, N. (2021). Wire arc additive manufacturing (WAAM) of Inconel 625 Alloy and its microstructure and mechanical properties International Research Journal of Engineering and Technology (IRJET) 8 1517–1528.
59. Ola, O. T., & Doern, F. E. (2024). A study of cold metal transfer clads in nickel-base Inconel 718 superalloy Materials & Design 57 51–59. https://doi.org/10.1016/j.matdes.2013.12.060
60. Xu, F. J., Lv, Y. H., Xu, B. S., Liu, Y. X., Shu, F. Y., & He, P. (2013). Effect of deposition strategy on the microstructure and mechanical properties of Inconel 625 superalloy fabricated by pulsed plasma arc deposition Materials & Design 45 446–455. https://doi.org/10.1016/j.matdes.2012.07.013

6 Smart Manufacturing in the Defence Sector
A Comprehensive Review and Analysis of Technological Advancements, Integration, and Challenges

Surendra C. Ghorpade and Sumati Sidharth

6.1 INTRODUCTION

The concept of "smart manufacturing" originated in the United States but is now widely utilised all over the world and has acquired considerable importance in business and academia recently. Numerous manufacturing systems are disguising themselves as SMSs. Networking data and information and communication technologies (ICTs) are used by SM (smart manufacturing), a set of manufacturing practices, to manage manufacturing activities [1]. Manufacturing has advanced, becoming more complicated, mechanised, and computerised. Smart manufacturing is the novel concept which uses various tools like sensors, communication technologies, data-driven modelling, simulation, and predictive analysis with manufacturing landscape of today and tomorrow. Once put into practice, these ideas and innovations would ensure that smart manufacturing will be defining the future of industrial revolution [2]. In the past, manufacturing was only defined as a single process or number of sub-activities that converted raw material into finished products. However, there is much more to manufacturing than is typically understood. Manufacturing today takes data-driven business operations into account at various levels, which has given birth to smart manufacturing technology. Future SMSs will have special capabilities for self-assembly to create intricate and personalised products to take advantage of both the present and new markets. Performance is continuously maintained and enhanced by SM using information.

6.2 SMART MANUFACTURING: DEFINED

Table 6.1 brings out definitions of smart manufacturing from different perspectives.

DOI: 10.1201/9781032725086-7

TABLE 6.1
Definitions of Smart Manufacturing

View	Definition	Reference
Engineering	Application of advanced and smart technologies which facilitate rapid and stable manufacturing of new products, dynamic adaptability to personalised product specifications, and optimisation of all stages of production in real time.	[3]
Networking	Integration of cyber-physical systems, IoT, and IIoT backed by sensors and communication technology to capture data at all stages of manufacturing a product.	[4]
Decision-making	Utilisation of big data analytics to aid manufacturing process, predict issues in the process, and improve productivity by optimising the operations including planning, diagnosis, and real-time assessment.	[5]

6.3 THE JOURNEY SO FAR

2005 Expansion of platforms for the Internet, e-commerce, social networking, and smartphones at a rate consistent with Moore's Law in terms of connectivity, data, and computing power. The term "cyberinfrastructure" began to be used.

2006 Smart manufacturing first used in National Science Foundation meeting on cyberinfrastructure.

2010 Smart Manufacturing Leadership Council (SMLC) coordinated with more than 50 industries for a workshop to develop infrastructure and capabilities in respect of smart manufacturing.

2014 The German Standardization Roadmap for Industry 4.0 (DKE/DIN Industry 4.0) Version 1.0 was released [6].

2010–2016 Leading manufacturers in the United States adopted the use of smart manufacturing processes. Various stakeholders were brought on common platform by organisations like the Manufacturing Enterprise Systems Association (MESA) and the SMLC to fast track implementation of smart manufacturing [7].

2016 The reports "Smart Manufacturing Landscape Explained" by MESA International and "Standards Landscape for Smart Manufacturing" by NIST were both released [8].

2016 The smart manufacturing institute of the United States (CESMII) was established as one of several US manufacturing institutes with the goal of bringing together business, academic, and government partners to boost manufacturing competitiveness in the United States.

2017 Release of the CESMII Roadmap for Smart Manufacturing. To assist manufacturers in evaluating their business practices and creating roadmaps for advancing the implementation of smart manufacturing, consulting firms issued guidelines like Singapore Smart Industry Readiness Index [9].

2017 National Network for Manufacturing Innovation (NNMI) accelerated R&D to promote adoption of smart manufacturing technologies.

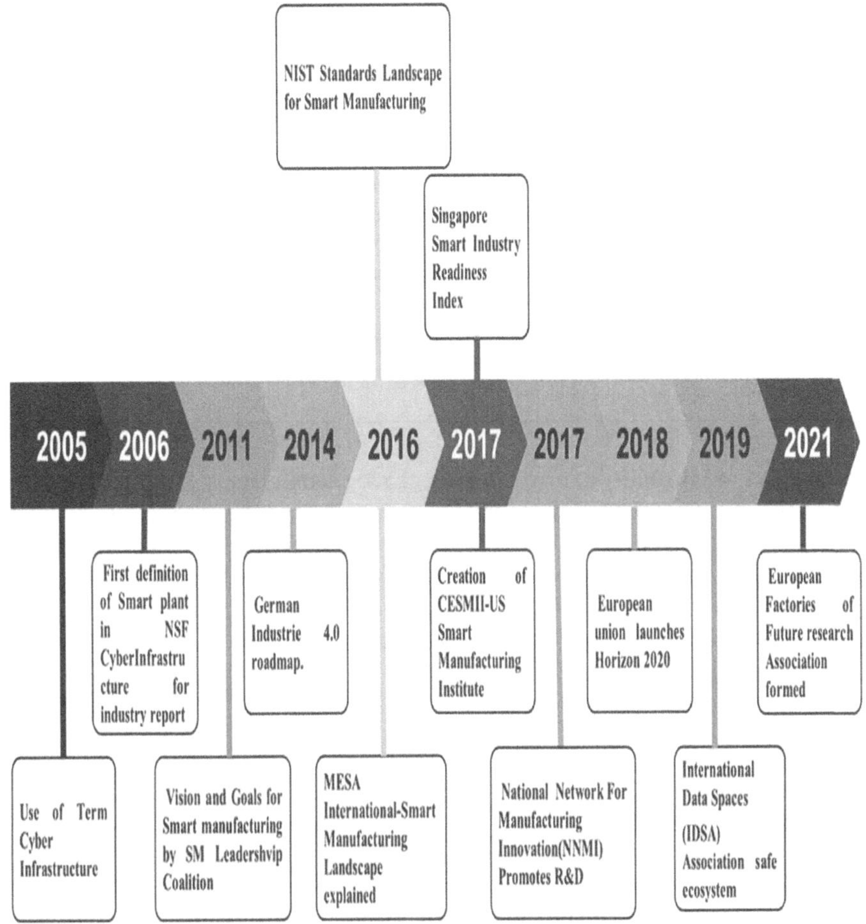

FIGURE 6.1 The journey of smart manufacturing.

2018 European Union launched Horizon 2020 initiative to fund projects for Industry 4.0.

2019 International Data Spaces Association (IDSA) secured a sovereign data sharing ecosystem.

2021 European Factories of Future Research Association facilitated collaboration between industry, academia, and policymakers for the promotion of smart manufacturing. Figure 6.1 depicts the progress.

6.4 TECHNOLOGICAL ADVANCEMENTS

The manufacturing sector has made significant progress from earlier machining techniques to present automation and entered the fourth generation of industrial revolution after going through various changes and advancements [10]. The introduction of smart materials presents many challenges which include performance

TABLE 6.2
Smart Factory

Components [14]	Objectives [15]
Smart production: which focuses on the efficient interaction of humans, machines, and tools	Data collection: the gathering of a variety of useful information/data from the equipment
Smart services: such as real-time production/ maintenance, monitoring, and predictive analysis	Data analysis: the data is analysed to produce information that is valuable for corporate and operational decision-making.
Smart machines: deals with high-performing systems, modern engineering shops, and integrated planning of material requirements with automatic waste segregation and reduction [16].	Implementation: automating processes, optimising systems, and guiding corporate strategy

aspects and sustainability of material through recyclability and durability. The smart material has a potential for enhancing product functionalities in sectors like aerospace, automotive, and construction [11]. To increase Germany's competitiveness in the manufacturing sector, a group of commercial, political, and academic specialists coined the phrase "Industry 4.0" in 2011. It places a strong emphasis on machine learning, interaction through the IoT, and the processing of real-time data. Industry 4.0 utilises Internet of Things (IoT) and Industrial IoT (IIoT) to communicate with machines, share real-time information, and make decisions [12]. The technological advancements in the field are brought out in subsequent paragraphs:

6.4.1 Concept of Smart Factory

Industry 4.0 uses cyber-physical systems (CPSs) to automate manufacturing processes and share information across various platforms. Data mining and control has been significantly enhanced by IoT and IIoT, machine learning, artificial intelligence (AI), and big data analytics [13]. The factories utilising this concept are called smart factories which operate on real-time interaction between smart machines, humans, and the entire manufacturing system. The objectives and components of smart factory are brought in Table 6.2.

6.4.2 Smart Manufacturing Ecosystem

The production, management, design, and engineering processes are all included in the manufacturing industry based on the concept of smart manufacturing ecosystem [8]. The framework of smart manufacturing ecosystem with its dimensions, functions, and constituents are brought out in Table 6.3.

6.5 DEFENCE MANUFACTURING IN INDIA

In 1962, the Department of Defence Production (DDP) was established with the goal of building an integrated infrastructure to facilitate the manufacture of weapons/ systems/products required by Defence Services. Over the past 60 years, DDP has

TABLE 6.3
Smart Manufacturing Ecosystem

Dimensions	Function	Constituents
Product	It involves the flow of information and controls of product from design to end of life cycle	Design Process planning Production engineering Service and use Recycling
Production	The dimension is responsible for entire management of production facility and caters for design, development, operation, and phasing out processes.	Design Build Commissioning O&M Decommissioning and recycling
Business	It is responsible for managing supplier and customer interactions and their resolution.	Source Plan Deliver and return

been successful in creating production facilities in collaboration with the Ordnance Board, Defence Public Sector undertakings, and private industry affiliated with the sector [17]. The Indian defence manufacturing industry makes a significant contribution to the Indian economy [18]. The Indian government has taken many initiatives to address the needs of its armed forces while also reducing imports of defence equipment from other countries [19]. As of 2021, India is third in the world's greatest defence budget. According to India's Budget for 2022–23, 25% of the defence research and development budget has been set aside for domestic industry and start-ups to foster innovation in new technologies. As of October 2022, 595 industrial licences had been given to 366 businesses engaged in the defence sector.

6.5.1 Defence Industrial Base

Is the Indian defence sector worthy of global attention, given its massive defence industrial base, huge defence budget, ongoing negotiations with global defence companies, and rush of legislative reforms? The answer is a clear positive. Despite a bleak global economic situation following the epidemic, the Indian economy has maintained a strong position, becoming the world's fifth largest economy with an estimated growth rate of 6% by 2024. According to the most recent economic survey, growth in 2023–24 is expected to continue at 6.0%–6.8%. Aside from this expansion, the economy has seen a credit growth of 30.5% in the micro, small, and medium enterprises ("MSME") sector [20].

India has a robust defence industry, including 41 ordnance factories ("OF"), 9 defence public sector undertakings ("DPSU"), and over 100 private companies. The DRDO, India's principal defence research organisation, oversees more than 50 laboratories. India's armed forces are the third largest in the world. The government has created two dedicated defence industrial corridors in Tamil Nadu and

Uttar Pradesh to cluster defence industries and use existing infrastructure and human talent. Government programmes like iDEX and DTIS promote innovation in the defence and aerospace industries.

India's defence expenditure has increased dramatically during the last two decades. From 2000 to 2010, India's defence expenditure nearly tripled, from INR 58,587 crore to INR 141,781 crore. In 2015, the defence budget allocation increased to INR 222,370 crore. Currently, India is the world's third largest military spender, with its defence expenditure accounting for 2.15% of total GDP (gross domestic product). Over the next five to seven years, the Indian government intends to spend USD 130 billion on fleet upgrading across the military services. The sector received INR 5.94 lakh crore in the 2023–24 budget, up 13% from the previous year.

6.5.2 Stakeholders of Defence Manufacturing

The network of numerous stakeholders linked with the Defence Manufacturing industry is stretched across the country, promoting production of Defence services' projected requirements. The key stakeholders are:

6.5.2.1 Defence Research and Development Organisation

In 1958, the Defence Science Organization and the Indian Army's Technical Development Establishments united to become the DRDO. The DRDO began with a network of 11 research institutions. It currently has around 53 research laboratories, institutes, and 15 centres of excellence. The Director General of DRDO is also the Secretary of the Department of Defence Research within the Ministry of Defence. Since its establishment, DRDO has shifted its focus from inspection to full-scale design, development, and manufacturing projects. It has built a wide range of goods, including unmanned aerial vehicles, combat vehicles, and electronic warfare devices.

6.5.2.2 New Public Sector Undertakings

In the year 2021, the Government of India completed the dissolution of Ordnance Factory Board (OFB) and consolidated 41 factories into seven new public sector undertakings (PSUs) to manufacture defence hardware comprising of ammunition to heavy weapons and military vehicles [21].

6.5.2.3 Old PSUs

The nine public sector undertakings were retained under administrative control of DDP, Ministry of Defence [22].

6.5.2.4 Defence Private Sector Undertakings

In 1998, the Government of India established a joint task force in collaboration with the Confederation of Indian Industry to encourage private sector participation. Furthermore, based on its suggestions, the defence sector was opened to private business in 2002. It was a major policy shift to support indigenous manufacturing, which was followed up by the Kelkar Committee in 2004. In the recent past, through policy reforms and with an intent to exploit technology and capability of the private sector, there has been steady rise in the participation and contribution of private companies to defence manufacture [23].

India's private sector has not only become one of the most active participants in the defence industrial complex, but it has also dramatically altered the industry's face. The sensitive and strategic character of the defence industry, as well as its direct impact on national security and foreign policy interests, has been mentioned as the reason to discourage private sector participation. In 2001, the Indian defence industry allowed 100% private sector participation due to the strong performance of the private sector in the decade following liberalisation in 1991, whereas the public sector struggled. Today, the private sector is an essential component of the defence industry.

6.6 INTEGRATION OF SMART MANUFACTURING TECHNOLOGIES WITH DEFENCE ECOSYSTEM

Next-generation industries are now supported by revolutionary digital and smart technologies which connect physical systems to the Internet. The significant announcements of the top technological nations are Industry 4.0 (Germany), Made in China 2025 (China), Industrial Internet (USA), and Society 5.0 (Japan). These technologies have the goal of integrating current manufacturing process with smart/digital technology [24]. For many of their operations, modern militaries rely heavily on electronic networks for communication and data. These networks, which perform both operational and sustenance tasks, are quickly developing into the pillars of military operation. Operational networks, logistical networks, administrative networks, etc. are examples of network-centric concepts used in the military. These networks are primarily focused on information exchange for the needs of the military, including situational awareness building, operational and routine communication, logistical management, imagery transmission, and administration [25]. In the future, as the military's reliance on technology develops, so will their reliance on data networking. In general, militaries now have shorter time frames available to them for assessing situations, making choices, carrying out tasks, reacting to events in the battlespace, etc. due to the speeding up of things brought about by technology. Without a doubt, this accelerating necessitates improving administrative and logistical processes in addition to operational ones. As a result of these specifications, future military technology will increasingly be networked. In the future, military networks would be connected to items like weapons, transportation platforms, logistical packages, and ordnance consignments. These items would all have electronics embedded in them that would generate and exchange data as well as be increasingly controlled and regulated without the need for traditional human oversight.

The smart manufacturing technologies are shown in Figure 6.2 and their applications in defence sector are discussed below:

6.6.1 Virtual Reality

The ability to experience computer-generated visuals and films that simulate real-world events is provided by virtual reality (VR). The main constituents are video and audio device, location enabler like GPS, and facility to communicate with external devices. The designing and testing costs can be significantly reduced. Additionally, it facilitates the digital manufacturing process by enabling consumers to envision and test items in a mock-up setting. This opens up more options for product modification,

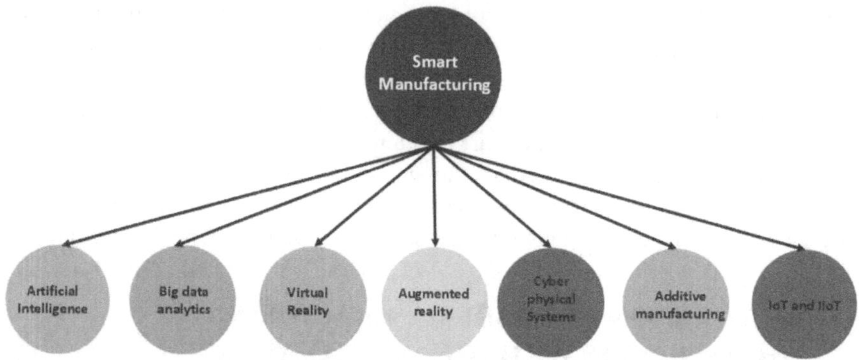

FIGURE 6.2 Smart manufacturing technologies.

redesign, and quick testing of product designs [26]. Some of the applications of virtual reality in the defence environment are explained in Figure 6.3 [27, 28].

6.6.2 Augmented Reality

Augmented reality concept simulates actual environment built through computer simulation of the real world. The wearable mobile devices activate the setting of the environment and permit interaction in real time for training, simulation, and validation purposes prior to actual manufacturing of a system or product. Using this interactive and cost-effective technology has been instrumental in enhancing the training and testing with varied scenarios and situations [29, 30]. Table 6.4 highlights the applications.

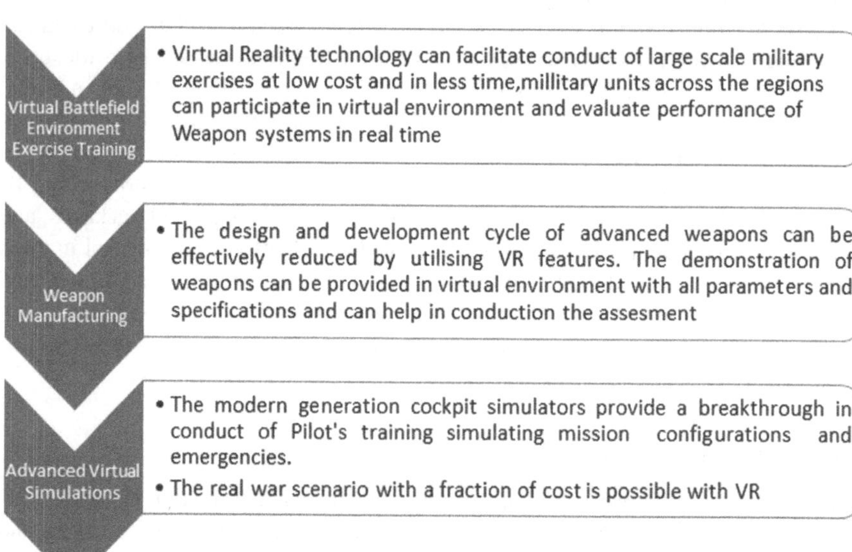

FIGURE 6.3 Virtual reality applications.

TABLE 6.4
Applications of Augmented Reality

Sl no	Application	Description
01	C-130 Hercules loadmaster training	A C-130 is a military transport plane that many nations frequently deploy for cargo and resupply missions. The loadmaster is the crew responsible for cargo handling and rigging specialist who airdrops troops and equipment out of the back of the C-130 to deliver cargo. The training generally is carried out on ground-mounted fuselages with necessary features of interiors without wings and tail. The quantity of training devices and the accuracy of the already available devices, however, were deemed insufficient to achieve the training objectives. Various components of loadmaster training can be carried out effectively using head-worn AR displays fitted to regular flying helmets [31]
02	Battlefield Augmented Reality Systems (BARS)	With the help of the Battlefield Augmented Reality System, artificial battlefields may be created in the real world and can be traversed without the restrictions seen in virtual environments. Typically, training facilities' walls are projected with a virtual environment and virtual actors to create training scenarios for soldiers. This is only applicable in indoor settings and requires substantial infrastructure. By overcoming these limitations, AR enables soldiers to engage in outdoor training activities where they can physically move across genuine landscapes [32]. For tactical situational awareness, the HUNTR (Heads-Up Navigation, Tracking, and Reporting) system has been created. With the use of AR technology, this system offers assistance data such as street names, building names, and historical details [33].
03	Technical guidance	It comprises the technologies for both virtual technical instructions and real-time remote support. Engineers in the US Army and Navy, Lockheed Martin, BAE Systems, and other organisations use this technology [33].

6.6.3 Cyber-Physical System

Industrial manufacturing engineers can utilise the Internet to directly monitor and control industrial operations, and cloud computing allows control engineers to access the industry's control system from anywhere. The CPS uses Internet-based data processing systems to connect with the real world [34].

6.6.4 Air Traffic Management System

In the near future, smart air vehicles are anticipated to predominate, particularly for military use. One of the most well-known instances of intelligent air vehicle is drone. CPSs are anticipated to have a significant impact on aviation and management of air traffic (ATM) in the future since awareness of physical location is a crucial issue for the next generation of air vehicles. Today, radar towers are used to control air traffic, and computing support systems have a limited understanding of the physical world. Therefore, for the next generation of air transportation systems, there is a critical need for close integration of the computational and physical capacities [35].

6.6.5 Additive Manufacturing

By creating parts from a variety of materials, additive manufacturing (AM) technologies give the manufacturing sector a number of advantages by enhancing design freedom and cutting production costs and lead times [36]. The technology of AM, usually known as 3D printing, opens up new possibilities for intelligent manufacturing technologies. The adaptability of AM technology allows for incorporating customer-specific requirements, rapid production, and effective prototyping resulting in reduced costs [37]. The reverse engineering (RE) method creates a digital copy of a product in which original blueprints, drawings, or other technical specifications are no longer available because the product technology has become outdated. RE produces two-dimensional cross-sectional pictures or point cloud data, which are then used to rebuild a computer-aided design (CAD) model [38]. Conversely, AM methods are becoming more and more well-liked since they can create a tangible object with intricate details by layering on materials. The main prerequisite for AM is a CAD model, which is produced by RE procedures and used to construct a product once the required processes are completed in the AM interface. The main contribution of AM to smart manufacturing is the ability to reverse engineer any parts or products using 3D scanning, to allow for design reconfiguration, and to quickly reproduce samples for testing and validation [39, 40]. As per researchers, the global 3D printing market is expected to expand at a compound annual growth rate of 23.3% from 2023 to 2030 [41]. The application of the technique is discussed below:

6.6.5.1 Complex Products Manufacturing

Real-time flaw detection and control as well as process efficiency improvement are now achievable thanks to improvements in AM, which span the prefabrication stage to the finished product [42]. These technologies appeal to aerospace firms because they may be used to produce extremely complicated and high-performing products. Although there is currently a thriving sector for producing polymer parts for high-performance military aircraft, direct metal fabrication technologies are probably where this industry is seeing the most attention. When producing complex parts, incremental sheet forming (ISF) lowers costs and eliminates the need for dies. High-quality components for the automotive, aerospace, and medical industries are produced by sheet metal forming [43]. Titanium is well suited for use in carbon composite aircraft designs because direct-metal AM methods can process it quite easily. When producing complex parts, ISF lowers costs and eliminates the need for dies.

6.6.6 Big Data Analytics

Big data analytics (BDA) enables real-time decision-making for smart manufacturing with the help of data collection and analysis from multiple sources. The manufacturing of products has been made more interactive by providing platform for interaction of customer and manufacturers and possibility of feedback at all stages of production. BDA assists in identifying and overcoming the problems of manufacturing and helps in avoiding product failures in the real world. Predictive manufacturing is possible due to features of data exploitation of BDA [44, 45]. Table 6.5 brings out applications and use cases of BDA.

TABLE 6.5
BDA Applications

Sl no.	Application	Description	Use Cases
01	Data farming	Data farming is a technique that creates landscapes of potential outcomes and uses high-performance computers to get inputs from trends. The fundamental concept is to input data into simulation through a variety of inputs and subsequently to reap substantial volumes of data as simulated outputs.	– The US Marine Corps' Albert project encouraged making decisions and put an emphasis on issues like "what if" and "what" factors are most important [46]. – PAXSEM is the integration of sensors of various types like optical or IR and effectors which are different types of weapons to provide description of technical system [47].
02	Course of action analysis	It is essential to use simulations to identify specific vital elements of the battle area in order to successfully complete mission planning. The military commander's evaluation of potential plans and various decision points is very crucial. The simulation-based course of action (COA) analysis method requires testing a variety of scenarios over a wide parameter value space [48].	Synthetic Theatre Operations Research Model (STORM) is used by the US Navy and Air Force to simulate a campaign level which can last for few minutes to hours based on complexity of the simulation [49].
03	Acquisition of new systems	In absence of real experiments or their exorbitant costs, modelling and simulation are extensively used to predict antiair, antisurface, and antisubmarine systems' warfare capabilities. M&S plays an important role in acquiring new platforms by using computer technology to support all phases from design to development. The simulation software has extremely accurate models that could be utilised for tactical-level constructive and virtual simulation [50].	Australia's acquisition of naval systems using Ship Air Defence Model [51].
04	Testing and evaluation of THAAD system	The Terminal High Altitude Area Defence (THAAD) system was designed to defend against MRBM on the US homeland and armed forces. Analysing the rapidly growing amount of data gathered from missile defence flight tests presented a problem for the THAAD system's testing and evaluation. The system was extremely complicated and software-intensive, and it included a variety of parts (such as radars, launchers, and interceptors). The simulation methodology was used by the THAAD programme allowing testing up to system level and testing in real time [50].	

6.6.7 Artificial Intelligence

The scientific term AI describes objects that can recognise conditions and take appropriate actions in response. The capacity to produce such enhanced artefacts has a greater influence on our culture [52]. AI is being effectively used in smart manufacturing systems to increase human-robot interaction, reduce humans in danger areas of operation, identify faults or defects, and improve maintenance aspects of system [53]. By harnessing the most recent advancements in AI and Industry 4.0, 3D printing has the ability to completely transform manufacturing capacities and organisations. Additionally, 3D printing will change where and how manufacturing is carried out in a variety of global markets and industrial sectors. The largest impact will be in early adopter industries like aerospace, automotive, life science, and military, where it will allow us to accomplish things with devices that have never been done before without the need for specialised machinery [54]. The features of AI are self-decision, optimisation, and immediate reaction to physical changes by altering production schedules, controlling machine operations, automatic changeover of tools, and prior warning of impeding problem [55]. Figure 6.4 depicts applications of AI.

Autonomous robots

For underwater examinations of marine/civil infrastructure, an AI-enabled ROV has been created with cutting-edge feature identification capabilities.

Block chain based automation

After their identities have been verified, the Defense forces can share information using the BCIDM (Block chain based Identity Management) technology.

C4ISR Systems

AI can assist in identifying opposing aircraft and even their strategy. This makes it simpler to mount a successful counterattack or take evasive action.

Cyber Security

Malware detection runs audits, exploits Android applications, and reports the found malware in a user-friendly manner.

Artificial Intelligence in Defence

Human Behavior Analysis

The onset of driver weariness and drowsiness can be precisely predicted and detected in real-time by a non-intrusive AI-based technology.

Intelligent Systems

A 360-degree, intelligent video monitoring system powered by AI helps with low-speed handling by giving the driver a real-time, 360-degree picture of the entire vehicle.

Internet of Battle Things

A soldier's helmet is equipped with a "Smart Helmet" that records 3D data of the surrounding area in real time & creates a 3D map of the same environment using an AI algorithm.

Autonomous Weapon Systems

Based on the inputs from the sensors and mission computer, SMART-CMDS automatically disperses the right kind and quantity of payloads in the right order against active threats.

FIGURE 6.4 AI applications.

6.6.8 IoT and IIoT

The IIoT is an application of IoT in industries, whereas IoT is mainly concerning domestic applications of Internet like smart homes, transportation applications, logistics functions and supply chain, agriculture, and healthcare. Industries can now work together through Internet platforms and achieve common goals [56]. These interactions among the many parts aid in the optimisation of resources, tools, and materials as well as production planning, fault localisation, predictive maintenance, and enhanced human-machine interaction (HMI). IIoT also enables the digital presentation of goods, procedures, and manufacturing facilities to clients for promotional and educational purposes [57, 58]. The applications of IoT in defence are in Figure 6.5.

6.7 CHALLENGES

Smart manufacturing is a futuristic technology with multiple advantages over traditional manufacturing. However, various challenges associated are discussed below [12].

6.7.1 Security Issues

Smart manufacturing system is based on integrated network to share and access information between manufacturing units, machines, and consumers. The backbone

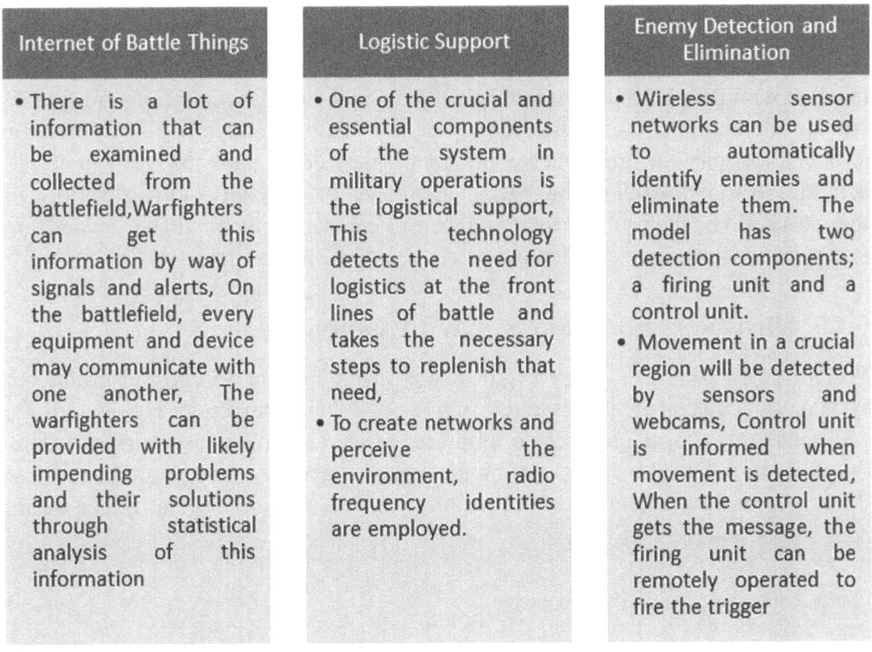

Internet of Battle Things

- There is a lot of information that can be examined and collected from the battlefield,Warfighters can get this information by way of signals and alerts, On the battlefield, every equipment and device may communicate with one another, The warfighters can be provided with likely impending problems and their solutions through statistical analysis of this information

Logistic Support

- One of the crucial and essential components of the system in military operations is the logistical support, This technology detects the need for logistics at the front lines of battle and takes the necessary steps to replenish that need,
- To create networks and perceive the environment, radio frequency identities are employed.

Enemy Detection and Elimination

- Wireless sensor networks can be used to automatically identify enemies and eliminate them. The model has two detection components; a firing unit and a control unit.
- Movement in a crucial region will be detected by sensors and webcams, Control unit is informed when movement is detected, When the control unit gets the message, the firing unit can be remotely operated to fire the trigger

FIGURE 6.5 Applications of IoT and IIoT.

of the system is Internet which calls for end-to-end data encryption and necessary protection. It is important to secure every node and element of the network from data theft, exploitation, and unwanted use. The security of the system as a whole is a guiding factor at all stages and needs to be prioritised [59].

6.7.2 System Integration

One of the major hurdles in successful implementation of technology is its integration with existing ones for seamless operation. The problem arises due to compatibility issues and protocols of different generation systems. The integration of such technologies at various levels of maturity and life cycle stages is a huge challenge to be addressed. Additionally, a stronger communication system is needed because of machine-to-machine communication and system interconnectivity [60].

6.7.3 Interoperability

The feature allows various systems to comprehend, understand, and exploit features of each other for maximising returns. With this facility, they can exchange data and information without importance to the origin of software/hardware. I4.0 Interoperability is further subdivided into Operational, Systematic, Technological, and Semantic. To use Interoperability effectively, it is important to adequately synchronise communication protocols. System can have constraints like bandwidth, frequency of operation, communication mode, and capabilities of hardware [61].

6.7.4 Safety in Human-Robot Collaboration

A cobot, also known as a lightweight robot, is a specific kind of robot that can collaborate with people in the workplace while introducing novel concepts from HMI. Human-robot cooperation is defined by the International Federation of Robotics as a robot's capacity to communicate and cooperate with humans to carry out certain activities in an industrial setting. Occupational health and safety of those working on the site should be the primary focus; dangerous environments should be avoided, and appropriate occupational health and safety should be maintained.

6.7.5 Return on Investment in New Technology

When transitioning to another sophisticated technology in an existing production system, the financial analysis and return on investment are carefully considered. The additional investment required to embrace newer technology is weighed against the production losses incurred during an upgrade, and the time required to recover the return on investment with the income from the present system influences the adoption of newer technology.

6.7.6 Multilingualism

Smart manufacturing systems should be capable of handling multilingual operations and interpreting any human-language instructions into machine language in order to command the machine to do the necessary operation. To make the phrase smart manufacturing more realistic and to include AI and modern technologies into manufacturing systems, it must be able to receive instructions directly from the operator, either in voice or text format.

6.8 CONCLUSION

The comprehensive review of cutting-edge technological developments and issues in smart manufacturing within the defence industry has illuminated the transformative potential of incorporating cutting-edge technology. The combination of IoT, data analytics, AI, robotics, and AM promises to transform the way defence systems are developed, produced, and maintained. The defence sector is on the verge of a technological revolution. The review highlights the variety of benefits that smart manufacturing offers the defence industry. Improved productivity, accuracy, and quality control are the key ingredients that can significantly shorten production schedules, cut costs, and use resources more effectively. Utilising data-driven insights enables decision-makers to make wise decisions, resulting in efficient operations and competitive benefits.

However, there are unique difficulties involved in integrating this technology into the defence sector. The threat of hostile assaults on networked systems makes cybersecurity a top priority. It is impossible to stress the importance of strong security measures, data encryption, and strict authentication procedures. In order to make sure that innovations adhere to set standards and do not jeopardise the integrity of the defence industry, regulatory compliance and standardisation also continue to be important areas that need careful consideration. Future research on the topic can explore the status of adoption of smart manufacturing techniques resulting in improvement in operational readiness and efficiency enhancement based on data analysis [62–66].

REFERENCES

1. S. Mittal, M. A. Khan, D. Romero, and T. Wuest, "Smart manufacturing: Characteristics, technologies and enabling factors," *Proceedings of the Institution of Mechanical Engineers, Part B: Journal of Engineering Manufacture*, vol. 233, no. 5, pp. 1342–1361, Apr. 2019. doi: 10.1177/0954405417736547
2. A. Kusiak, "Smart manufacturing," *International Journal of Production Research*, vol. 56, no. 1–2, pp. 508–517, Jan. 2018. doi: 10.1080/00207543.2017.1351644.
3. Kumar, Ajay, Parveen Kumar, and Yang Liu. "Industry 4.0 Driven Manufacturing Technologies." 2024. https://doi.org/10.1007/978-3-031-68271-1
4. P. Zheng *et al.*, "Smart manufacturing systems for industry 4.0: Conceptual framework, scenarios, and future perspectives," *Frontiers of Mechanical Engineering*, vol. 13, no. 2, pp. 137–150, Jun. 2018. doi: 10.1007/s11465-018-0499-5
5. J. Lee, E. Lapira, B. Bagheri, and H. Kao, "Recent advances and trends in predictive manufacturing systems in big data environment," *Manufacturing Letters*, vol. 1, no. 1, pp. 38–41, Oct. 2013. doi: 10.1016/j.mfglet.2013.09.005
6. Kumar, Ajay, Yang Liu, and Rakesh Kumar, eds. Handbook of Intelligent and Sustainable Manufacturing: Tools, Principles, and Strategies. CRC Press, 2024. https://doi.org/10.1201/9781003405870

7. C. Leiva, "On the Journey to a Smart Manufacturing Revolution," IndustryWeek. Accessed: Aug. 10, 2023 [Online]. Available: https://www.industryweek.com/technology-and-iiot/systems-integration/article/21967056/on-the-journey-to-a-smart-manufacturing-revolution
8. Y. Lu, K. Morris, and S. Frechette, "Current standards landscape for smart manufacturing systems," *National Institute of Standards and Technology*, NIST IR 8107, Feb. 2016. doi: 10.6028/NIST.IR.8107
9. "The Smart Industry Readiness Index." Accessed: Aug. 10, 2023 [Online]. Available: https://www.edb.gov.sg/en/about-edb/media-releases-publications/advanced-manufacturing-release.html
10. M. Hammad, A. Badshah, G. Abbas, H. Alasmary, M. Waqas, and W. Khan, "A provable secure and efficient authentication framework for smart manufacturing industry," *IEEE Access*, vol. 11, pp. 67626–67639, Jun. 2023. doi: 10.1109/ACCESS.2023.3290913
11. A. Kumar, P Kumar, A. K. Srivastava, and V. Goyat, *Modeling, Characterization, and Processing of Smart Materials.* IGI Global, 1AD. Accessed: Oct. 20, 2023. [Online]. Available: https://www.igi-global.com/gateway/book/318402
12. S. Phuyal, D. Bista, and R. Bista, "Challenges, opportunities and future directions of smart manufacturing: A state of art review," *Sustainable Futures*, vol. 2, p. 100023, Jan. 2020. doi: 10.1016/j.sftr.2020.100023
13. H. E. Negulescu, "A configuration model of the smart manufacturing company," *Bulletin of the Transilvania University of Brasov. Series V: Economic Sciences*, pp. 79–88, Jun. 2023. doi: 10.31926/but.es.2023.16.65.1.9
14. U. M. Dilberoglu, B. Gharehpapagh, U. Yaman, and M. Dolen, "The role of additive manufacturing in the era of Industry 4.0," *Procedia Manufacturing*, vol. 11, pp. 545–554, Jan. 2017. doi: 10.1016/j.promfg.2017.07.148
15. L. Damiani, M. Demartini, G. Guizzi, R. Revetria, and F. Tonelli, "Augmented and virtual reality applications in industrial systems: A qualitative review towards the industry 4.0 era," *IFAC-PapersOnLine*, vol. 51, no. 11, pp. 624–630, Jan. 2018. doi: 10.1016/j.ifacol.2018.08.388
16. A. Parveen, A. Kumar, R. K. Mittal, and R. Goel, Ed., *Waste Recovery and Management: An Approach toward Sustainable Development Goals.* Boca Raton: CRC Press, 2023. doi: 10.1201/9781003359784
17. Paul, Aditi, Somnath Sinha, Parveen Kumar, Shirasthi Choudhary, Krishna Samdani, and Saumya Mishra. "Overview of Cyber Security in Intelligent and Sustainable Manufacturing." In *Handbook of Intelligent and Sustainable Manufacturing*, pp. 95–118. CRC Press, 2025.
18. A. K. Jain, "Modernization of Indian defence forces: Challenges & prospects," *International Journal of Science and Research (IJSR)*, vol. 7, no. 5, 2016.
19. Defence Reports, "Defence Manufacturing Industry in India – IBEF," India Brand Equity Foundation. Accessed: Jul. 14, 2023 [Online]. Available: https://www.ibef.org/industry/defence-manufacturing
20. "Indian-Defence-Industry-Redefining_Frontiers-web.pdf." Accessed: Jul. 08, 2023 [Online]. Available: https://nishithdesai.com/Content/document/pdf/ResearchPapers/Indian-Defence-Industry-Redefining_Frontiers-web.pdf
21. "Seven New Defence Public Sector Units (DPSUs)," Drishti IAS. Accessed: Jul. 25, 2023 [Online]. Available: https://www.drishtiias.com/daily-updates/daily-news-analysis/seven-new-defence-public-sector-units-dpsus
22. "Defence Public Sector Undertakings | Department of Defence Production." Accessed: Jul. 25, 2023 [Online]. Available: https://www.ddpmod.gov.in/defence-public-sector-undertakings
23. "Indian Defense Industry -Top Private Sector Players," Defense Industry. Accessed: Jul. 25, 2023 [Online]. Available: http://defenseindustry.in/top-private-sector-players/

24. J. M. Müller, and K.-I. Voigt, "Sustainable industrial value creation in SMEs: A comparison between Industry 4.0 and made in China 2025," *International Journal of Precision Engineering and Manufacturing-Green Technology*, vol. 5, no. 5, pp. 659–670, Oct. 2018. doi: 10.1007/s40684-018-0056-z
25. Lavanya, Raja, G. Bhavani, C. Santhiya, G. Vinoth Chakkaravarthy, and Parveen Kumar, "Digital Twins and IoT for Sustainable Manufacturing: A Survey of Current Practices and Future Directions." In *Handbook of Intelligent and Sustainable Manufacturing*, pp. 80–94. CRC Press, 2025.
26. Kumar, Mukul, Sourabh Anand, Pushpendra S. Bharti, Manoj Kumar Satyarthi, Parveen Kumar, and Ajay Kumar. "Progressive Automation: Mapping the Horizon of Smart Manufacturing with RoboDK Workstations and Industry 4.0." In *Industry 4.0 Driven Manufacturing Technologies*, pp. 335–354. Cham: Springer Nature Switzerland, 2024.
27. J. Kovar, K. Mouralova, F. Ksica, J. Kroupa, O. Andrs, and Z. Hadas, "Virtual reality in context of Industry 4.0 proposed projects at Brno University of Technology," in *2016 17th International Conference on Mechatronics – Mechatronika (ME)*, Dec. 2016, pp. 1–7.
28. Bhatia, Vineet, Ajay Kumar, Sumati Sidharth, Sanjeev Kumar Khare, Surendra Chandrakant Ghorpade, Parveen Kumar, and Gaydaa AlZohbi. "Industry 4.0 in Aircraft Manufacturing: Innovative Use Cases and Patent Landscape."" In *Industry 4.0 Driven Manufacturing Technologies*, pp. 103–137. Cham: Springer Nature Switzerland, 2024.
29. P. Fraga-Lamas, T. M. FernáNdez-CaraméS, Ós. Blanco-Novoa, and M. A. Vilar-Montesinos, "A review on industrial augmented reality systems for the industry 4.0 shipyard," *IEEE Access*, vol. 6, pp. 13358–13375, 2018. doi: 10.1109/ACCESS.2018.2808326
30. Anand, Sourabh, Manoj Kumar Satyarthi, Pushpendra S. Bharti, Parveen Kumar, and Ajay Kumar. "Digital Twin Integration for Enhanced Control in FDM 3D Printing." In *Industry 4.0 Driven Manufacturing Technologies*, pp. 373–388. Cham: Springer Nature Switzerland, 2024.
31. Tege, Saurabh, and Parveen Kumar. "Intelligent and Sustainable Manufacturing Applications in the Automotive Industry." In *Handbook of Intelligent and Sustainable Manufacturing*, pp. 305–335. CRC Press.
32. F. Manuri, and A. Sanna, "A survey on applications of augmented reality," *Advances in Computer Science: An International Journal*, vol. 5, no. 1, pp. 18–27, 2016.
33. D. Aslan, B. B. Çetin, and I. G. Özbilgin, "An innovative technology: Augmented reality based information systems," *Procedia Computer Science*, vol. 158, pp. 407–414, Jan. 2019. doi: 10.1016/j.procs.2019.09.069
34. E. A. Lee, "The past, present and future of cyber-physical systems: A focus on models," *Sensors*, vol. 15, no. 3, Art. no. 3, Mar. 2015. doi: 10.3390/s150304837
35. V. Gunes, S. Peter, T. Givargis, and F. Vahid, "A survey on concepts, applications, and challenges in cyber-physical systems," *KSII Transactions on Internet and Information Systems*, vol. 8, pp. 4242–4268, Dec. 2014. doi: 10.3837/tiis.2014.12.001
36. A. Kumar, P. Kumar, R. K. Mittal, and V. Gambhir, "Materials Processed by Additive Manufacturing Techniques," in *Advances in Additive Manufacturing*, Elsevier, 2023, pp. 217–233. Accessed: Mar. 22, 2024 [Online]. Available: https://www.sciencedirect.com/science/article/pii/B9780323918343000144
37. A. Kumar, R. K. Mittal, and A. Haleem, Eds., "Additive Manufacturing," in Advances in Additive Manufacturing, in Additive Manufacturing Materials and Technologies, Elsevier, 2023, pp. v–xiii. doi: 10.1016/B978-0-323-91834-3.00033-8
38. A. Kumar, P. Kumar, H. Singh, A. Haleem, and R. K. Mittal, "Integration of Reverse Engineering with Additive Manufacturing," in *Advances in Additive Manufacturing*, Elsevier, 2023, pp. 43–65. Accessed: Mar. 24, 2024 [Online]. Available: https://www.sciencedirect.com/science/article/pii/B9780323918343000284

39. M. Mehrpouya, A. Dehghanghadikolaei, B. Fotovvati, A. Vosooghnia, S. S. Emamian, and A. Gisario, "The potential of additive manufacturing in the smart factory industrial 4.0: A review," *Applied Sciences*, vol. 9, no. 18, Art. no. 18, Jan. 2019. doi: 10.3390/app9183865
40. A. Gisario, M. Kazarian, F. Martina, and M. Mehrpouya, "Metal additive manufacturing in the commercial aviation industry: A review," *Journal of Manufacturing Systems*, vol. 53, pp. 124–149, Oct. 2019. doi: 10.1016/j.jmsy.2019.08.005
41. G. Goyal, A. Kumar, and D. Sharma, "12 Recent Applications of Rapid Prototyping with 3D Printing: A Review," in*3D Printing Technologies: Digital Manufacturing, Artificial Intelligence, Industry 4.0.* Walter De Gruyter, 2024, p. 245.
42. A. Kumar, R. K. Mittal, and A. Haleem, *Advances in Additive Manufacturing: Artificial Intelligence, Nature-Inspired, and Biomanufacturing.* Elsevier, 2022. Accessed: Mar. 22, 2024 [Online]. Available: https://books.google.com/books?hl=en&lr=&id=c8h6EAAAQBAJ&oi=fnd&pg=PP1&dq=info:2DPEeVdY_UsJ:scholar.google.com&ots=s_cRplKIa-&sig=x7sdzcpDj2J2Wul4Hd3zVPYnIMc
43. A. Mittal Ravi Kant, *Incremental Sheet Forming Technologies: Principles, Merits, Limitations, and Applications.* Boca Raton, FL: CRC Press, 2020. doi: 10.1201/9780429298905
44. M. M. Mabkhot, A. M. Al-Ahmari, B. Salah, and H. Alkhalefah, "Requirements of the smart factory system: A survey and perspective," *Machines*, vol. 6, no. 2, Art. no. 2, Jun. 2018. doi: 10.3390/machines6020023
45. S. Ren, Y. Zhang, Y. Liu, T. Sakao, D. Huisingh, and C. M. V. B. Almeida, "A comprehensive review of big data analytics throughout product lifecycle to support sustainable smart manufacturing: A framework, challenges and future research directions," *Journal of Cleaner Production*, vol. 210, pp. 1343–1365, Feb. 2019. doi: 10.1016/j.jclepro.2018.11.025
46. G. Horne, S. Sanchez, S. Seichter, K. Haymann, and D. Nitsch, "Data Farming in Support of Military Decision Makers," Oct. 2009. Accessed: Aug. 17, 2023 [Online]. Available: https://www.semanticscholar.org/paper/Data-Farming-in-Support-of-Military-Decision-Makers-Horne-Sanchez/4857a4e90e57bd293a8cca21aafbea87916b9846
47. D. Kallfass, and T. Schlaak, "NATO MSG-088 case study results to demonstrate the benefit of using Data Farming for military decision support," in *Proceedings of the 2012 Winter Simulation Conference (WSC)*, Dec. 2012, pp. 1–12. doi: 10.1109/WSC.2012.6465132
48. U. Z. Yildirim, I. Sabuncuoglu, B. Tansel, and A. Balcioglu, "A design of experiments approach to military deployment planning problem," in *2008 Winter Simulation Conference*, Dec. 2008, pp. 1234–1241. doi: 10.1109/WSC.2008.4736195
49. M. L. McDonald *et al.*, "Enhancing the Analytic Utility of the Synthetic Theater Operations Research Model (STORM)," Dec. 2014. Accessed: Aug. 17, 2023 [Online]. Available: https://www.semanticscholar.org/paper/ENHANCING-THE-ANALYTIC-UTILITY-OF-THE-SYNTHETIC-McDonald-Upton/d4f5015cce7ed1dfb6a2c3ed7795139c65958363
50. X. Song, Y. Wu, Y. Ma, Y. Cui, and G. Gong, "Military simulation big data: Background, state of the art, and challenges," *Mathematical Problems in Engineering*, vol. 2015, p. e298356, Nov. 2015. doi: 10.1155/2015/298356
51. S. Boinepalli, and G. Brown, "Simulation Studies of Naval Warships using the Ship Air Defence Model (SADM)," 2010. Accessed: Aug. 17, 2023 [Online]. Available: https://www.semanticscholar.org/paper/Simulation-Studies-of-Naval-Warships-using-the-Ship-Boinepalli-Brown/acac603dd5db627334edde1254551829794ef0ff
52. P. Rajendra *et al.*, "Impact of artificial intelligence on civilization: Future perspectives," *Materials Today: Proceedings*, vol. 56, pp. 252–256, 2022.

53. Ajay, H. Singh, Parveen Kumar, and B. AlMangour, Ed., *Handbook of Smart Manufacturing: Forecasting the Future of Industry 4.0*. Boca Raton: CRC Press, 2023. doi: 10.1201/9781003333760.
54. A. Kumar, P. Kumar, N. Sharma, and A. K. Srivastava, *3D Printing Technologies: Digital Manufacturing, Artificial Intelligence, Industry 4.0*. Walter de Gruyter GmbH & Co KG, 2024. Accessed: Mar. 24, 2024 [Online]. Available: https://books.google.com/books?hl=en&lr=&id=-e_tEAAAQBAJ&oi=fnd&pg=PR5&dq=info:AK8zDpUKoWsJ:scholar.google.com&ots=3LM-o8eSag&sig=gn54o804nky1kDe34DDEZ5zhF5g
55. J. Lee, H. Davari, J. Singh, and V. Pandhare, "Industrial artificial intelligence for Industry 4.0-based manufacturing systems," *Manufacturing Letters*, vol. 18, pp. 20–23, Oct. 2018. doi: 10.1016/j.mfglet.2018.09.002
56. M. K. G. Biswas Tarun Gupta, D. Mangal, and P. Thapliyal, Don, "Study and Analysis of IoT (Industry 4.0): A Review," in *Handbook of Smart Manufacturing*, CRC Press, 2023.
57. C. Perera, A. Zaslavsky, P. Christen, and D. Georgakopoulos, "Context aware computing for the internet of things: A survey," *IEEE Communications Surveys & Tutorials*, vol. 16, no. 1, pp. 414–454, 2014. doi: 10.1109/SURV.2013.042313.00197
58. L. Atzori, A. Iera, and G. Morabito, "The internet of things: A survey," *Computer Networks*, vol. 54, no. 15, pp. 2787–2805, Oct. 2010. doi: 10.1016/j.comnet.2010.05.010
59. K.-D. Thoben, S. Wiesner, and T. Wuest, "'Industrie 4.0' and smart manufacturing – A review of research issues and application examples," *International Journal of Automation Technology*, vol. 11, pp. 4–19, Jan. 2017. doi: 10.20965/ijat.2017.p0004
60. "smart-manufacturing-report2.pdf." Accessed: Aug. 10, 2023 [Online]. Available: https://www.sme.org/globalassets/sme.org/media/white-papers-and-reports/smart-manufacturing-report2.pdf
61. D. Chen, G. Doumeingts, and F. Vernadat, "Architectures for enterprise integration and interoperability: Past, present and future," *Computers in Industry*, vol. 59, no. 7, pp. 647–659, Sep. 2008. doi: 10.1016/j.compind.2007.12.016
62. C. Mahatme, J. Giri, F. Mohammad, M. S. Ali, T. Sathish, N. Sunheriya, and R. Chadge, "Experimental and numerical investigation of PLA based different lattice topologies and unit cell configurations for additive manufacturing," *The International Journal, Advanced Manufacturing Technology*, 2024. doi: 10.1007/s00170-024-13882-4
63. J. Giri, T. Sathish, T. Sheikh, N. Sunehriya, P. Giri, R. Chadge, C. Mahatme, and A. Parthiban, "Automatic liver segmentation using U-Net deep learning architecture for additive manufacturing," *Interactions*, vol. 245, no. 1, 2024. doi: 10.1007/s10751-024-01927-9
64. M. S. Tufail, J. Giri, E. Makki, T. Sathish, R. Chadge, and N Sunheriya, "Machinability of different cutting tool materials for electric discharge machining: A review and future prospects," *AIP Advances*, vol. 14, no. 4, 2024. doi: 10.1063/5.0201614
65. K. L. Narasimhamu, M. Natarajan, P. Thejasree, E. Makki, J. Giri, N. Sunheriya, R. Chadge, C. Mahatme, P. Giri, and T. Sathish "Development of hybrid optimization model using Grey-ANFIS-Jaya algorithm for CNC drilling of aluminium alloy," *Journal of Engineering*, vol. 2024, pp. 1–12, 2024. doi: 10.1155/2024/1476770
66. M. Natarajan, T. Pasupuleti, J. Giri, H. A. Al-Lohedan, L. N. Katta, F. Mohammad, N. Sunheriya, R. Chadge, C. Mahatme, P. Giri, S. Mallik, and T. Sathish, "Optimization of wire spark erosion machining of Grade 9 titanium alloy (Grade 9) using a hybrid learning algorithm," *AIP Advances*, vol. 14, no. 1, 2024. doi: 10.1063/5.0177658

7 Design and Development of Indexing Jig Using CAD Software

Pankaj Shakkarwal and Nitin Kumar Waghmare

7.1 INTRODUCTION

The demand for practically all commodities and services is rising due to the rapid growth of the human population. Businesses must improve their production levels to meet this demand. However, many businesses cannot compromise on their profit margins in order to increase their production level. Companies must therefore match consumer expectations while maintaining the quality of their products and their own profit margins. The only way to hit this goal is to keep the production costs as low as possible. Companies must ensure proper resource utilization for this reason. They need to boost productivity and minimize losses during production [1].

The basic setup for drilling process via jig is illustrated in Figure 7.1. A jig is a tool used to hold work pieces primarily for positioning and supporting the work piece. Additionally, it directs the cutting tool to carry out a specific task. Drilling, reaming, milling, and tapping all operation are performed by jigs.

These are specialized design instruments that make industrial processes like machining, assembly, and inspection easier [2]. The main industrial uses of jigs and fixtures are to reduce the cost of manufacturing, increase the productivity, higher accuracy of parts, interchangeability, reduce quality control expenses.

A new method was proposed to enhance error efficiency by utilizing a two-dimensional manifold that reduces the dimensionality of the workspace and detailed contribution as mentioned in Table 7.1. Simulations were performed in DELMIA and MATLAB to verify the effectiveness of proposed method [3]. A variation analysis method based on small-sample data for aero-structure assembly was suggested. The technique is a useful addition to more conventional variation analysis techniques that use probability distributions of variation sources [4]. Drill jig, then, boosts production by doing away with the need for individual placement, marking, and frequent checking. Fixtures can also aid in streamlining the metalworking processes carried out on specialized machinery [5, 6]. The design and simulation of a beam and clamping fixture aim to reduce idle time, minimize human labor requirements, and maximize output rates [1]. Studied about the design of jig for Weight Lever, which includes things like decrease production times, better and more accurate dimensioning, less rejection, and cost effectiveness [7, 8]. Designing and manufacturing an acyclic jig to mitigate issues associated with manual marking processes. Reducing the production cycle time with implemented combinations [2]. Solidworks was used to create the fixtures and jigs. Increased worker safety and facilitating mass productions [9].

DOI: 10.1201/9781032725086-8

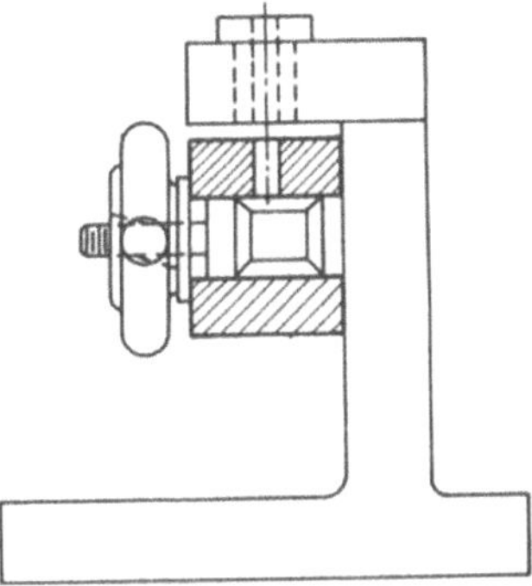

FIGURE 7.1 Basic structure of a drill jig.

TABLE 7.1
Summary of the Literature Studies and Highlights the Contribution of the Work

Author's (Year)	Objective Function	Methods	Analysis
Zhu et al. [3]	Reduce the dimensionality of the workspace	Two-dimensional manifold	DELMIA and MATLAB
Mei et al. [4]	For the assembly of compliant aerospace structures	Hybrid assembly variation model	Variation Analysis method
Kundu and Patel [5]	To increase the productivity and the accuracy	Drill jig designed	Reducing the set up cost and manual fatigue
Babu and Reddy [1]	Simulation and design of the clamping fixture and beam	Pneumatic Clamping Fixture	Idle time was reduce and Maximize output rate
Ali and Mahalle [7]	Decrease production times, better and more accurate dimensioning	Jig for Weight Lever	Less rejection, and cost effectiveness
Momin et al. [2]	Design and manufacturing	Acrylic jig the cost of jig manufacturing is reduced	Reducing production cycle time
Shrivastava and Verma [9]	Worker safety and facilitating mass productions	Manufacturing of paddy weeder	Saving 30% of time elapsed
Sangamesh and Kulkarni [10]	Design and fabrication of jig	Shaft	Cycle times for hydraulic and manual indexing have also been compared
Leonarld et al. [12]	To create a drill bush	Design of tools	Manufacture high-quality components
Dongre et al. [13]	Reduction of idle time and weight if jigs and fixture	Manufacturing of chassis bracket	Determining the stress exerted on jigs, fixtures, and brackets
Chaskar et al. [14]	Reduce the fatigue of the worker	Drilling operation of different hole sizes on one jig itself	Improved efficiency

The shaft that, while indexing, must support the weight of the work piece and the drill jig. Cycle times for hydraulic and manual indexing have also been compared [10, 11]. To create a drill bush that is adjustable so that it can better hold work pieces of different sizes. Tool design that reliably yields high-quality components [12]. For manufacturing the chassis bracket, design jigs and fixtures and analyze the stress and strain that occur in both the jigs/fixtures and the chassis bracket. Stresses were calculated which acting on jigs and fixtures and bracket [13]. Creating a drilling jig that can perform drilling operations of various hole sizes on a single jig to eliminate part rejections [14]. Some other materials also play an important role in jigs and fixture design and manufacturing process [15–20].

The objectives of the present study to improve the methodology of conventional techniques by designing and development of indexing jig using Creo Parametric.

7.1.1 Problem Statement

Meeting the modern industry's demand for maximum productivity at minimal cost poses a challenge to tool designers. In current scenario, as small-scale industries are performing machining operation, i.e. marking, punching, drilling, reaming, and tapping have been perform manually. Due to which machining technique has slowed down the productivity.

In the current study, Cero Parametric software is used to design and fabricate an indexing drilling jig to increase the productivity and reduce the labor cost. Time required without jig is listed in Table 7.2.

7.1.1.1 Objective

The objectives of the present study to improve the methodology of conventional techniques by designing and development of indexing jig using Creo Parametric second, software was used to design each component, including the base plate, locator, clamping devices, bushes, jig plate, and indexing mechanism. The whole design procedure was completed with the help of Cero Parametric software which helps for designing, drafting assembly and useful for customized applications and manipulations. Calculations are made for specific parameters like thrust, clamping force, and rpm.

7.2 DESIGN CONSIDERATION

All the tool design begins in the mind of the tool designer. To translate tooling concepts into usable hardware, extensive preparation and study are required. As with

TABLE 7.2
Time Required before Jig

S. No.	Process Name	No. of Workers Required	Time Required (S)
1	Marking + Punching	1	108 + 72 = 180
2	Job Setting + Drilling	1	20 + 45 = 65
	Total	2	245

TABLE 7.3
Specifications of Indexing Drilling Jig

S. No.	Components of Jig	Dimensions (mm)	Material
1	Base Plate	100 * 100	Mild Steel
2	Support Plate	100 * 100	Mild Steel
3	Jig Plate	100 * 10, R = 10	EN31
4	Jig Bush	Outer = 8, Inner = 5	EN353
5	Springs	Outer = 8, Inner = 5, L = 36	Standard
6	lever	L = 135	EN31
7	Indexing Plate	Cutting Edge = 45°	EN31
8	Knob	R = 100	Mild Steel
9	Locater	L = 73, Max. Dia. = 12	Mild Steel
10	Clamper	D = 38	Mild Steel

any tool design, the first step is to arrange all relevant data. Production schedules and part drawings are closely examined to determine the precise tool needed. The second step involved developing the dimensions and individual elements, including the base plate, locator, clamping devices, bushes, jig plate, and indexing mechanism listed in Table 7.3. All these parts have been designed, modeled, drafted with the help of Creo Parametric [21, 22].

7.2.1 Component Description – Housing of Indexing Jig

Machining components is made up of mild steel. The primary advantages of mild steel include stability, reduced machining time, and even material distribution. The angular holes inclined at 25° in equally spaced as shown in Figure 7.2 [23, 24].

7.2.1.1 Design of Base Plate and Support Plate

The Base plate is an essential component in the design of the jig. Typically, the base is used to support the complete assembly, such as clamps, supports, locators, and other information required to reference, locate, and hold the parts while working is

FIGURE 7.2 3D model of the component.

FIGURE 7.3 3D model of base plate and support plate.

performed. Generally, the size of the part and the task to be completed dictate the shape and size of the base plate and support plate. We used both vertical and horizontal plates in our model, with the horizontal plate serving as the base and the vertical plate as the support as shown in Figure 7.3. To avoid structural distortion during the machining and non-machining processes, in design, the base plate is shaped to fit the work piece's measurements.

7.2.1.1.1 Design of Jig Plate

The component of a jig drill that sets the drill bushing's holes is called the plate. Typically, the size of the bushing used determines the thickness of the jig plate. The bushing should have sufficient length to effectively support and guide the tool as specified. Generally, precision can be prevented with a length that is one to two times the tool diameter. In order to ensure tool accuracy, the bushing's wall thickness should be able to easily bear all cutting forces. Material used for jig plate is EN31 as shown in Figure 7.4 [21, 22].

7.3 DESIGN OF LOCATION OF COMPONENTS

7.3.1 Design of Jig Bush

Drill bushings are employed to precisely locate and guide drills, reamers, taps, counter bores, countersinks, spot-facing tools, and other rotating tools commonly

FIGURE 7.4 3D model of jig plate.

FIGURE 7.5 3D model of jig bush.

used for creating or modifying holes shown in Figure 7.5. Drill bushings are typically hardened and ground to exact sizes to ensure the necessary repeatability in the jig.

7.3.1.1 Design of Locator

It is crucial that the work piece is positioned correctly before beginning any machining operation to ensure accuracy and precision. To make sure that the work piece can be loaded and unloaded simply, locators must be chosen as shown in Figure 7.6. Additionally, it makes sure that work pieces are solidly held and properly positioned shown in Figure 7.6 [21, 22].

7.3.1.1.1 Design of Indexing Plate

A circular plate that has been graduated or that has circular rows of holes that are spaced differently and is used in machines (such as for cutting gear teeth or graduating circles).To facilitate the precise positioning of holes or other machining operations on the work piece, the plate can be rotated while the work piece is in place as shown in Figure 7.7 [21, 22].

7.3.1.1.1.1 Design of C-Washer Using inward pressure, a clamp is a fastening tool that holds or secures things firmly together to prevent movement or separation as shown in Figure 7.8. The jig design is safe since, according to the calculation, the clamping force is greater than the drilling force (machining force), and the clamp selection is further supported by its formability.

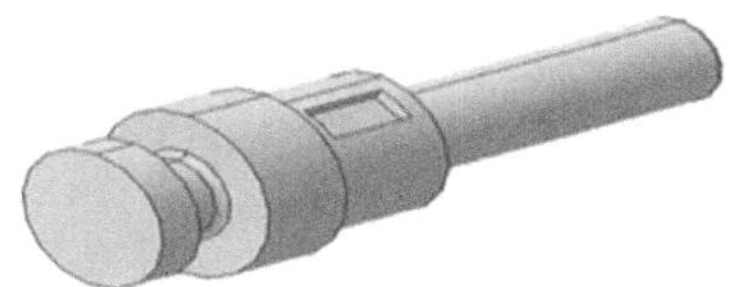

FIGURE 7.6 3D model of locator.

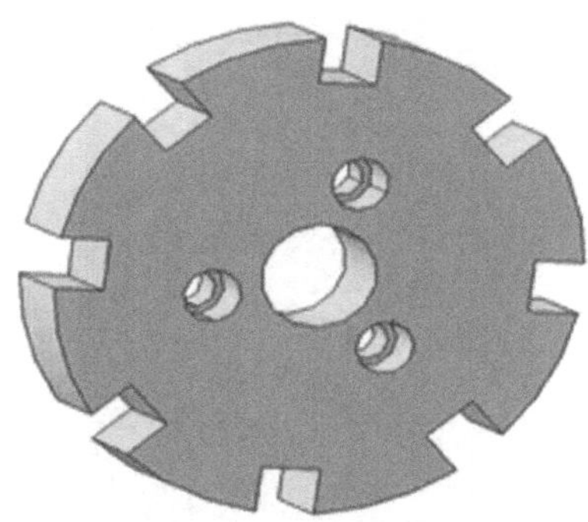

FIGURE 7.7 3D model of indexing plate.

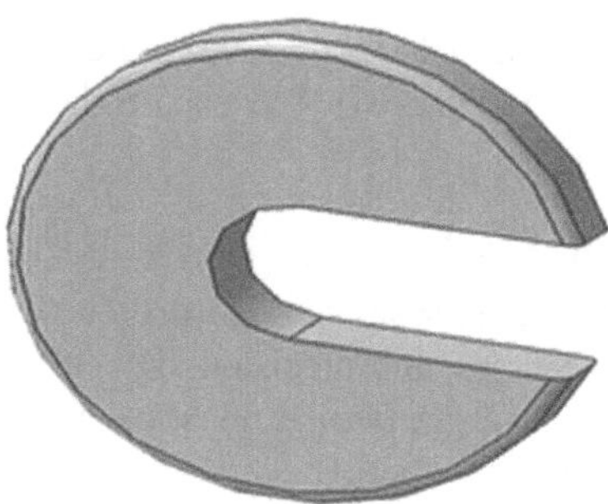

FIGURE 7.8 3D model of c-clamp.

7.3.1.1.1.1.1 Design of Tapper Bush The Taper Lock bush, sometimes referred to as a Taper bush or Taper Fit bush, is commonly used in power transmission drives to mount pulleys, sprockets, and couplings on shafts as shown in Figure 7.9. The relevant shaft and key way sizes are pre-bored and keyed into the Taper Lock bush. The bush's exterior is tapered to fit the bore of the component that will be mounted on the shaft.

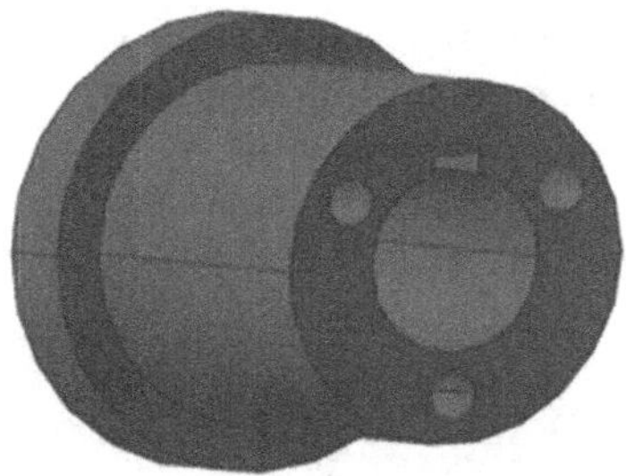

FIGURE 7.9 3D model of tapper bush.

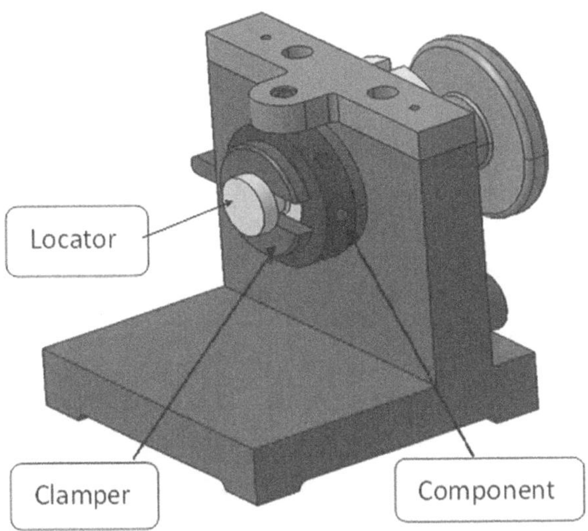

FIGURE 7.10 3D model of complete assembly of indexing jig in isometric front view.

7.4 INDEXING DRILL JIG ASSEMBLY

Complete assembly of indexing jig is shown in Figure 7.10.

7.5 INDEXING MECHANISM

The "Indexing Plate and Lever Arrangement" serves as the mechanism for indexing, functioning as follows: lowering the lever disengages the spring, causing the locator and indexing plate to rotate. This action aligns the taper bush with the next required hole position, and the lever locks into the precise hole location at the corner of the indexing plate, completing the indexing process.

This is a typical indexing method that is simple to use, has little operational time, and doesn't require skilled personnel. Here, it is easy to see that the indexing plate has cut lines at 45°, and a lever ensured the component would be positioned correctly for the following hole as shown in Figure 7.11.

The locator pin is the most important component in the design of the drill jig because it is crucial to maintain the relationship; as a result, a locator pin is used. In this design, the locator pin is parallel to the jig base plate and the component and locator axes are inclined perpendicularly to one another as shown in Figure 7.12.

Therefore, the notion of a secure design is upheld, ensuring that even unskilled operators can effortlessly operate the jig, as the component is securely held in the desired position, leaving no room for positional errors.

7.6 DRILLING FORCE CALCULATION

For several parameters, such as thrust, clamping force, and rpm, the following computations are carried out and compared with the allowable limits when designing for the selection of the necessary pieces of the indexing type of drill jig [25].

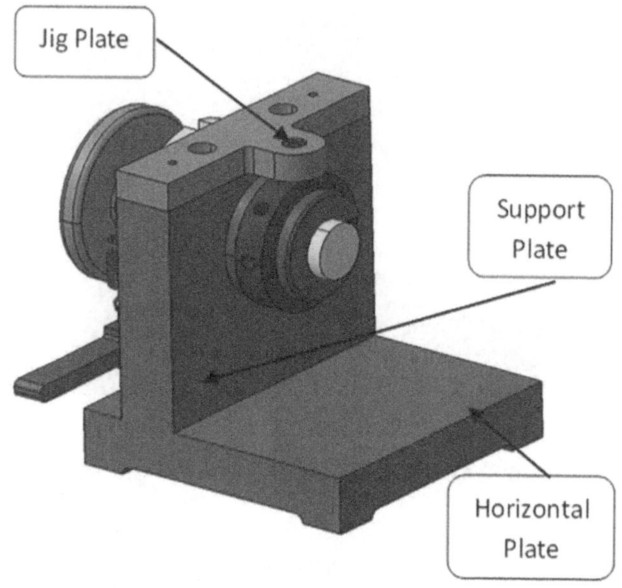

FIGURE 7.11 3D model of complete assembly of indexing jig in isometric back view.

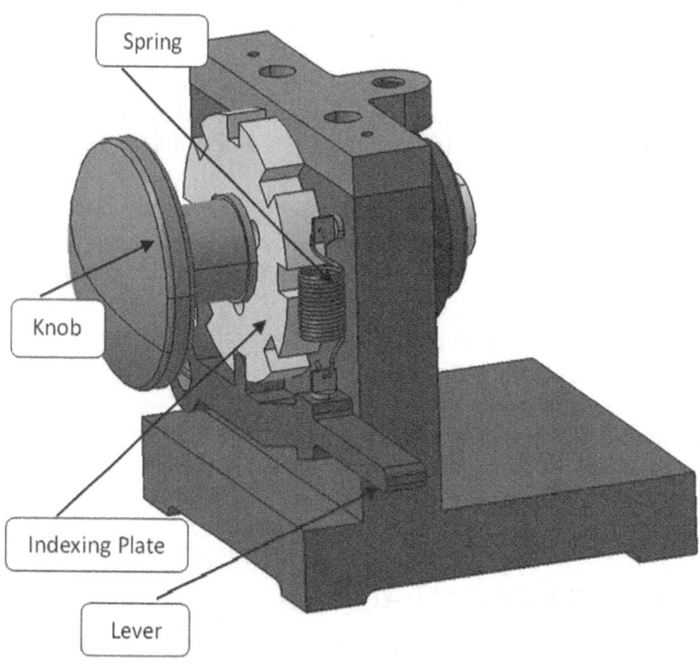

FIGURE 7.12 3D model of complete assembly of indexing jig in front view.

7.6.1 Calculations of Cutting Forces

Drilling Thrust

The axial force a drill bit applies to the work piece or substance it is drilling is referred to as the "drilling thrust". The following formula can be used to estimate the axial thrust F (N) [26]:

$$F = (K * K_c * f * d)/2, \tag{7.1}$$

K_c = Specific cutting force, which depends primarily on the material being machined = 650 N/mm^2

f = Feed per rotation = 0.5 mm, d = Tool diameter = 25 mm

K = The coefficient depends on the geometry of the tip of the tool, we can consider an average value of 0.6

$$F = (0.6 * 650 * 0.5 * 25)/2.$$

Thrust Force (F) = 2437 N

Drilling Torque

It is the force that the drill motor applies to the drill bit in order to rotate it and cause it to interact with the material being drilled.

T = Drilling torque in Nm
C = Constant depend upon the material
f = Drill feed in mm/rev.
D = Tool diameter in mm

$$\begin{aligned} T1 &= C * D^{1.8} f^{0.75} \\ &= 0.36 * 25^{1.8} * 0.5^{0.75} \\ &= 70.27\,\text{N-m}. \end{aligned} \tag{7.2}$$

$$\text{Torque} = 70.27\,\text{N-m}$$

Spindle Speeds

The term "spindle speed" describes the spindle's rotational speed in a machine tool or drilling apparatus, which is commonly expressed in revolutions per minute (RPM).

$$\begin{aligned} V &= (3.14 * D * N)/(1000) \\ 28\,\text{m/min} &= 3.14 * 25 * N/1000. \\ N &= 357\,\text{rpm} \end{aligned} \tag{7.3}$$

Power of Drilling

The quantity of mechanical energy used or necessary for drilling is referred to as the drilling power. What is meant by "power of drilling" is the amount of energy needed to drill successfully and efficiently.

$$\begin{aligned} P &= (2*3.14*\mathrm{N}*\mathrm{T1})/60{,}000 \\ &= (2*3.14*357*70.27)/60{,}000. \\ \text{Cutting Power}(P) &= 3.82\,\mathrm{kW} \end{aligned} \quad (7.4)$$

7.6.2 Calculation of Marching Parameters

Machining Time

The productivity and efficiency of machining processes are reflected in machining time, which is a crucial parameter in manufacturing activities. It includes all time-related elements of transforming raw materials into completed parts, impacting industrial production schedules and expenses overall.

$$\begin{aligned} \mathrm{T} &= \mathrm{L}/(\mathrm{N}*\mathrm{f}) \\ &= 30/(0.5*357)*60. \\ &= 10.08\ \text{seconds} \end{aligned} \quad (7.5)$$

Volume of Metal Removed/min

The "Volume of Metal Removed per Minute" (VMR/min) is a metric used in machining and manufacturing processes to quantify the amount of material that is cut away from a workpiece in a given period of time, typically 1 minute.

$$\begin{aligned} &= \text{Area of hole} * \text{Feed} * \text{Speed} \\ &= 3.14*12.5*12.5*0.5*357 \\ &= 87576.56\,\mathrm{mm}^3 \end{aligned} \quad (7.6)$$

Energy Consumption

It is the amount of energy used by a system, device, or process over a specified period.

$$\begin{aligned} &= \text{Vol. of metal removed} / \text{min} / \text{Power} \\ &= 87{,}576.56/3.82 \\ &= 22.93\,\mathrm{mm}^2/\text{watt min.} \end{aligned} \quad (7.7)$$

The above result can be summarized in Table 7.4 which allows us to easily compare and observed the required forces during the drilling process by the designed setup that is known as indexing drill jig.

TABLE 7.4
Summary of Calculated Values

Parameters	Calculated Values
Drilling Thrust	2437 N
Drill Torque	70.27 N-m
Machining Time	10.08 seconds
Power of Drilling	3.82 kW
Volume of Metal Removed/min	87576.56 mm^3
Energy Consumption	22.93 mm^2/watt min
Clamping Force	3500 N

TABLE 7.5
Comparison

S. No.	Existing Solution (Industry Data)	Proposed Solution (Design Jig)
1	Operation Time = 245 seconds (Marking + Setting + Drilling)	Operation Time = 60 seconds (Loading + Drilling + Unloading)
2	Number of Workers Required = 2	Number of Workers Required = 2
3	Rate of Production = 116 Jobs/Day (Working Hours = 8 hours)	Rate of Production = 480 Jobs/Day (Working Hours = 8 hours)
4	Labor Cost = Rs 2.4/-	Labor Cost = Rs 0.30/-

It is evident from Table 7.4 that the clamping force exceeds the drilling force, indicating that the design is safe for machining. Therefore, the design is evaluated and meets the requirements of the interchangeable part idea. It is also error-free [26–30].

Drilling results, such as bore and surface finish, are found to be within tolerance. The advantage of our Indexing jig model can well define by Table 7.5.

7.7 CONCLUSION

Designing and manufacturing of indexing using material is mild steel which ultimately result into:

1. Since the clamping force is more than the estimated drilling forces, the design is secure.
2. There is a reduction in production idle time and a proper assembly of the indexing type drill jig.
3. Operation time, no. of workers, labor cost was reduced with compare an existing solution in Table 7.5.
4. Drilling results, including as bore and surface finish, were found to be within tolerance.

7.8 FUTURE SCOPE

Indexing jigs are expected to grow significantly in the future in a number of industries. Indexing jigs are expected to become increasingly important as technology develops because they help increase manufacturing processes' accuracy, adaptability, and efficiency. The integration of indexing jigs with sophisticated automation and robotics systems is a critical component of their future development. The ability to make modifications quickly and easily thanks to this connection will shorten setup times and boost output overall.

Moreover, improvements in manufacturing processes and materials science will probably result in the creation of indexing jig designs that are both more robust and lightweight. These developments will increase manufacturing processes' overall cost effectiveness in addition to improving mobility. Future indexing jigs might also include energy-efficient mechanisms and environmentally friendly materials in response to the increased demand for sustainability, which would be in line with international environmental standards.

Essentially, by fusing cutting-edge technology, sustainability principles, and seamless integration into smart manufacturing settings, indexing jigs have the potential to completely transform manufacturing procedures in the future.

REFERENCES

1. Babu, T. Anji, & Reddy, D. Vidyasagar. "Design and simulation of pneumatic clamping fixture to minimize the idle time and maximize the production rate in welding process". *International Journal & Magazine of Engineering, Technology, Management and Research* 7(2015): 288–291.
2. Momin, Mujaffar, Lokhande, Sanket, Gunavant, Pradip, & Kokil, Narendra. "Design and manufacturing of acrylic jig". *International Research Journal of Engineering and Technology* 3(5)(2016): 1571–1574.
3. Zhu, Weidong, Li, Guanhua, Dong, Huiyue, & Ke, Yinglin. "Positioning error compensation on two-dimensional manifold for robotic machining". *Robotics and Computer Integrated Manufacturing* 59(2019): 394–405.
4. Mei, Biao, Zhu, Weidong, Ke, Yinglin, & Zheng, Pengyu. "Variation analysis driven by small-sample data for compliant aero-structure assembly". *Assembly Automation* 39(2019): 101–112.
5. Kundu, Pawan Kumar, & Patel, Savan Kumar. "Design and fabrication of work holding device for a drilling machine". *International Journal of Pure and Applied Research in Engineering and. Technology* 1(8)(2013): 177–186.
6. Muniappan, A., & Thiagarajan, C. "Design and fabrication of stab arm weldment for drilling jig". *International Journal of Applied Research and Studies* 7(2014).
7. Ali, Mohsin, & Mahalle, Ganesh. "Design and analysis of weight lever drilling jig". *International Journal of Pure and Applied Research in Engineering and Technology* 8(2013): 177–186.
8. Abouhenidi, Hamad Mohammed. "Jig and fixture design". *International Journal of Scientific & Engineering Research* 5(2)(2014): 142–153.
9. Shrivastava, Ankit, & Verma, Ajay. "Implementation of improved jigs and fixtures in the production of non-active rotary paddy weeder". *Science Technology and Arts Research. Journal* 4(2014): 152–157.

10. Sangamesh, & Kulkarni, V. V. "Design and fabrication of indexing set up for drill jig". *International Research Journal of Engineering and Technology* 03(06)(2016).
11. Midhun, R, & Vignesh, A. "Design and fabrication of jigs and fixtures for drilling operation". *International Journal for Scientific Research & Development* 3(12)(2016): 853–859.
12. Leonarld, Premodh, Gladson, G. Jerom Nithin, & Sunil, Kumar. "Design and optimization of drill jig". *International Journal of Scientific Research Engineering & Technology* 5(8)(2016): 458–463.
13. Dongre, Sawita D., Gulhane, U. D., & Kuttarmare, H.C. "Design and finite element analysis of jigs and fixtures for manufacturing of chassis bracket". *International Journal of Research in Advent Technology* 2(2)(2014).
14. Chaskar, Keta A., Ingle, Arjun D., Phadale, Litesh A., Patil, Sumit S., & Ingole, P. R. "Design & manufacturing of drilling jig". *International Journal for Engineering Applications and Technology* 6(3)(2016): 27–30.
15. Ajay, Praveen, Mittal, Ravi, Kant, & Goel, Rajesh. Waste Recovery and Management: An Approach toward Sustainable Development Goals. CRC Press (2023).
16. Kumar, Ajay, Mittal, Ravi Kant, & Haleem, Abid. Advances in Additive Manufacturing: Artificial Intelligence, Nature-Inspired, and Bio Manufacturing. Elsevier (2022).
17. Waghmare, Nitin Kumar, & Khan, Sabah. "Extraction and characterization of nanocellulose fibrils from Indian sugarcane bagasse – An agro waste". *Journal of Natural Fibers* 19(2022).
18. Singh, Suraj Kumar, Waghmare, Nitin Kumar, Khan, Sabah, & Mishra, R. K. "Mechanical properties of epoxy hybrid composites reinforced with agave fiber and zinc powder," *AIP Conference Proceedings* 2317(2021).
19. Prabin, Sahana, Waghmare, Nitin Kumar, Singh, Suraj Kumar, & Khan, Sabah. "Mechanical properties of natural fiber reinforced epoxy composites: A review". *Procedia Computer Science* 152(2019): 375–379.
20. Kamble, P. D., Giri, J., Makki, E., Sunheriya, N., Sahare, S. B., Chadge, R., Mahatme, C., Giri, P., T, S., & Panchal, H. "An application of hybrid Taguchi-ANN to predict tool wear for turning EN24 material". *AIP Advances* 14(1)(2024). https://doi.org/10.1063/5.0186432
21. Ajay, Mittal, & Kant, Ravi. Incremental Sheet Forming Technologies: Principles, Merits, Limitations, and Applications. CRC Press (2020).
22. Kumar, Ajay, Kumar, Praveen, Srivastava, Ashish Kumar, & Goyal, Vikas. Modeling, Characterization, and Processing of Smart Materials. IGI Global (2023).
23. Kumar, Love, & Sharma, Rajiv Kumar. Smart Manufacturing and Industry 4.0: State-of-the-Art Review. CRC Press (2023).
24. Kumar, Ajay, Singh, Hari Singh., Kumar, Praveen, & AlMangour, Bandar. (Eds.). Handbook of Smart Manufacturing: Forecasting the Future of Industry 4.0. CRC Press (2023).
25. Joshi, Prakash H., & Paramly, Robert O. Jigs and Fixtures Design Manual. 2nd Ed. McGraw-Hill (2013).
26. Narsaiah, S. R. V., Radhakrishna, S. L., Yerra, Lakshmipathi, & Prasanna, V. "Design and analysis of indexing type drill jig for a missile component". *International Journal of Mechanical Engineering and Technology* 8(2017): 889–898.
27. Giri, J., Sathish, T., Sheikh, T., Sunehriya, N., Giri, P., Chadge, R., Mahatme, C., & Parthiban, A. "Automatic liver segmentation using U-Net deep learning architecture for additive manufacturing". *Interactions* 245(1)(2024). https://doi.org/10.1007/s10751-024-01927-9
28. Praveen, N., Siddesh Kumar, N. G., Prasad, C., Giri, J., Albaijan, I., Mallik, U., & Sathish, NT. "Effect of pulse time (Ton), pause time (Toff), peak current (Ip) on MRR

and surface roughness of Cu–Al–Mn ternary shape memory alloy using wire EDM". *Journal of Materials Research and Technology/Journal of Materials Research and Technology* (2024). https://doi.org/10.1016/j.jmrt.2024.03.122

29. Adakane, R., Washimkar, P. V., Chaudhari, S. S., Giri, J., Sathish, T., Parthiban, A., & Mahatme, C. "Prediction and analysis of material removal rate and tool wear for electric discharge machining of H16 material using ANN and ANOVA". *Interactions* 245(1)(2024). https://doi.org/10.1007/s10751-024-01933-x
30. Natarajan, M., Pasupuleti, T., Abdullah, M. M. S., Mohammad, F., Giri, J., Chadge, R., Sunheriya, N., Mahatme, C., Giri, P., & Soleiman, A. A. "Assessment of machining of hastelloy using WEDM by a multi-objective approach". *Sustainability* 15(13)(2023): 10105. https://doi.org/10.3390/su151310105

8 Formability and Fracture Criteria Study of Sintered Al-Fe_3C-Cu Composites under Cold Upsetting

Rajeshkannan Ananthanarayanan, A. K. Jeevanantham, Devi Seenivasagam, Ravichandran Rajeeth Kumar, and Rajkumar Shashank Rajkumar

8.1 INTRODUCTION

Aluminium (Al) matrix composites have evolved over 40 years and have grown significantly due to their preferable and potential applications in various fields [1, 2]. It attracted vast attention due to their combined properties of matrix and reinforcement that amalgamate toughness with high mechanical strength and thermal and wear resistance [3, 4]. As far as reinforcement material is concerned, the vast majority of oxides, carbides, silicides and nitrides were tested; specifically, SiO_2, Al_2O_3, SiC-TiC, TiN, ashes and graphites are the common materials being used [5, 6]. Several combinations have brought out significant achievements in specific properties. Especially low-specific density, high strength and stiffness and good wear resistance are the paramount categories to establish in research [7]. The most widely employed technique to produce Al-Composite is stir casting and powder metallurgy (P/M) in industries. While these techniques enable net-shape processing, it still faces issues with porosity that compromise structural applications, highlighting areas that require further research [8–10].

Among alloying elements, copper (Cu) is considered the optimal addition to Al, with its incorporation ranging from 1 to 6% due to the significant enhancement in mechanical properties within this range [11]. The workability of various carbide-reinforced materials was examined using P/M processing. It was found that titanium carbide (TiC) performed best in terms of densification and formability at 2 wt% and 4 wt%, with iron carbide (Fe_3C) being the second best. These findings were true irrespective of preform geometries as well [12]. Therefore, in the present investigation, Cu inclusions of 2 and 4% are considered, while for the carbide reinforcement, Fe_3C of 4% is considered as there is limited research available in the literature, unlike TiC.

As reported, P/M is one of the cost-effective net-shape forming and can embed isotropic properties [13, 14]. When conventional P/M processing combines with traditional forging, it gets the benefit of both net-shape forming and the load-bearing

DOI: 10.1201/9781032725086-9

capacity for the products [15]. Despite these advantages, presence of porosity seems to be the main detrimental. Therefore, synthesising various combinations of alloys/composites necessitates delineating the forming limit and fracture criteria. The formability depends on the deformation process, though the material and other process parameters that largely govern its characteristics [16].

The formability characteristics are determined using the stress/strain formability index, which considers the impact of hydrostatic stress relative to effective stress. Given that P/M combined with forging is classified as bulk metal forming, known as powder forging, it is essential to examine the deformation characteristics under triaxial conditions [17]. The fracture occurs in the deforming preform due to tensile stress in the free-deforming surface. Though the primary stress is compressive in powder forging, it is inevitable that the appearance of tensile stress at the free-deforming surface stops further formability of the preform [18–23].

Therefore, the study is carried out predominantly on the effective strain at fracture in the triaxial condition against the stress formability index for the chosen composite. Using Oyane's fracture criterion, an attempt has been made to formulate a fracture model (FM) for the composites under investigation with the influence of compositional and lubricant variations and compared with the experimentally observed values. Further, an analysis of variance (ANOVA) study is performed to reveal the significance of the influential parameter and its quantitative contributions. Meanwhile, a further attempt has been made to detail the FM calculation, which is explained in the 'theoretical study' chapter.

8.2 MATERIALS AND METHODS

The study aims to prepare specimens with the selected compositions and subject it to deformation processes to evaluate formability and FMs under triaxial stress and strain conditions. Figure 8.1 presents a detailed flow diagram illustrating the entire research process, from stating the objective to drawing conclusions. While Chapter 3 covers the theoretical study, this chapter focuses on the experimental methods.

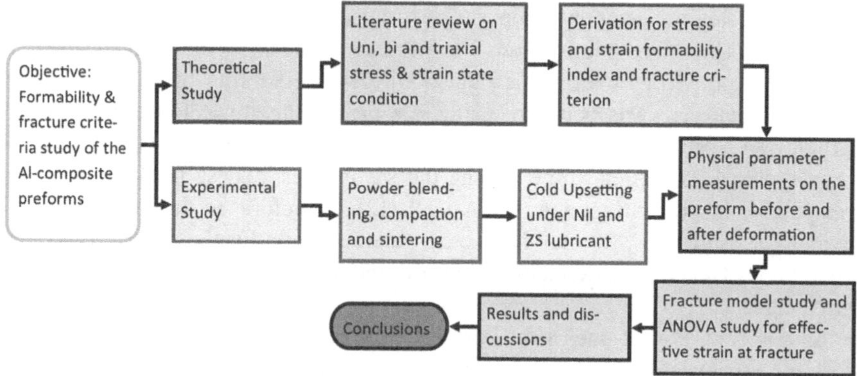

FIGURE 8.1 Research methodology.

8.2.1 Experimental Study

8.2.1.1 Powder Characterisation and Preparation

Al powders of the required amount were measured using a digital balancing machine of 10^{-4} g accuracy to prepare the compacts of 0.2 aspect ratio. The compacts are cylindrical in shape with a diameter of 28 mm and height of 7 mm to produce a constant initial relative density of 86%. Likewise, to prepare a powder blend corresponding to Al-4Fe_3C, Al-4Fe_3C-2Cu and Al-4Fe_3C-4Cu, the elemental powders of Al, Fe_3C and Cu of the required amount was carefully measured and blended using RETSCH planetary ball mill (PM400MA) until homogeneity was ensured [11]. The powder characterisation, such as Al powders sieve analysis, and other powder blends apparent density and flow rate measurements were carried out using a hall flow meter given in Table 8.1. Similarly, the compressibility test has also been performed to determine the required pressure to be applied to obtain constant relative density, as shown in Table 8.1. Table 8.1 further affirms that the flow rate decreases when any elements are added to pure Al. This is due to the use of Fe_3C and Cu powders with particle sizes smaller than 47 μm. The reduced particle size increases the surface contact between particles, ultimately lowering the flow rate [24]. The purity of Al powder through chemical analysis reveals 99.5%, and the remaining were insoluble impurities, while the purity of Cu and Fe_3C powders are 99.5 and 99.7%, respectively.

8.2.1.2 Powder Compaction and Sintering

The pressure required for compacting the mixture of Al powder and other powder blends was determined to be 156±10 MPa. A ram speed of 1 mm/sec was set using a 100-tonne capacity hydraulic press. The amount of powder needed to achieve a 0.2 aspect ratio compact was measured and poured into a compaction die set assembly, where the necessary pressure was applied to produce the compacts. These compacts were placed inside an atmospheric-controlled electric muffle furnace, with argon gas used during sintering to maintain an inert atmosphere and prevent oxidation. The sintering temperature was set at 561°C for 60 min, and the specimens were left to cool in the furnace after the sintering process.

TABLE 8.1
Powder Characterisation

Composition	Flow Rate (s/50g)		Apparent Density (g/cc)		Compressibility (g/cc) at the Pressure of 156±10 MPa		
Al	87.562		1.171		2.298		
Al-4Fe_3C	80.637		1.308		2.354		
Al-4Fe_3C-2Cu	80.257		1.354		2.391		
Al-4Fe3C-4Cu	79.419		1.370		2.427		
Sieve size (mm) for Al	500	+250	+200	+150	+100	+75	−75
Wt% retained	Trace	0.4	2.2	15.2	56.4	10.6	14.5

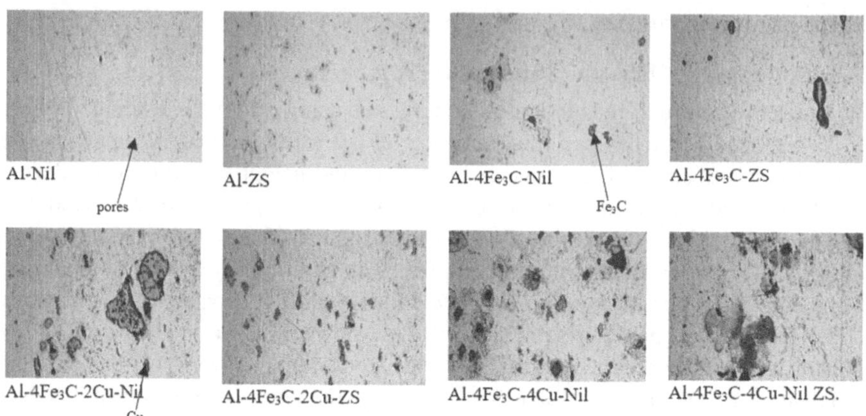

FIGURE 8.2 Micrographs of fully deformed specimens at 1000× magnification.

8.2.1.3 Powder Forging and Micrographs

The sintered and furnace-cooled specimens were wiped to free of any dirt and made initial measurements, including diameters, height and density. These specimens are then subjected to cold upsetting using the same hydraulic press in an incremental loading of 2 tonnes until a visible crack appears on the free-deforming surface. The specimens are deformed under nil lubricant between two mirror-polished flat dies, and the remaining specimens are deformed under zinc stearate (ZS) lubricant conditions. After each incremental step of loading, the deformed diameters, height and density are measured using a digital vernier calliper and Archimedes principle [11], respectively. The obtained results are then analysed for their workability characteristics.

Meanwhile, the fully deformed specimens under each composition and frictional condition have been cut into two halves diametrically. One-half of the cut specimen is prepared to view the microstructure using an Inverted Microscope of model ZEISS Axio Vert.A1 Mat with AxioVision LE software for image acquisition with 1000× magnification. The micrographs are shown in Figure 8.2, which are taken in the middle of the cut portion. It can be seen that the existence of micropores in all the categories of specimens. But these pores are almost round or spherical in shape, particularly under nil lubricant deformed specimens, due to the hydrostatic pressure caused by the dry friction conditions.

On the other hand, the preforms deformed under ZS exhibit some elongation of pores in the diametric direction. Simultaneously, the Al and Cu particles are significantly deformed perpendicular to the applied load (i.e., in the diametric direction). However, Fe_3C particles show relatively negligible deformation due to their hardness.

8.3 THEORETICAL ANALYSIS

As the preforms undergo plastic deformation, it experiences extensive strain, accompanied by induced stresses, complicating further deformation [25]. These stresses and strains are derived in order to establish the workability phenomenon in all three different states, respectively, uniaxial, biaxial and triaxial, which are detailed below.

8.3.1 Uniaxial Condition

The cylindrical preforms are deformed axially, and so height reduces while diameter increases; thus, the axial stress, σ_z and axial strain, ε_z can be given as

$$\sigma_z = P/A_i, \tag{8.1}$$

$$\varepsilon_z = ln\left(h_f/h_o\right), \tag{8.2}$$

where P is the applied load, A_i is the instantaneous area of cross-section, h_f is the final height of the preform and h_o is the initial height of the preform. In the uniaxial condition, radial stress, σ_r, and hoop or circumferential stress, σ_θ, are not applicable; therefore, the effective stress, σ_{eff}, can be written as follows [26]:

$$\sigma_{eff} = -\sigma_z, \tag{8.3}$$

whereas

$$\sigma_r = \sigma_\theta = 0. \tag{8.4}$$

Likewise, the effective strain, ε_{eff}, is

$$\varepsilon_{eff} = -\varepsilon_z = ln\left(h_o/h_f\right). \tag{8.5}$$

On the other hand, the preforms deformation in the lateral direction is evident. So, the diametrical strain shall be either stated as radial, ε_r, or hoop strain, ε_θ, which can be expressed as

$$\varepsilon_\theta = \varepsilon_r = ln\left(D_f/D_o\right), \tag{8.6}$$

where D_f is the final diameter, and the D_o is the initial diameter. In the uniaxial, it is assumed that the bulge diameter, D_b, is of no effect or its effect is neglected [27]. So, the average or hydrostatic or mean stress, σ_m, and mean strain, ε_m, in uniaxial can thus be calculated using the following expression:

$$\sigma_m = \sigma_z/3 \text{ and } \varepsilon_m = \varepsilon_z/3. \tag{8.7}$$

8.3.2 Biaxial Condition

In biaxial or plane conditions, both axial, z and hoop direction, θ for the cylindrical geometry preform is considered to develop the corresponding stresses and strains with the assumption that radial component, $r = 0$ [28]. It can be noticed that the bulging is quite evident under nil lubricant deformation, and to some extent, there will be variation in the bulge portion of the deforming diameter even under lubricant condition, and so, the derivation for the hoop strain, including bulge diameter, D_b is important to consider, which is explained elsewhere, and it is given as [29]

$$\varepsilon_\theta = \left(2D_b^2 + D_c^2\right)/\left(3D_0^2\right), \tag{8.8}$$

and that of hoop stress,

$$\sigma_\theta = \left[(1+2\alpha)/(2+\alpha)\right]\sigma_z, \tag{8.9}$$

where

$$\alpha = \varepsilon_\theta/2\varepsilon_z. \tag{8.10}$$

Eq. (8.10) can also be termed new Poisson's ratio, α, and D_c is the contact diameter. Therefore, the mean stress and mean strain for the plane condition can be written as

$$\sigma_m = \left(\sigma_z + \sigma_\theta\right)/3 \text{ and } \varepsilon_m = \left(\varepsilon_z + \varepsilon_\theta\right)/3. \tag{8.11}$$

As detailed elsewhere [30] for the expression of effective stress in the biaxial state using α, and σ_z is given as

$$\sigma_{eff} = (0.5+\alpha)\sqrt{\left[3\left(1+\alpha+\alpha^2\right)\right]}\sigma_z. \tag{8.12}$$

The effective strain gradient, $d\varepsilon_{eff}$, can be written using relative density, R, and differences in the strain as given in [31]

$$d\varepsilon_{eff} = \left\{ \begin{array}{l} \left[2\left\langle 3\left(2+R^2\right)\right\rangle\right]\left[\left(d\varepsilon_1 - d\varepsilon_2\right)^2 + \left(d\varepsilon_2 - d\varepsilon_3\right)^2 + \left(d\varepsilon_3 - d\varepsilon_1\right)^2\right] \\ +\left[\left(\left\langle d\varepsilon_1 + d\varepsilon_2 + d\varepsilon_3\right\rangle^2/3\right)\left(1-R^2\right)\right] \end{array} \right\}^{0.5}. \tag{8.13}$$

Applying in Eq. (8.13), the principal axial strain into a cylindrical coordinate, with the assumption of proportional straining with a constant ratio of $d\varepsilon_1 : d\varepsilon_2 : d\varepsilon_3$ or $d\varepsilon_z : d\varepsilon_r : d\varepsilon_\theta$ and for the plain strain condition, $\varepsilon_r = 0$, results into

$$\varepsilon_{eff} = \left\{\left[2/\left\langle 3\left(2+R^2\right)\right\rangle\right]\left[\left(\varepsilon_z\right)^2 + \left(-\varepsilon_\theta\right)^2 + \left(\varepsilon_\theta - \varepsilon_z\right)^2\right] + \left[\left(\left\langle \varepsilon_z + \varepsilon_\theta\right\rangle^2/3\right)\left(1-R^2\right)\right]\right\}^{0.5}. \tag{8.14}$$

Eq. (8.14) can further be simplified into

$$\varepsilon_{eff} = \left\{\left[4/\left\langle 3\left(2+R^2\right)\right\rangle\right]\left[\varepsilon_z^2 + \varepsilon_\theta^2 - \varepsilon_z\varepsilon_\theta\right] + \left[\left(\left\langle \varepsilon_z + \varepsilon_\theta\right\rangle^2/3\right)\left(1-R^2\right)\right]\right\}^{0.5}. \tag{8.15}$$

8.3.3 Triaxial Condition

The radial component is equated to hoop component in triaxial condition; as a result, the hoop stress can be determined from the expression detailed elsewhere [32], and it is given as

$$\alpha = \left[\left(2+R^2\right)\sigma_\theta - R^2\left(\sigma_z + 2\sigma_\theta\right)\right] \Big/ \left[\left(2+R^2\right)\sigma_z - R^2\left(\sigma_z + 2\sigma_\theta\right)\right]. \quad (8.16)$$

Solving Eq. (8.16) for σ_θ gives

$$\sigma_\theta = \left[\left(2\alpha + R^2\right)\Big/\left(2 - R^2 + 2R^2\alpha\right)\right]\sigma_z. \quad (8.17)$$

Likewise, Eq. (8.11) shall be rewritten for the triaxial condition to determine hydrostatic stress and strain,

$$\sigma_m = \left(\sigma_z + 2\sigma_\theta\right)/3 \text{ and } \varepsilon_m = \left(\varepsilon_z + 2\varepsilon_\theta\right)/3. \quad (8.18)$$

The σ_{eff} for the triaxial component can be calculated using the following expression as described elsewhere [33]:

$$\sigma_{eff} = \left\{\left[\sigma_z^2 + 2\sigma_\theta^2 - R^2\sigma_\theta^2 + 2\sigma_z\sigma_\theta\right] \Big/ \left[2R^2 - 1\right]\right\}^{0.5}. \quad (8.19)$$

Eq. (8.13) can be modified by substituting cylindrical strain coordinates with $\varepsilon_r = \varepsilon_\theta$ for triaxial state, results into the following expression:

$$\varepsilon_{eff} = \left\{\left[2\Big/\left\langle 3\left(2+R^2\right)\right\rangle\right]\left[2\varepsilon_z^2 + 2\varepsilon_\theta^2 - 4\varepsilon_z\varepsilon_\theta\right] + \left[\left(\left\langle 2\varepsilon_\theta - \varepsilon_z\right\rangle^2/3\right)\left(1 - R^2\right)\right]\right\}^{0.5}. \quad (8.20)$$

8.3.4 Formability Factor and Fracture Model

The extent of compressive deformation on the cylindrical preform experiences a compressive stress and strain, axially. However, the radial or hoop component experiences severe tensile stress or strain that results in fracture initiation in the hoop region. Thus, the strain corresponding to the fracture can be termed the forming limit strain, which can be determined using the effective strain at fracture. During fracture, the hydrostatic or the three components of stress or strain takes importance. Consequently, a stress or strain formability factor, β_σ or β_ε, respectively was proposed by Vujovic and Shabaik, 1986, encompassing the hydrostatic components with respect to the effective component [27]. The relationship between stress and strain formability factor can be given as

$$\beta_\sigma = 3\sigma_m/\sigma_{eff}, \text{ and } \beta_\varepsilon = 3\varepsilon_m/\varepsilon_{eff}. \quad (8.21)$$

Shima and Oyane, 1976, proposed a FM for porous metals as [31]

$$\int_0^{\varepsilon_{eff}^f} \left[1 + \left(\sigma_m / \sigma_{eff}\right)\right] d\varepsilon_{eff} = C_2. \quad (8.22)$$

Substituting Eq. (8.21) in Eq. (8.23) and solving the integration for effective strain at fracture, ε^{f}_{eff}, and replacing stress formability factor at fracture, β^{f}_{σ}, results into

$$\varepsilon^{f}_{eff} = 3C_1C_2 / \left(3C_1 + \beta^{f}_{\sigma}\right). \quad (8.23)$$

Eq. (8.24) can be modified to determine the constants, C_1 and C_2:

$$1/\varepsilon^{f}_{eff} = \left(3C_1 + \beta^{f}_{\sigma}\right)/3C_1C_2. \quad (8.24)$$

$$1/\varepsilon^{f}_{eff} = (1/C_2) + \left(\beta^{f}_{\sigma}/3C_1C_2\right). \quad (8.25)$$

Eq. (8.26) is of the typical linear equation form $(Y = a + bX)$
Therefore,

$$Y = 1/\varepsilon^{f}_{eff},\ X = \beta^{f}_{\sigma},\ b = 1/3C_1C_2 \text{ and } a = 1/C_2. \quad (8.26)$$

The constants, C_1 and C_2 defines the formability limit in the form of the hyperbola that can be determined using the linear regression analysis on the experimentally obtained data for ε^{f}_{eff} and β^{f}_{σ}. Such as the following expression shall be used to determine the constants, b and a; using these constants, C_1 and C_2 can be determined.

$$b = \left[n\Sigma(XY) - \Sigma Y \Sigma X\right] / \left[n\Sigma X^2 - (\Sigma X)^2\right], \quad (8.27)$$

$$a = \left[\Sigma Y - b\Sigma X\right]/n. \quad (8.28)$$

8.4 RESULTS AND DISCUSSION

Figure 8.3a and b is drawn between effective strain at fracture and stress formability factor at fracture in triaxial condition. Particularly, in Figure 8.3a, the influence of lubricants and in Figure 8.3b, the influence of compositions are portrayed. In these figures, the term used, such as Comps, refers to compositions, Exp refers to experimental values and FM refers to the fracture model. In Figure 8.3a, the terms Al, $4Fe_3C$, 2Cu and 4Cu, respectively, refer to pure Al, Al-$4Fe_3C$, Al-$4Fe_3C$-2Cu and Al-$4Fe_3C$-4Cu. Therefore, the general trend of this figure (Figure 8.3a) is that as the stress formability factor at fracture increases, the effective strain at fracture decreases; this behaviour is profound for the compositions that are deformed under ZS lubricant conditions. But for nil lubricant, the curve is almost flat, which indicates the addition of $4Fe_3C$ in Al and the further addition of 2Cu and 4Cu in Al-$4Fe_3C$ composite advances effective strain at fracture, nevertheless, the trend is not substantial against low-stress formability index at fracture, unlike in ZS lubricant. This reveals that the addition of iron carbide and further incremental addition of Cu profoundly increases effective strain to fracture at the lower stress formability index for ZS lubricant. However, in the case of pure Al, effective strain at fracture achieved the peak against the highest stress formability index at fracture as compared to any other composite preforms under concern, irrespective of lubricant, this could be due to its excellent malleability property. The curves drawn for each category in

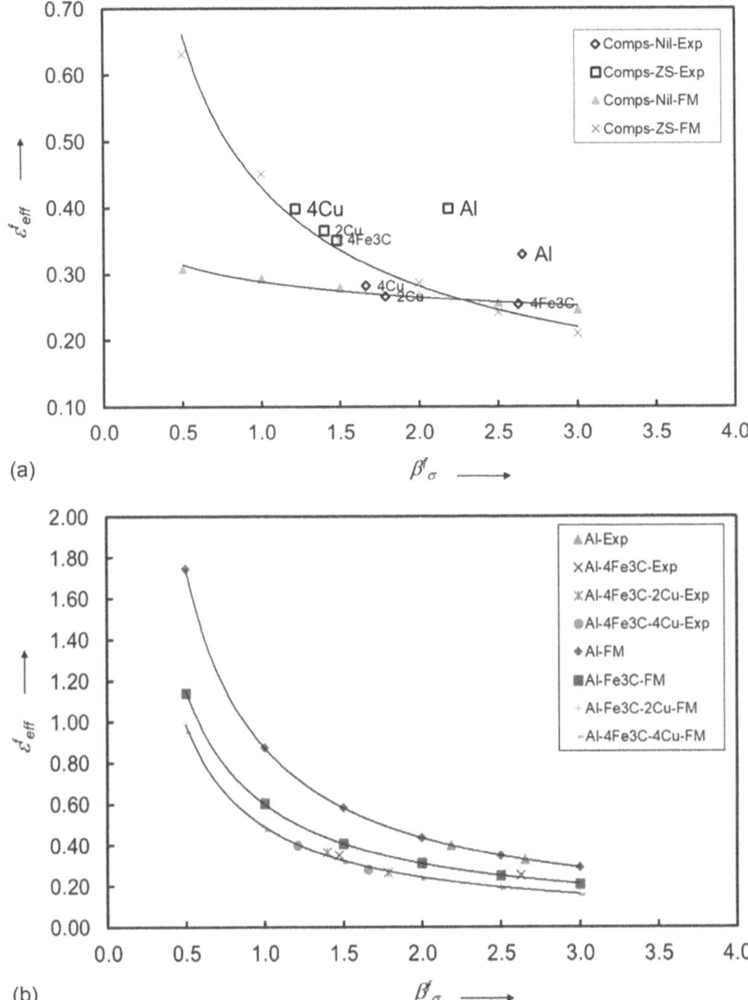

FIGURE 8.3 (a) ε^f_{eff} vs β^f_σ in both experimental and fracture model under lubricants influence. (b) ε^f_{eff} vs β^f_σ in both experimental and fracture model under the compositional influence.

Figure 8.3a and b uses the FM Eq. (8.23), where the experimental values for each category are embossed. The experimental values for ε^f_{eff} and β^f_σ under each category are shown in Table 8.2. These values are utilised to predict the constant values, C_1 and C_2; eventually, the FM for compositions and lubricants are formulated, and it is tabulated in Table 8.3. Tables 8.2 and 8.3 show the FM and the experimental values (Exp) on each category that are well synchronised except in Figure 8.3a, where Al preforms are a little away from the FM.

Figure 8.3b further ascertains that pure Al possesses the highest effective strain at fracture for any given stress formability index, which reveals that the straining of components is substantial for any given stress formability index at fracture. This

TABLE 8.2
ε^f_{eff} and β^f_σ Values

Composition	Lubricant	ε^f_{eff}	β^f_σ
Al	Nil	0.3299	2.6558
Al-4Fe_3C		0.2545	2.6297
Al-4Fe_3C-2Cu		0.2657	1.7864
Al-4Fe_3C-4Cu		0.2824	1.6613
Al	ZS	0.3988	2.1834
Al-4Fe_3C		0.3517	1.4744
Al-4Fe_3C-2Cu		0.3661	1.3973
Al-4Fe3C-4Cu		0.3990	1.2129

behaviour is substantially retarded when 4%Fe_3C is added with pure Al and is further decreased when 2Cu or 4Cu is added. Another interesting fact to notice is that between 2Cu and 4Cu addition, there is no separation between the curves; in other words, effective strain at fracture is in line with any given stress formability factor at fracture for both 2Cu and 4Cu added composites.

Figure 8.4 is drawn for the strain formability index against the stress formability index. It is evident that the increase of stress formability increases strain formability linearly. Interestingly, the influence of compositions under ZS lubricant is practically negligible, however, in nil lubricant conditions, the addition of 4Fe_3C in Al decreases the rate of strain formability slightly. On the contrary, adding 2Cu in Al-4Fe_3C little improves the strain formability, likewise, a similar trend of mild improvement in strain formability is visible by adding further 2Cu with Al-4F_3C-2Cu composite. However, the comparison between the lubricants reveals that preforms deformed under ZS lubricant showed higher strain formability indexes invariable to the compositions. Similarly, regardless of the compositions, all compacts deformed under ZS lubricant experience earlier fracture compared to those deformed under nil lubricant

TABLE 8.3
Constants, C_1 and C_2 and the Fracture Model

Compositions	Lubricant	C_1	C_2	Fracture Model
Al-4Fe_3C; Al-4Fe_3C-2Cu and Al-4Fe_3C-4Cu	Nil	−7.226	0.252	$\varepsilon^f_{eff} = 3.063/(9.444 + \beta^f_\sigma)$
Al-4Fe_3C; Al-4Fe_3C-2Cu and Al-4Fe_3C-4Cu	ZS	−7.497	0.351	$\varepsilon^f_{eff} = 0.783/(0.748 + \beta^f_\sigma)$
Al	Nil & ZS	2.371×10^{-4}	1228.8	$\varepsilon^f_{eff} = 0.874/(7.1 \times 10^{-4} + \beta^f_\sigma)$
Al-4Fe_3C	Nil & ZS	0.0214	10.03	$\varepsilon^f_{eff} = 0.644/(0.064 + \beta^f_\sigma)$
Al-4Fe_3C-2Cu	Nil & ZS	-2.29×10^{-3}	−70.71	$\varepsilon^f_{eff} = 0.486/(-0.007 + \beta^f_\sigma)$
Al-4Fe_3C-4Cu	Nil & ZS	-1.07×10^{-3}	−147.13	$\varepsilon^f_{eff} = 0.473/(-0.003 + \beta^f_\sigma)$

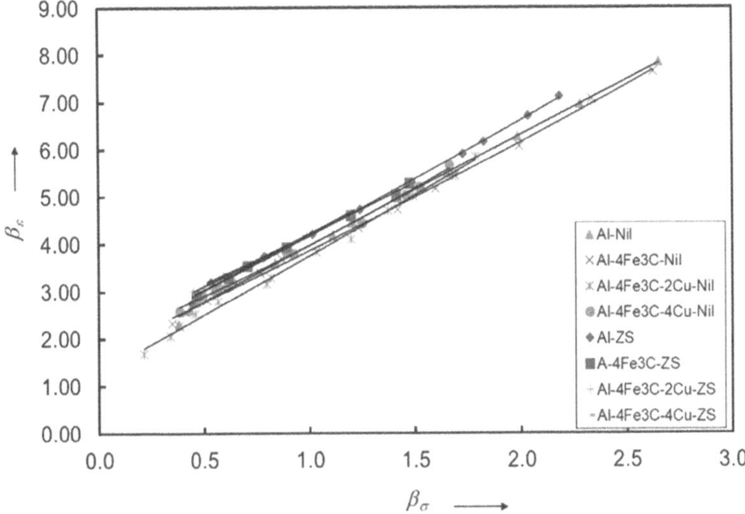

FIGURE 8.4 Strain formability index, β_ε, against stress formability index, β_σ.

conditions. This could be attributed to the higher strain formability, potentially leading to premature failure.

An ANOVA study was performed on effective strain at fracture across different compositions concerning stress formability index at fracture. The ANOVA test is valuable for evaluating the significance of parameter influences based on the P-value. Additionally, the combined assessment of degrees of freedom (df) and the F-value helps identify the most critical factor, as detailed elsewhere [11]. Hence, this study helps to quantify the influence of stress formability index and compositions. Table 8.4 shows the ANOVA study for effective strain at fracture. It can be observed from the P-value in Table 8.4 that it is less than 0.05, which portrays the significance of the two variables. Upon determining the percentage contribution (%C), nearly 82% of the stress formability factor influences the effective strain at fracture, while the influence of the composition is 12%.

On the other hand, the error percentage is 6%, which is relatively higher. It can be due to the influence of lubricants within the composition that is subjected to deformation and the interaction effect within the variables. Using the ANOVA results,

TABLE 8.4
ANOVA for ε^f_{eff}

Source of Variation	SS	df	MS	F	P-Value	F_{Crit}	%C
β^f_σ	2.83002	5	0.566004	42.09024	2.4E-08	2.90129	82.6
Composition	0.40849	3	0.136164	10.12569	0.000676	3.28738	11.9
Error	0.20171	15	0.013447				5.9
Total	3.44022	23					100

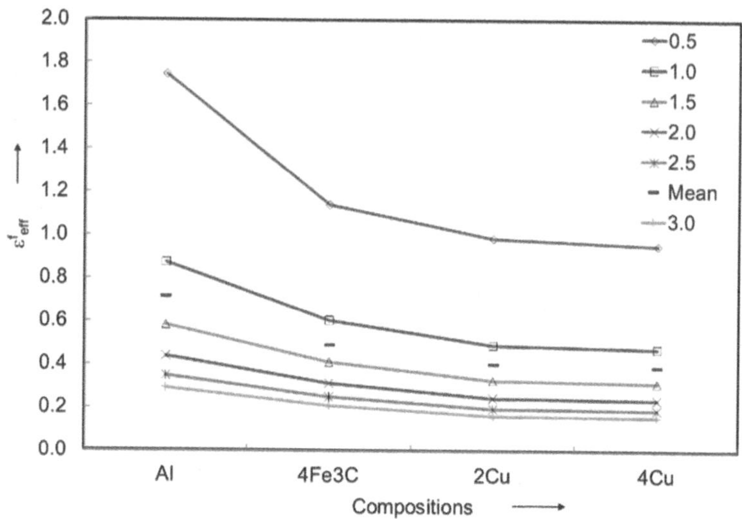

FIGURE 8.5 Influence of compositions in β^f_σ against ε^f_{eff}.

a Figure 8.5 is constructed to study the effect of composition in stress formability index at fracture, β^f_σ, against effective strain at fracture, ε^f_{eff}. In Figure 8.5, the points 0.5, 1, 1.5, 2, 2.5 and 3 refer to the various stress formability factor at fracture. Likewise, the $4Fe_3C$, 2Cu and 4Cu in x-axis referring to Al-$4Fe_3C$, Al-$4Fe_3C$-2Cu and Al-$4Fe_3C$-4Cu, respectively. Figure 8.5 reveals that iron carbide added with Al reduces effective strain at fracture, which means early failure is evident. This behaviour is particularly profound for the lower values of stress formability factor at fracture. But further addition of 2Cu or 4Cu reduces the rate of retardation in the effective strain at fracture; however, this effect is mild.

8.5 FUTURE SCOPE AND DIRECTIONS

Future research should focus on extending the study of formability and fracture criteria of Al-Fe_3C-Cu composites by incorporating additional alloying elements and varying their concentrations to enhance the mechanical properties and workability of these composites. Investigating the effect of different sintering techniques, such as spark plasma sintering or hot isostatic pressing, on the microstructure and fracture behaviour could yield valuable insights. Additionally, exploring the impact of advanced lubrication techniques and surface treatments on the cold upsetting process may help reduce friction and improve the formability of the composites.

Implementing computational modelling and simulation techniques, such as finite element analysis (FEA), can provide a deeper understanding of the stress-strain distribution during deformation and predict fracture initiation more accurately. Future studies should also focus on the real-world application and performance of these composites in various industries, such as automotive and aerospace, to validate the laboratory findings and ensure practical viability. Furthermore, investigating the recyclability and environmental impact of these composites can contribute to sustainable material development.

8.6 CONCLUSIONS

Following are the major finding from the formability and FM criteria in triaxial stress and strain state conditions of sintered Al-Fe_3C-Cu composites subjected to cold upsetting.

The FM determined between effective strain and stress formability index at fracture for the compositional and lubricant effect using Oyane's fracture criteria is well synchronised with the experimental values. For lower stress formability values substantial increase in effective strain at fracture is observed. The lubricant effect in Al-4Fe_3C-4Cu is nil as it perfectly synchronising with Al-4Fe_3C-2Cu, so any addition of Cu in Al-4Fe_3C-2Cu makes no difference in the fracture strain path. Likewise, compositional impact in nil lubricant is negligible for all practical purposes, but in ZS lubricant, it is substantial. Nevertheless, pure Al stand aloof for both the lubricant condition.

Further, an ANOVA study revealed the significance of the main effects, such as stress formability index and composition in the effective strain at fracture, which helped to quantify its impact to 82% and 12%, respectively.

The strain and stress formability index revealed a linear relationship between them. The preforms deformed under nil lubricant extended the strain formability, but in ZS lubricant, the effect is quite the opposite, on the other hand, its strain formability is high for any given stress formability factor. In addition, the influence of compositions in nil lubricant is evident; on the contrary, ZS lubricant has no effect.

REFERENCES

1. Singh, A., A. K. Srivastava, A. Kumar, et al. 2023. "Design and Development of Plentiful Fly Ash-Based Glass Powder-Reinforced Plastic Composite Bricks for Low Water Absorption and High Compressive and Flexural Strength." International Journal on Interactive Design and Manufacturing. https://doi.org/10.1007/s12008-023-01580-6
2. Srivastava, A. K., S. Tiwari, P. Pachauri, N. Gupta, B. Sunil, and A. Kumar. 2024. "Bonding Strength and Microstructural Features of Al5083-AZ31B Alloys Laminated Sheet through Friction Stir Additive Manufacturing." Journal of Adhesion Science and Technology 38 (4): 583–96. https://doi.org/10.1080/01694243.2023.2240637
3. Tripathi, A., N. K. Jha, R. N. Hota, A. Kumar, and R. Tyagi. 2024. "Green Sound-Absorbing Material Prepared by Using Natural Fiber for Building Acoustics." Proceedings of the Institution of Mechanical Engineers, Part E: Journal of Process Mechanical Engineering. https://doi.org/10.1177/09544089241253973
4. Kumar, A., V. Kumar Shrivastava, P. Kumar, A. Kumar, and V. Gulati. 2024. "Predictive and Experimental Analysis of Forces in Die-Less Forming Using Artificial Intelligence Techniques." Proceedings of the Institution of Mechanical Engineers, Part E: Journal of Process Mechanical Engineering. https://doi.org/10.1177/09544089241235473
5. Saxena, A., T. K. Gupta, R. Srivastava, A. K. Srivastava, and A. Kumar. 2024. "A Comparison of Different Experimental Approaches in the Determination of Dynamic Fracture Toughness (J1d) Behavior for RHA Steel Cost Sustainable MMAW

Weldments." Journal of Adhesion Science and Technology: 1–22. https://doi.org/10.1080/01694243.2024.2334267

6. Kumar Murmu, S., S. Chattopadhayaya, R. Cep, A. Kumar, A. Kumar, S. Kumar Mahato, A. Kumar, P. Ranjan Sethy, and K. Logesh. 2024. "Exploring Tribological Properties in the Design and Manufacturing of Metal Matrix Composites: An Investigation into the AL6061-SiC-Fly ASH Alloy Fabricated via Stir Casting Process." Frontiers in Materials 11. https://doi.org/10.3389/fmats.2024.1415907
7. Ajay, P., S. Ahmad, J. Sharma, and V. Gambhir (Eds.). 2023. Handbook of Sustainable Materials: Modelling, Characterization, and Optimization (1st ed.). CRC Press. https://doi.org/10.1201/9781003297772
8. Singh, A., H. Parveen, and B. AlMangour (Eds.). December 2023. Handbook of Smart Manufacturing: Forecasting the Future of Industry 4.0 (1st ed.). CRC Press. https://doi.org/10.1201/9781003333760
9. Srivastava, A. K., A. Kumar, P. Kumar, P. Gautam, and N. Dogra. 2023. "Research Progress in Metal Additive Manufacturing: Challenges and Opportunities." International Journal on Interactive Design and Manufacturing (IJIDeM). https://doi.org/10.1007/s12008-023-01661-6
10. Kumar, A., P. Kumar, R. K. Mittal, and V. Gambhir. 2023. "Chapter 12 – Materials Processed by Additive Manufacturing Techniques." In Advances in Additive Manufacturing, edited by Ajay Kumar, Ravi Kant Mittal, and Abid Haleem, 217–33. Additive Manufacturing Materials and Technologies. Elsevier. https://doi.org/10.1016/B978-0-323-91834-3.00014-4
11. Jeevanantham, A. K., D. R. Seenivasagam, and R. Ananthanarayanan. 2020. "Strain Hardening Analysis and Modelling for Sintered Al-Cu-TiC Preforms with Varying Process Parameters during Cold Upsetting." Journal of Materials Research and Technology 9 (5): 12007–18. https://doi.org/10.1016/j.jmrt.2020.08.065
12. Narayan, S., and A. Rajeshkannan. 2017). "Hardness, Tensile and Impact Behaviour of Hot Forged Aluminium Metal Matrix Composites." Journal of Materials Research and Technology 6 (3): 213–19. https://doi.org/10.1016/j.jmrt.2016.09.006
13. Kumar, A., P. Kumar, A. K. Srivastava, and V. Goyat (Eds.). 2023. Modeling, Characterization, and Processing of Smart Materials. In Advances in Chemical and Materials Engineering. IGI Global. https://doi.org/10.4018/978-1-6684-9224-6.
14. Kumar, A., P. Kumar, R. K. Mittal, and H. Singh. 2023. "Chapter 9 – Preprocessing and Postprocessing in Additive Manufacturing." In Advances in Additive Manufacturing, edited by Ajay Kumar, Ravi Kant Mittal, and Abid Haleem, 141–65. Additive Manufacturing Materials and Technologies. Elsevier. https://doi.org/10.1016/B978-0-323-91834-3.00005-3
15. Huang, C.-C., and J.-H. Cheng. 2005). "A New Forming-Limit Criterion for Fracture Prediction in a Powder Forging Application." International Journal of Mechanical Sciences 47 (7): 1123–45. https://doi.org/10.1016/j.ijmecsci.2005.02.014
16. Rezayat, M. 2022."Deformation Mechanism in Particulate Metal Matrix Composites." Journal of Alloys and Compounds 890: 161512. https://doi.org/10.1016/j.jallcom.2021.161512
17. Borra, N. V. S., and V. V. K. P. Davuluri. 2022. "Experimental Investigations of Al-Cr3C2 Composite Preform Densification and Deformation." Annales de Chimie – Science des Matériaux 46 (4): 185–92. https://doi.org/10.18280/acsm.460403
18. Tharmaraj, R., M. Joseph Davidson, and S. Richard. 2022. "Role of Localized Heating on the Workability of Powder Metallurgical Al–4% Ti Components in Cold Compression." Proceedings of the Institution of Mechanical Engineers, Part C: Journal of Mechanical Engineering Science 236 (5): 2428–46. https://doi.org/10.1177/09544062211030305
19. Sathish, T., R. Saravanan, S. J. Arunachalam, A. Parthiban, and J. Giri. 2024. "Comparing Kevlar Fiber Influence on PP/sisal/SiO2 Nanoparticle Fillers/Kevlar

Hybrid Nanocomposites with Plain Nanocomposite for Enhanced Mechanical Properties." Interactions 245 (1). https://doi.org/10.1007/s10751-024-01935-9
20. Thakare, P. A., N. Kumar, V. B. Ugale, J. Giri, N. Sunheriya, and H. A. Al-Lohedan. 2024. "Effect of Impact and Flexural Loading on Hybrid Composite Made of Kevlar and Natural Fibers." AIP Advances 14 (4). https://doi.org/10.1063/5.0195907
21. Natarajan, M., T. Pasupuleti, M. M. S. Abdullah, F. Mohammad, J. Giri, R. Chadge, N. Sunheriya, C. Mahatme, P. Giri, and A. A. Soleiman. 2023. "Assessment of Machining of Hastelloy Using WEDM by a Multi-Objective Approach." Sustainability 15 (13): 10105. https://doi.org/10.3390/su151310105
22. Kamble, P. D., J. Giri, E. Makki, N. Sunheriya, S. B. Sahare, R. Chadge, C. Mahatme, P. Giri, S. T., and H. Panchal. 2024. "An Application of Hybrid Taguchi-ANN to Predict Tool Wear for Turning EN24 Material." AIP Advances 14 (1). https://doi.org/10.1063/5.0186432
23. Tufail, M. S., J. Giri, E. Makki, T. Sathish, R. Chadge, and N. Sunheriya. 2024. "Machinability of Different Cutting Tool Materials for Electric Discharge Machining: A Review and Future Prospects." AIP Advances 14 (4). https://doi.org/10.1063/5.0201614
24. Duggirala, R., and R. Shivpuri. 1992. "Effects of Processing Parameters in P/M Steel Forging on Part Properties: A Review Part I Powder Preparation, Compaction, and Sintering." Journal of Materials Engineering and Performance 1 (4): 495–503. https://doi.org/10.1007/BF02682687
25. Kaku, S. M. Y., A. K. Khanra, and M. J. Davidson. 2018. "Effect of Deformation on Properties of Al/Al-Alloy ZrB_2 Powder Metallurgy Composite." Journal of Alloys and Compounds 747: 666–75. https://doi.org/10.1016/j.jallcom.2018.03.088
26. Abdel-Rahman, M., and M. N. El-Sheikh. 1995. "Workability in Forging of Powder Metallurgy Compacts." Journal of Materials Processing Technology 54 (1): 97–102. https://doi.org/10.1016/0924-0136(95)01926-X
27. Vujovic, V., and A. H. Shabaik. 1986. "A New Workability Criterion for Ductile Metals." Journal of Engineering Materials and Technology 108 (3): 245–49. https://doi.org/10.1115/1.3225876
28. Rajeshkannan, A. 2010. "Workability Studies on Cold Upsetting of Sintered Copper Alloy Preforms." Materials Research 13: 457–64. https://doi.org/10.1590/S1516-14392010000400006
29. Narayan, S., and A., Rajeshkannan. 2016. "Workability Studies of Sintered Aluminium Composites during Hot Deformation." Proceedings of the Institution of Mechanical Engineers, Part B: Journal of Engineering Manufacture 230 (3): 494–504. https://doi.org/10.1177/0954405414556497
30. Ananthanarayanan, R., N. S. Rai, M. Chand, N. Prasad, R. Ram, and M. Naicker. 2014. "Densification Behaviour of Sintered-Forged Aluminium Composite Preforms." Proceedings of the Institution of Mechanical Engineers, Part B: Journal of Engineering Manufacture 228 (3): 441–49. https://doi.org/10.1177/0954405413501811
31. Shima, S., and M. Oyane. 1976. "Plasticity Theory for Porous Metals." International Journal of Mechanical Sciences 18 (6): 285–91. https://doi.org/10.1016/0020-7403(76)90030-8
32. Rajkumar, P. R., C. Kailasanathan, A. Senthilkumar, N. Selvakumar, and A. JohnRajan. 2020. "Study on Formability and Strain Hardening Index: Influence of Particle Size of Boron Carbide (B_4C) in Magnesium Matrix Composites Fabricated by Powder Metallurgy Technique." Materials Research Express 7 (1): 016597. https://doi.org/10.1088/2053-1591/ab6c0b
33. Ravichandran, M., A. Naveen Sait, and V. Anandakrishnan. 2014. "Densification and Deformation Studies on Powder Metallurgy Al–TiO_2–Gr Composite during Cold Upsetting." Journal of Materials Research 29 (13): 1480–87. https://doi.org/10.1557/jmr.2014.143

Section II

Advancements in Manufacturing Methods and Techniques

9 Revolutionizing Manufacturing Operations in the Era of Industry 4.0

The Role of Industrial Internet of Things

Firdoos Afzal Bhat, Saad Parvez, and Ajay Kumar

9.1 INTRODUCTION

The industrial industry is witnessing a significant shift with the introduction of the Industrial Internet of Things (IIoT), which heralds a period of unparalleled collaboration, data-driven decision-making, and digitization. IIoT makes use of cutting-edge technologies to provide real-time monitoring, control, and process improvement for industrial processes. These technologies include actuators, sensors, wireless connectivity, and cloud-based computing. IIoT has fully dominated the manufacturing sector due to the employment of sensors, actuators, and other allied system integration devices [1].

A large amount of data generated via various industrial processes is the utmost challenge that has been conveyed by the IIoT [2]. This large amount of data is generated via sensors, equipment's, devices, and other systems that produce, generate, and transmit information [3]. Thus, such a huge amount of data generated needs to be effectively processed and analysed and hence requires advanced computational abilities and tools for analysis. Variety of data in the IIoT is again an issue that needs to be addressed [4]. Also, organizations use different data formats and protocols for the integration of sensors with the machinery. Thus, digital investigations must be incorporated so as to ensure the correctness, authenticity, and integrity of the large amount of data. IIoT is also recognized as a distributed and heterogeneous reservoir of large amounts of data.

It consists of various components, such as controllers, sensors, actuators, and network integration devices. And these components are delivered by many organizations that work on different software's and operation systems [5–7]. Hence, it's important to develop proper techniques/instruments for real-time analysis of the data [8, 9]. IIoT also raises the challenge of privacy, data breaches/security of the integrated systems, resulting in the proneness to data breaches of systems and data

DOI: 10.1201/9781032725086-11

sensitivity. Thus, proper security measures should be implemented in order to have data confidentiality and integrity.

The study goal is to provide an overview of IIoT, its significance, evolution, and role. It also includes the various components and architecture of IIoT, data management, and analytics in the IIoT, as well as future trends and challenges in the field of IIoT for manufacturing techniques. Because of the technological upheaval brought about by the Industry 4.0 age, this study will aid in the unification of the narrative surrounding the IIoT in the manufacturing sector. Additionally, potential recommendations for next advances and problems are presented.

9.1.1 Significance of Industrial Internet of Things

Given that the industrial industry gains a great deal from the application of IIoT. Here are a few of them:

Operational Efficiency: IIoT boosts productivity by continuously monitoring product developments, supply chains, and equipment operations; it helps producers increase operational efficiency by detecting inefficiencies, streamlining procedures, and cutting down on downtime through the analysis of sensor-gathered data.

Predictive Maintenance: The manufacturing company's capacity to create predictive maintenance schedules and regularly assess system performance is aided by IIoT. Therefore, by analysing data trends, IIoT aids in anticipating potential system failures and enabling necessary maintenance schedule measures that may otherwise present the system failure and other catastrophic losses. To present a manufacturing system view in real time, a process of descriptive analytics and data visualizations is used.

Predictive maintenance can be corrective ("when a fault is detected in equipment"), preventive ("focuses on reducing the likelihood of equipment breakdowns"), or risk-based ("to address risk-sensitive systems and machinery").

Quality Control: Through effective monitoring and administration of every industrial process, IIoT contributes to the improvement of product quality and standards. Thus, having access to real-time information makes it easier to identify deviations from requirements for quality and defects and to quickly conduct corrective actions to guarantee that the market only receives the best products [10]. Process control, control charts, acceptance sampling, and product QC are a few examples of IIoT quality control.

Agile Manufacturing: IIoT enables industrial companies to react quickly to the ever-changing needs of their consumers and the dynamic nature of the market. Manufacturing companies can enhance their system processes' flexibility and adaptability and enable quick changes, alterations, and expansions as needed by using IIoT-enabled systems and devices.

Optimized Supply Chain Management: The Internet of Things (IIoT) improves supply chain management through data accessibility and access, demand forecasting, improved inventory management, and increased logistical efficiency. As a result, through supply chain management, accurate real-time monitoring of manufacturing, transportation, and inventory status enables better informed decision-making, shorter lead times, and increased efficiency.

9.1.2 Evolution of Manufacturing Skills with IIoT

The first ten years of the 19th century saw the transition from an agricultural to an industrial and urban economy, which gave rise to the industrial revolution. The primary fuels for the massive steam engines utilized in the manufacturing industries were coal, water, and steam. The first industrial revolution is thought to have started in Britain and spread to other countries [11]. The second industrial revolution, which took place in the late 1800s and early 1900s, was typified by the broad use of methodically developed inventions. The groundwork for the 20th century's growth was laid, and the shortcomings of the aforementioned revolutions were linked to hazardous working conditions [12].

The invention of transistors and microprocessors in the 1950s marked the start of the third industrial revolution by enabling fully automated manufacturing powered by a wide range of electrical equipment. Therefore, the integration of computers and digital sensors onto industrial floors led to a significant improvement in working conditions; nonetheless, persistent issues such as pollution, urban overcrowding, worker exploitation, and environmental degradation continued. The fourth industrial revolution (I4.0) was built upon the third industrial revolution, which focused on transistors, sensors, and microelectronics for data creation. In order to increase technological advancement, profitability, and operational efficiency, I4.0, or industrial automation, integrates cutting-edge digital technologies such as "Artificial Intelligence (AI), IoT, Machine Learning (ML), Cyber-Physical Systems (CPS), Cloud computing, Additive Manufacturing (AM) & Cybersecurity" with manufacturing machinery [13].

With the advancement of technology and rising customer demands, manufacturing capabilities have kept up with the times, leading to the development of increasingly complicated and creative manufacturing models over time. Below is a quick summary of key phases in the manufacturing system's evolution (Figure 9.1):

Craft Production: The manufacturing method was historically thought of as a craft that was learnt through production with the assistance of knowledgeable artists utilizing their hands, simple tools, and procedures. Craft production required a lot of labour because each product was created specifically to satisfy specific customer needs.

Mass Production: Through the introduction of assembly lines and standardized parts, the industrial revolution aided in the mass manufacture of goods. As a result, this assisted in the large-scale manufacture of goods by bringing them down in price and facilitating easy access for a wider range of consumers [14].

Lean Manufacturing: The late 20th century saw the emergence of lean manufacturing concepts as a result of mass production's inefficiencies. The goal

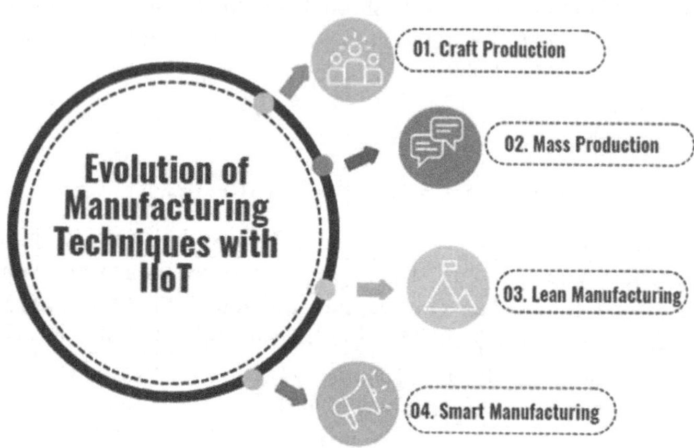

FIGURE 9.1 Evolution of manufacturing skills with IIoT. (Source: By Author.)

of lean manufacturing is to increase productivity and overall efficiency by concentrating on increasing value, reducing waste, and streamlining production processes.

Industry 4.0 and Smart Manufacturing: With the advent of Industry 4.0, industrial systems' productivity, efficiency, and performance were improved by the automation and integration of cutting-edge, new, and innovative technologies. Therefore, smart manufacturing uses cutting-edge technologies such as "robotics, AI, big data analytics, & IoT" [14] to construct an intelligent and networked production system.

9.1.3 Role of IIoT

With the introduction of I4.0, IIoT—the combination of networked devices, real-time data analytics, and automation—significantly enhances industrial processes [15]. An outline of how IIoT improves manufacturing operations is provided as follows:

Real-Time Monitoring: IIoT enables prompt replies for appropriate machine and industrial process monitoring and management with the use of embedded sensors. Thus, by having access to real-time data, production operations can be seen and controlled better, leading to increased productivity and efficiency.

Predictive Maintenance: IIoT makes it possible for manufacturing companies to create predictive maintenance plans by consistently tracking system performance. In order to anticipate system problems before they happen and to enable the appropriate maintenance schedule procedures to prevent system breakdowns and other catastrophic losses, IIoT analyses data trends.

Quality Control and Assurance: By guaranteeing constant product quality and standards, the Industrial Internet of Things enables precise monitoring and management of industrial operations. The real-time data insights allow for quick remedial actions to be taken in order to maintain standards of quality and early error identification.

Agile Manufacturing: IIoT enables industrial companies to react quickly to the dynamic shifts in consumer demand and production needs. The incorporation of IIoT-enabled devices and systems into production processes, thus, enhances flexibility and adaptability by permitting prompt modifications, customization, and expansion as needed.

Supply Chain Optimization: By enhancing communication and transparency throughout the supply chain, IIoT aids in the effective administration of forecasting, logistics, and inventory management systems. Manufacturers may reduce lead times, improve overall supply chain performance, and make well-informed decisions by having real-time access to information on inventory levels, production progress, and shipment routes.

9.2 LITERATURE REVIEW

Due to the global wave of the Covid-19 pandemic, people had to reevaluate and rearrange their industrial plans, socioeconomic policies, and sociopolitical institutions [16]. This demonstrated how important it is to have manufacturing technology that is ethical, sustainable, and mission-driven in order to enable long-term growth [17]. The adoption of cutting-edge technologies like "blockchain, artificial intelligence, big data, reconfigurable manufacturing systems, virtual manufacturing, blockchain, additive manufacturing & IIoT," which led to the creation of AMS (Advanced Manufacturing Strategy), has recently altered the landscape of manufacturing strategy.

IoT has many different effects. For example, the self-governing automobile is a modern IoT device under development [18]. In connection to the Internet of Things, the industry's operational procedures are likewise evolving. The Internet of Things created especially to meet the needs of industry is known as the IIoT. Yang et al. [19] address many IoT applications in industrial settings. The IoT gives industry new perspectives. IoT makes it possible to actively control equipment power utilization. In 2014, Shrouf et al. put up a plan to raise energy proficiency [20]. The technique involves using sensors and probes to track and analyse energy usage over time. It is possible to define new practices and strategies following data analysis. Similar to this, Wang et al. suggested an architecture to raise the energy efficiency of IIoT [21]. They recommend an architecture to boost output. Notable advancements have been made in energy efficiency. Another issue is downtime limitation, particularly for the industry. Preventive maintenance is one of the options.

On the other hand, it requires to be finished at a precise period. In 2012, Xu and colleagues introduced a technique that utilizes the Internet of Things to predict malfunctions [22]. The information is gathered by sensors or cyber-physical systems. To construct the forecast, the authors of [23] suggested using big data technologies. These tools work well because they can handle large volumes of data with ease. Automation is one specific sector need that the IoT helps with. Industry 4.0 emphasizes flexibility, which calls for reconfigurable production processes. Robots capable of withstanding operational software updates are needed for this kind of task. IoT makes this extra intelligence possible. Cobots were created with the fusion of robotics and IoT [24].

These systems need to be connected to networks. The "world of networks" thus faces new technological tests that require attention [25]. Researchers and developers have been interested in blockchain integration with Internet of Things devices lately. However, there hasn't been much research done on how blockchain technology might help with Internet of Things security requirements. IoTs rigorously govern data flow management, processing power, available bandwidth, data throughput, storage, QoS (Quality of Service) prioritization, and processing power [26]. If handled improperly, authentication is a crucial component of IoT security and can quickly result in numerous breaches [27]. One of the key elements of dependable data transmission is trust among smart object communication [28]. On these objects, almost 90% of data exchanges are not properly encrypted [29]. The issue seems to be that a lot of businesses have a lot of consumer devices connected to their networks.

Also, a lot of devices are linked with overall enterprise systems, which puts the devices and the data on them in danger in the case that this network is breached [30, 31]. To ascertain the reliability of these applications and their producers, regulations and legislation are necessary. In today's intelligent and networked society, zero trust is becoming more and more prevalent [32]. Most of the current blockchain platforms are unworkable due to the amount of data created by smart objects. By 2025, 41.6 billion objects will be connected to the Internet, meaning that a huge amount of information will increase. As things stand, blockchain systems will not be able to hold that much of the enormous amount produced by IoT devices without becoming unresponsive [33]. One advantage of fusing blockchain technology with IoT is defence against cyber-attacks [34]. Blockchain, which aims to serve as the basis for Internet of Things applications, including interactions and transactions, includes smart contracts as a fundamental element [35, 36]. Many other studies are also performed integrating manufacturing and software technologies [37–41].

9.3 IIoT COMPONENTS AND ARCHITECTURE

9.3.1 IIoT Components

Handling IIoT communication with physical components, such as sensing, actuating, or digitally labelling real objects, is known as "Real-World Interaction." Because the interfacing task significantly depends on the application, a specific understanding of the technical and physical systems to be dealt with is required. The different crucial IIoT elements depicted in Figure 9.2 are as follows:

Sensors are the devices that recognize physical alterations and convert into digital signals [42]. The changes that are to be recorded are physical quantity, sensor behaviours and its materials, and, most crucially, nominal changes like range, stability, motion detection, resolution, or reaction time [43], which help differentiate among the many types of sensors. Commonly monitored parameters include temperature, force, velocity, acceleration,

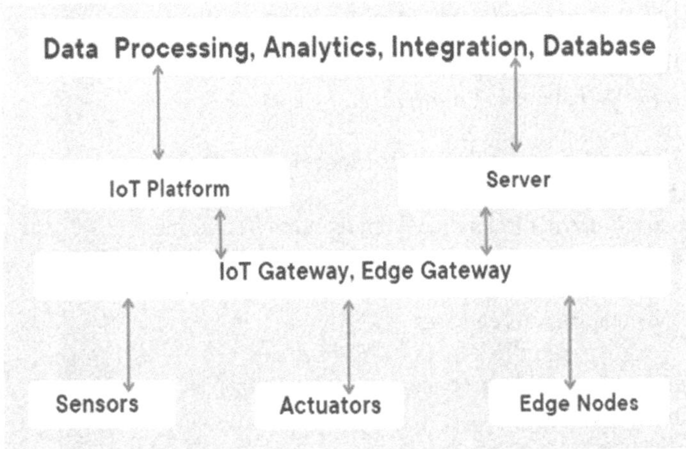

FIGURE 9.2 IIoT infrastructure. (Source: By Author.)

light, noise, flow, level, and human-machine interfaces [44, 45]. There are numerous methods for choosing a sensor.

Actuators are used for altering the state or turning on and off the functionality of physical entities [46]. They convert energy (such as electric, hydraulic, or pneumatic) or data into "motion" in order to accomplish this [47]. Determine the "maximum force, frequency range/speed, and stroke"—the three primary actuator parameters that place severe limitations on actuator performance. Examples of secondary features are response time, precision (with regard to force, position, and operation duration), minimum thrust, endurance, temperature stability, resistance to "shock and vibration," size, and weight [48]. Other aspects of the "integration of an actuator into an existing system include energy input type, input/output characteristics, interfaces (e.g., digital connection to field buses or analogue inputs), aspects of constructive integration (e.g., via screws, vs. specialized designs), and the degree of intelligence" (e.g., whether more complex control algorithms are required) [49].

Edge Devices: These parts, which are also called computing devices or edge gateways, serve as a bridge among sensors, actuators, and the main IIoT infrastructure. They are in charge of the preliminary data processing, filtering, and compilation near the source, which minimizes latency and lowers the demand on network capacity.

Gateways: In the IIoT ecosystem, these devices facilitate communication and compatibility among diverse network zones, protocols, and devices. Gateways guarantee that data is seamlessly integrated and exchanged among edge devices, actuators, sensors, and the main IIoT network.

Cloud Platforms: Whether in dispersed or centralized systems, cloud platforms offer a stable and scalable basis for IIoT data processing, analysis, and storing. They provide the tools and analytical skills required to manage, interpret, and display large datasets in real time as well as in the past.

9.3.2 IIoT Architectural Framework

From the literature, it was found that there are four works classified as IIoT architectural frameworks that are as follows:

- The IoT Architectural Reference Model (IoT ARM),
- IEEE standard 2413–2019,
- Industrial Internet Reference Architecture (IIRA), and
- Reference Architecture Model Industry 4.0 (RAMI4.0) [48].

IoT ARM comprises three parts:

- A reference model (IoT concepts and definitions),
- Architecture, and
- Application guidance.

The Industrial Internet Consortium recognized IIRA [49], which involves four perspectives: "commercial, practice, purposeful, & application." From a commercial point of view, IoT organization sponsors the organization's vision and ambitions, whereas the practice perspective demonstrates predicted utilization of the system. The functional view includes operational elements, and it's both internal and external relations. Lastly, the application perspective includes skills required to accomplish the defined, well-designed mechanisms, as well as configuration, connectivity architecture, and nominal portrayals [49].

German "Plattform Industrie 4.0" created RAMI4.0. It is a domain-driven architecture that integrates IIoT concerns while additionally concentrating on production and operational specifics [49]. The main IIoT architectural frameworks are shown in Figure 9.3.

Edge Computing: This involves the local processing and analysis of data at or near where it is generated (end of network). Edge computing in IIoT setups performs immediate data pre-processing, filtering, and analysis, reducing the need for bandwidth and enabling decisions to be made in real-time.

Communication Protocols: IIoT architectures rely on diverse communication protocols to facilitate the transfer of data among devices, edge nodes, gateways, and cloud services. Thus, the protocols, including CoAP (Constrained Application Protocol), OPC UA (Open Platform Communication Unified Architecture), MQTT (Message Queuing Telemetry) and Modbus, are frequently used for the optimization and increase the reliability, scalability, and proper usage of bandwidth.

Data Storage: The IIoT helps in processing the vast volume of real-time and historical data collected through the sensors and devices, which needs to be properly handled. These storage devices include databases, data lakes, and distributed file systems that can be located either onsite or on cloud platforms.

Analytics: In order to derive meaningful insights from a huge pile of collected data via sensors and devices, IIoT plays a significant role in analysing the vast amount of data. Manufacturing processes can enhance efficiency,

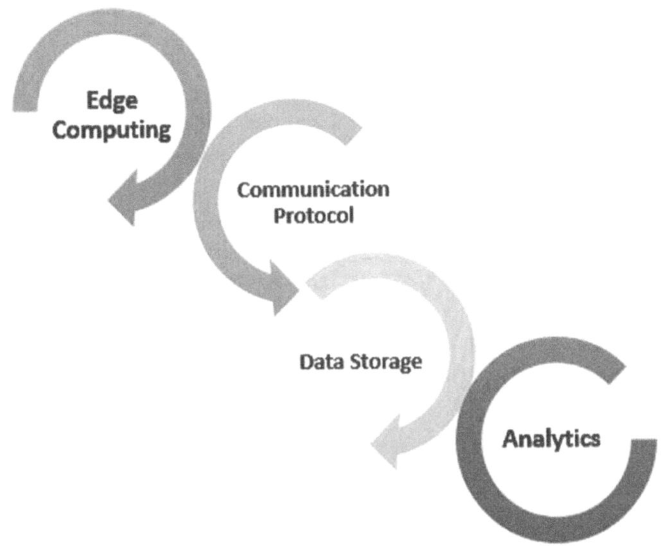

FIGURE 9.3 IIoT architectural framework. (Source: By Author.)

identify problems, and adjust operation processes by employing techniques, such as pattern recognition, predictive modelling, real-time analytics, and anomaly detection.

9.4 DATA MANAGEMENT AND ANALYTICS IN IIoT

Data Collection: A large amount of information collected via the use of sensors and devices. IIoT helps in effective data collection by means of properly, efficiently, and reliably gathering data from various sources, including edge devices, gateways, and cloud services [50].

Data Storage: The larger amount of data manufactured by sensors involves scalable and dependable storage solutions. The variety of data would be stored using a variety of approaches, including distributed file systems, databases, and data lakes.

Data Processing: Cleaning, filtering, and aggregating raw sensor data are examples of data processing procedures used to extract valuable insights. Batch processing and stream processing are two methods for successfully handling both historical and real-time data streams.

Data Integration: In IIoT systems, the integration of data collected from different bases, like sensors, business systems, and external databases, is essential. This integration helps to acquire the compatibility and coherence of the data across various platforms.

Data Security: Data security of both essential and non-essential data is considered to be essential in all aspects. In order to avoid unwanted access or data breaches, secure measures should be followed, such as encryption, secure communication protocols, and access controls.

9.5 DATA ANALYTICS TOOLS AND ALGORITHMS

Descriptive Analytics: In descriptive analytics, insights are provided by prior events and trends, by supporting manufacturing firms in order to have a better empathetic of traditional data and performance metrics. The various essential tools used in identifying critical metrics and areas for development are data visualization, reporting, and dashboards.

Predictive Analytics: Predictive analytics employees' machine learning algorithms in order to forecast future trends based on previous datasets. This includes predictive maintenance algorithms in order to anticipate machine-driven disasters by optimizing maintenance schedules and decreasing system disruptions. Also, demand forecasting algorithms are employed to predict market trends and optimize manufacturing strategies.

Prescriptive Analytics: Prescriptive analytics predicts the necessary actions to be taken in order to improve industrial processes. Optimization algorithms are thus employed to find the most appropriate manufacturing strategy, proper allocations of resources and supply chain strategies so as to increase efficiency and decrease cost.

9.6 CHALLENGES IN CYBERSECURITY AND RISK MANAGEMENT WITH IIoT

Due to the interconnection of different systems, devices in IIoT integration in the manufacturing system, different cybersecurity challenges arise due to this [50]. Some of these challenges are as follows:

Data Breaches: The transmission of critical data in the IIoT devices which are interconnected is prone to cybersecurity attacks with an aim to steal the confidential information of the manufacturing organization in order to disrupt their operation.

Malware Threats: The various devices/networks connected through the IIoT are vulnerable to contagions via malicious software such as ransomware that results in data theft, system failures, and unauthorized access to the attackers, causing substantial manufacturing process delays and financial damages to the manufacturing organization.

Unauthorized Access: Due to the absence of robust authentication approaches and poor access controls, unauthorized entry into IIoT systems loop and thus posing potential risks to important data and manufacturing process safety.

Supply Chain Vulnerabilities: Due to the dependency on third-party vendors/suppliers in IIoT manufacturing systems, this results in the introduction of potential risks to the supply chain. These peripheral entities attack the main IIoT infrastructure and disrupt operation processes [36].

9.7 APPROACHES TO ENHANCING CYBERSECURITY IN IIoT

The security of the IIoT manufacturing system is a great concern for the organization in the age of Industry 4.0. With the aim of strengthening the security of IIoT manufacturing organizations, industrial units can implement essential cybersecurity practices such as "encryption, authentication, controlled access, network division, ongoing surveillance, and workforce education." Inspecting cybersecurity—continuously evaluating, updating, and enhancing strategies—is essential to diminish emerging hazards and exposures within the IIoT manufacturing domain [50]. The various effective processes in order to minimize cybersecurity risks in IIoT manufacturing processes are as follows (Figure 9.4):

Encryption: To defend sensitive/sensible data against unauthorized access/interception and manipulation, encryption helps us to transmit data along IIoT devices, edge components, gateways, and cloud services. Thus, employing strong encryption standards ensures data privacy and reliability.

Authentication: Establishing authentication protocols, including multi-level authentication and digital certificates, authenticates the identities of handlers and devices, thus preventing illegal access and shielding against identity stealing and scams.

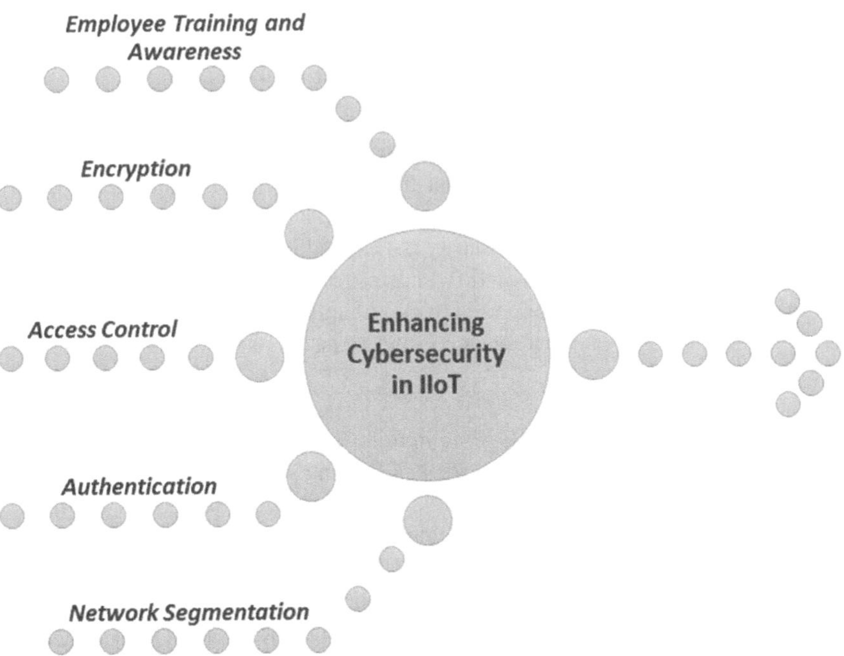

FIGURE 9.4 Enhancing cybersecurity in IIoT. (Source: By Author.)

Access Control: In order to prevent the illegal access in IIoT manufacturing system, this can be achieved by implementing detailed access control measures that restrict entry based on users' roles, permissions, and privileges. Thus, using role-based access control (RBAC) and the principle of least privilege aids unauthorized access and diminishes the influence of security breaches.

Network Segmentation: In order to isolate and defend security breaches and reduce the impact of cyber-attacks, network segmentation is essential. Thus, by separating networks by function, department and security level, manufacturing firms can defend themselves from attackers and protect critical resources.

Continuous Monitoring: The usage of real-time monitoring and alarm systems increases initial detection and retort to security issues in IIoT manufacturing systems. The technologies, including anomaly detection algorithms, intrusion detection systems and security information, and event management tools, are critical for detecting doubtful activity and probable threats.

Employee Training and Awareness: The education given to the employee on cybersecurity as well as improving awareness of prevalent dangers and attack vectors will aid in decreasing the risk of human error and insider threats. Regular training on data handling, security awareness programmes, and simulated phishing exercises will assist in strengthening security and foster a cybersecurity-conscious culture in manufacturing organizations.

9.8 FUTURE TRENDS AND CHALLENGES IN IIoT

With the advancement of manufacturing processes in the I4.0, probably in a new wave of industrialized insurgency called Society 5.0, new advanced technologies will be developed in order to persist in the highly dynamic manufacturing environment [51]. Nonetheless, overcoming hurdles related to costs, system complexity, skill shortages in the workforce, security issues, and interoperability is essential for harnessing the full capabilities of IIoT and fostering digital transformation within regions. For the IIoT to be widely used in manufacturing, coordination amongst academic institutions, government agencies, and industry players is essential in overcoming these obstacles.

Edge Computing: Current cloud computing technology struggles with processing large volumes of data swiftly, affecting service quality and the efficiency of networks and IoT devices. Edge computing emerges as a remedy by offering decentralized computing and storage, allowing data from IoT devices to be processed locally before being sent to the cloud [51]. This approach enhances response times and reduces network congestion. Edge computing and fog computing share the principle of local data processing; however, they differ in the processing location. Edge computing occurs directly on devices or near the network's edge, while fog computing operates within local network equipment, further from the data source. Despite cloud computing's role in supporting IoT, its latency issues in high-volume

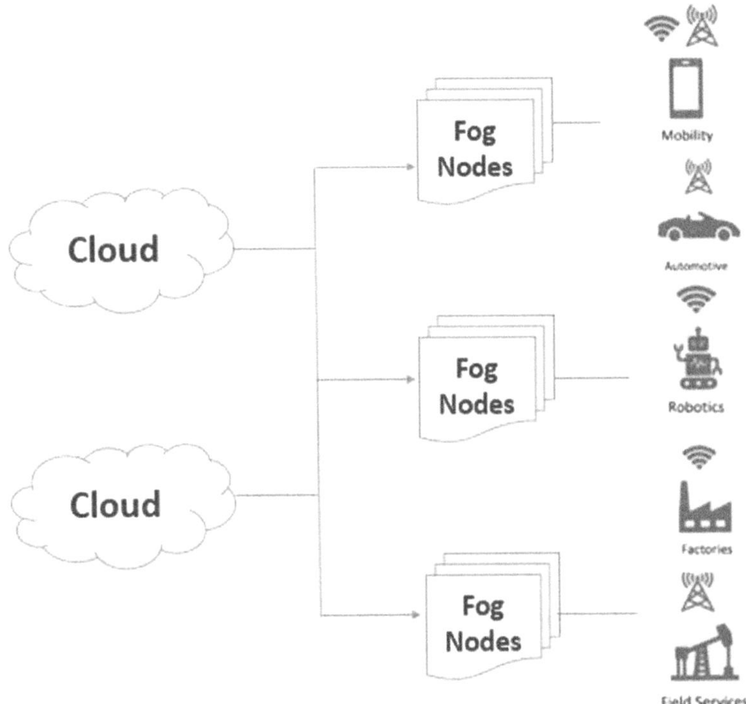

FIGURE 9.5 Edge computing. (Source: By Author.)

data scenarios are mitigated by edge and fog computing solutions, crucial for instantaneous decision-making (Figure 9.5).

6G and Beyond: The quest for a fully interconnected digital ecosystem necessitates advancements in communication technologies alongside digital and computational innovations. Telecommunication firms are exploring 6G technologies, expected to outperform 5G by offering higher frequencies, increased capacity, reduced latency, and better energy efficiency. 6G aims to enable a vast array of applications, from AI-driven logistics to digital healthcare and cooperative robots, promising significant technological breakthroughs by 2030.

Augmented Reality (AR): AR overlaps with the digital information into practical environments, making immersive experiences which blend physical and virtual elements. AR enriches the practical world with virtual, contrastingly virtual reality (VR), which completely immerses users in a digital environment [52]. AR's development traces back to the 1960s, evolving into mixed reality (MR) that merges real and virtual worlds, allowing interactive experiences with advanced sensing and imaging technologies.

Internet of Everything (IoE): Introduced by Cisco, IoE perception expands beyond Internet of Things (IoT) to encompass widespread connectivity among people, objects, data, and processes. This interconnectedness fosters

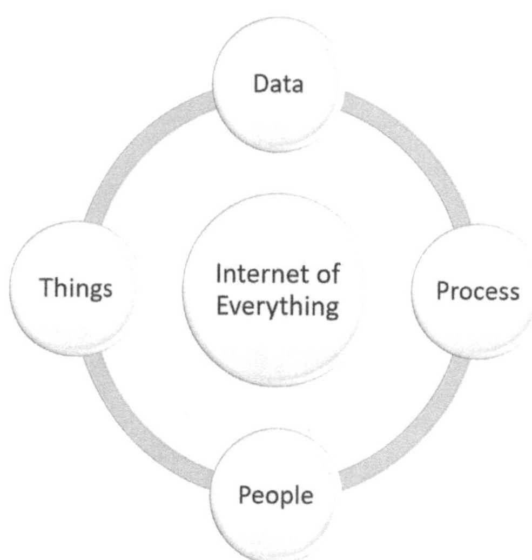

FIGURE 9.6 Internet of everything. (Source: By Author.)

new capabilities, enhanced experiences, and a multitude of opportunities, promising to intelligently connect billions of items worldwide. IoE operates on a closed-loop system where AI algorithms analyse data from various sources to deliver personalized user experiences, demonstrating the extensive potential and data flow within the IoE framework (Figure 9.6).

9.9 CHALLENGES FACING IIoT IMPLEMENTATION IN THE MANUFACTURING SECTOR

- The initial and ongoing costs associated with developing and maintaining an IIoT infrastructure pose a significant obstacle to its widespread adoption in manufacturing. The financial burden of acquiring sensors, edge devices, networking systems, and analytics platforms can be particularly daunting for small- to medium-sized enterprises with restricted budgets.
- The deployment of IIoT technologies demands expertise in areas, such as networking, cybersecurity, data analysis, and system integration. The inherent complexity of these systems and the necessity for them to be interoperable and scalable may present difficulties for manufacturers, especially those lacking robust technical teams.
- Skilled workforce is important for the integration of IIoT within manufacturing systems. However, the manufacturing industry has an abundance shortage of well-skilled and professional persons trained in cybersecurity and IIoT systems.
- The security of essential IIoT components (devices, networks, and cloud infrastructures) is more susceptible to cyber risk.
- The continuous challenge in implementation of IIoT and use of wide range devices, protocols, and platforms remains still issue for the manufacturing

systems. The lack of standardized protocols and the prevalence of proprietary solutions complicate the integration process and limit the scalability and adaptability of IIoT systems.

9.10 CONCLUSION

Experts in the field claim that currently the fourth industrial revolution is in progress, which is described by the incorporation of a number of disruptive digital advances into various production processes and equipment. This integration aims to make data collection easier by providing smooth communication channels among diverse components, resulting in more cost-effective product offerings. This study presents a complete review of IIoT and its impact on the manufacturing sector. In this study, the author stressed how IIoT is transforming the conventional manufacturing process in order to enhance the efficiency, productivity, and overall competitiveness of the manufacturing sector. Also, the evolution of the manufacturing sector as a result of IIoT integration was also highlighted. From the technical point of view, the author stressed the various components and architectural framework of IIoT and emphasized the interconnection of devices, sensors, and networks that helps in data exchange, and decision-making. Also, this study elucidated the significance of data management and analytics in binding the benefits of IIoT data generation and helping the organizations with suitable insights in order to optimize their processes and have scheduled maintenance. This study furthermore explains the essential role of advanced analytics techniques that will help in extracting useful data from the vast data stream.

Furthermore, this study acknowledged the challenges possessed by cybersecurity and risk management by interconnected industrial systems. By citing these challenges in the IIoT implementation, this study also provides proactive solutions, including encryption, authentication, and intrusion detection, in order to effectively curtail cyber-attacks. Also, this study will provide the detailed anticipated trends and issues that are evolving in IIoT implementation and highlight the prominence of recurrent innovation and adaptability in order to cope with emerging prospects and diminish possible threats.

Thus, this study's novelty arises in the unified narrative of the IIoT in the arena of the manufacturing sector due to the technological transformation during the Industry 4.0 era, and probable prescriptive is provided on forthcoming developments and challenges. This study will function as an introductory source for understanding the complex characteristics of IIoT and its implications in the current manufacturing sector. At the end, the author expects to provide readers with the understanding and perspective they need to navigate the complicated landscape of IIoT implementation and leverage its potential for transformation in driving long-term development and creative thinking in the manufacturing industry through tackling key concepts, technical considerations, and future perspectives.

The main challenges that will serve as the foundation for the other study are that more in-depth analysis on how IIoT will revolutionize manufacturing and more on-field, survey-based case studies and the various regions of economies views regarding the implementation of IIoT in the manufacturing processes should be addressed.

Further investigation is necessary into the challenges associated with the Internet of Things (IIoT), big data management, and supply chain operational and system coordination. In conclusion, from a theoretical standpoint, scholars should provide a structure that enables businesses to transition quickly among the industrial revolutions.

REFERENCES

1. Saigopal, V. V. R. G., and V. Raju. "IIoT Digital Forensics and Major Security Issues." In *2020 International Conference on Computational Intelligence (ICCI)*, 233–236, 2020.
2. Singh, Hari, Kumar Ajay, Kumar Parveen, and Bandar AlMangour (Eds.). *Handbook of Smart Manufacturing: Forecasting the Future of Industry 4.0* (1st ed.). CRC Press, 2023. https://doi.org/10.1201/9781003333760
3. Rani, S., K. Tripathi, and A. Kumar. "Machine Learning Aided Malware Detection for Secure and Smart Manufacturing: A Comprehensive Analysis of the State of the Art." *International Journal on Interactive Design and Manufacturing* (2023). https://doi.org/10.1007/s12008-023-01578-0
4. Wu, T., G. Jourjon, K. Thilakarathna, and P. L. Yeoh. "Mapchain-D: A Distributed Blockchain for IIoT Data Storage and Communications." *IEEE Transactions on Industrial Informatics* 19, no. 9 (2023): 9766–9776.
5. Sun, S., and J. Yu. "An Indoor Location Algorithm for Heterogeneous Devices and Environmental Changes." *Journal of Jilin University (Science Edition)* 61, no. 4 (2023).
6. Wang, H., C. Bi, Z. Shen, and P. Liu. "Two-Stage Location Privacy Protection Method for Mobile Crowd Sensing." *Journal of Jilin University (Science Edition)* 61, no. 5 (2023).
7. Zhang, P., X. Yu, X. Bai, C. Wang, J. Zheng, and X. Ning. "Joint Discriminative Representation Learning for End-to-End Person Search." *Pattern Recognition* 147 (2024): 110053.
8. Ning, E., C. Wang, H. Zhang, X. Ning, and P. Tiwari. "Occluded Person Re-Identification with Deep Learning: A Survey and Perspectives." *Expert Systems with Applications* 239 (2024): 122419.
9. Tian, S., L. Li, W. Li, H. Ran, X. Ning, and P. Tiwari. "A Survey on Few-Shot Class-Incremental Learning." *Neural Networks* 169 (2024): 307–324.
10. S., E. F., P. L. O., J. A. A., and S. P. G. "Integration of an IIoT Platform with A Deep Learning Based Computer Vision System for Seedling Quality Control Automation." In *2021 IEEE 3rd Eurasia Conference on IOT, Communication and Engineering (ECICE)*. 2021. https://doi.org/10.1109/ecice52819.2021.9645700
11. Fields, G. "Urbanization and the Transition from Agrarian to Industrial Society." *Berkeley Planning Journal* 13, no. 1 (1999). https://doi.org/10.5070/BP313113032
12. Tadesse, H., et al. "Bottlenecks and Productivity Analysis in Soft Drink Industry: A Case Study of East Africa Bottling Share Company." *International Journal on Interactive Design and Manufacturing (IJIDeM)* (2024): 1–13.
13. Li, D. X. "Industry 4.0—Frontiers of Fourth Industrial Revolution." *Systems Research and Behavioral Science* 37, no. 4 (2020): 531–534. https://doi.org/10.1002/sres.2719
14. Kumar, A., V. K. Shrivastava, P. Kumar, A. Kumar, and V. Gulati. "Predictive and Experimental Analysis of Forces in Die-Less Forming Using Artificial Intelligence Techniques." *Proceedings of the Institution of Mechanical Engineers, Part E: Journal of Process Mechanical Engineering* (2024). https://doi.org/10.1177/09544089241235473
15. Kumar, A., S. Rani, S. Rathee, and S. Bhatia (Eds.). *Security and Risk Analysis for Intelligent Cloud Computing: Methods, Applications, and Preventions* (1st ed.). CRC Press, 2023. https://doi.org/10.1201/9781003329947

16. Schwab, K., and T. Malleret. *COVID-19: The Great Reset*. Forum Publishing, 2020.
17. Mazzucato, M. *Mission Economy: A Moonshot Guide to Changing Capitalism*. Penguin, 2021.
18. Kavitha, D., and S. Ravikumar. “Designing an IoT Based Autonomous Vehicle Meant for Detecting Speed Bumps and Lanes on Roads.” *Journal of Ambient Intelligence and Humanized Computing* 12, no. 10 (2021).
19. Yang, C., W. Shen, and X. Wang. “Applications of Internet of Things in Manufacturing.” In *2016 IEEE 20th International Conference on Computer Supported Cooperative Work in Design (CSCWD)*, 670–675. May 2016.
20. Shrouf, F., J. Ordieres, and G. Miragliotta. “Smart Factories in Industry 4.0: A Review of the Concept and of Energy Management Approaches in Production Based on the Internet of Things Paradigm.” In *2014 IEEE International Conference on Industrial Engineering and Engineering Management*, 697–701. December 2014.
21. Wang, K., et al. “Green Industrial Internet of Things Architecture: An Energy Efficient Perspective.” *IEEE Communications Magazine* 54, no. 12 (2016): 48–54.
22. Xu, X., T. Chen, and M. Minami. “Intelligent Fault Prediction System Based on Internet of Things.” *Computational Mathematics and Applications in Advanced Technology and Consumer Control* 64, no. 5 (2012): 833–839.
23. Wan, J., et al. “A Manufacturing Big Data Solution for Active Preventive Maintenance.” *IEEE Transactions on Industrial Informatics* 13, no. 4 (2017): 2039–2047.
24. Jazdi, N. “Cyber Physical Systems in the Context of Industry 4.0.” In *IEEE International Conference on Automation, Quality and Testing, Robotics*, 1–4. May 2014.
25. Wollschlaeger, M., T. Sauter, and J. Jasperneite. “The Future of Industrial Communication: Automation Networks in the Era of the Internet of Things and Industry 4.0.” *IEEE Industrial Electronics Magazine* 11, no. 1 (2017): 17–27.
26. Al-Turjman, F., and A. Radwan. “Data Delivery in Wireless Multimedia Sensor Networks: Challenging and Defying in the IoT Era.” *IEEE Wireless Communications* 24 (2017): 126–131.
27. Wu, F., X. Li, S. Kumari, X. Li, J. Shen, K.-K. R. Choo, M. Wazid, and A. K. Das. “An Efficient Authentication and Key Agreement Scheme for Multi-Gateway Wireless Sensor Networks in IoT Deployment.” *Journal of Network and Computer Applications* 89 (2017): 72–85.
28. Altaf, A., H. Abbas, F. Iqbal, and A. Derhab. “Trust Models of Internet of Smart Things: A Survey, Open Issues, and Future Directions.” *Journal of Network and Computer Applications* 137 (2019): 93–111.
29. Rajendra, P., M. Kumari, S. Rani, N. Dogra, R. Boadh, A. Kumar, and M. Dahiya. “Impact of Artificial Intelligence on Civilization: Future Perspectives.” *Materials Today: Proceedings* 56 (2022): 252–256. https://doi.org/10.1016/j.matpr.2022.01.113
30. Bansal, A., A. Bansal, M. Kumar, and S. Bajwa. “Applications of Artificial Intelligence in Marketing.” In *Balancing Automation and Human Interaction in Modern Marketing*, 59–72. IGI Global, 2024.
31. Khan, F., A. U. Rehman, J. Zheng, M. A. Jan, and M. Alam. “Mobile Crowdsensing: A Survey on Privacy-Preservation, Task Management, Assignment Models, and Incentives Mechanisms.” *Future Generation Computer Systems* 100 (2019): 456–472.
32. Samaniego, M., and R. Deters. “Zero-Trust Hierarchical Management in IoT.” In *2018 IEEE International Congress on Internet of Things (ICIOT)*, 88–95. IEEE, 2018.
33. Novo, O. “Blockchain Meets IoT: An Architecture for Scalable Access Management in IT.” *IEEE Internet of Things Journal* 5 (2018): 1184–1195.
34. Kshetri, N. “Can Blockchain Strengthen the Internet of Things?” *IT Professional* 19 (2017): 68–72.

35. Reyna, A., C. Martín, J. Chen, E. Soler, and M. Díaz. "On Blockchain and Its Integration with IoT: Challenges and Opportunities." *Future Generation Computer Systems* 88 (2018): 173–190.
36. Kumar, A., P. Kumar, N. Sharma, and A. K. Srivastava (Eds.) *3D Printing Technologies: Digital Manufacturing, Artificial Intelligence, Industry 4.0.* Walter de Gruyter GmbH & Co KG, 2024.
37. Mahatme, C., J. Giri, F. Mohammad, M. S. Ali, T. Sathish, N. Sunheriya, and R. Chadge. "Experimental and Numerical Investigation of PLA Based Different Lattice Topologies and Unit Cell Configurations for Additive Manufacturing." *The International Journal of Advanced Manufacturing Technology/International Journal, Advanced Manufacturing Technology* (2024). https://doi.org/10.1007/s00170-024-13882-4
38. Natarajan, M., T. Pasupuleti, J. Giri, H. A. Al-Lohedan, L. N. Katta, F. Mohammad, N. Sunheriya, R. Chadge, C. Mahatme, P. Giri, S. Mallik, and T. Sathish. "Optimization of Wire Spark Erosion Machining of Grade 9 Titanium Alloy (Grade 9) Using a Hybrid Learning Algorithm." *AIP Advances*, 14, no. 1 (2024). https://doi.org/10.1063/5.0177658
39. Narasimhamu, K. L., M. Natarajan, P. Thejasree, E. Makki, J. Giri, N. Sunheriya, R. Chadge, C. Mahatme, P. Giri, and T. Sathish. "Development of Hybrid Optimization Model Using Grey-ANFIS-Jaya Algorithm for CNC Drilling of Aluminium Alloy." *Journal of Engineering*, 2024 (2024): 1–12. https://doi.org/10.1155/2024/1476770
40. Tufail, M. S., J. Giri, E. Makki, T. Sathish, R. Chadge, and N. Sunheriya. "Machinability of Different Cutting Tool Materials for Electric Discharge Machining: A Review and Future Prospects." *AIP Advances*, 14, no. 4 (2024). https://doi.org/10.1063/5.0201614
41. Kamble, P. D., J. Giri, E. Makki, N. Sunheriya, S. B. Sahare, R. Chadge, C. Mahatme, P. Giri, S. T., and H. Panchal. "An Application of Hybrid Taguchi-ANN to Predict Tool Wear for Turning EN24 Material." *AIP Advances*, 14, no. 1 (2024). https://doi.org/10.1063/5.0186432
42. International Organization for Standardization. *Internet of Things (IoT)—Reference Architecture.* Standard ISO/IEC 30141:2018, Geneva, CH; 2018. https://www.iso.org/standard/65695.html
43. AIOTI WG Standardisation. *High Level Architecture (HLA) – Release 5.0.* Technical specification, Alliance for Internet of Things Innovation; 2020.
44. White, R. M. "A Sensor Classification Scheme." *IEEE Transactions on Ultrasonics, Ferroelectrics, and Frequency Control* 34, no. 2 (1987): 124–126. https://doi.org/10.1109/T-UFFC.1987.26922
45. VDMA and wbk. *Leitfaden Sensorik für Industrie 4.0: Wege zu Kostengünstigen Sensorsystemen.* VDMA Forum Industrie 4.0 and Karlsruhe Institute of Technology (KIT) wbk Institute of Production Science; 2018.
46. Gerber, A., and J. Romeo. "Choosing the Best Hardware for Your Next IoT Project." 2017. https://developer.ibm.com/articles/iot-lp101-best-hardware-devices-iot-project/. [Accessed December 2, 2021].
47. Bauer, M., et al. "IoT Reference Model." In *Enabling Things to Talk*, 113–162. Springer, 2013.
48. Pivoto, D. G., L. F. de Almeida, R. da Rosa Righi, J. J. Rodrigues, A. B. Lugli, and A. M. Alberti. "Cyber-Physical Systems Architectures for Industrial Internet of Things Applications in Industry 4.0: A Literature Review." *Journal of Manufacturing Systems* 58 (2021): 176–192.
49. Deutsches Institut für Normung. *Referenzarchitekturmodell Industrie 4.0 (RAMI 4.0).* Standard DIN SPEC 91345:2016–04, 2016.
50. Bakhshi, Z., A. Balador, and J. Mustafa. "Industrial IoT Security Threats and Concerns by Considering Cisco and Microsoft IoT Reference Models." In *2018 IEEE Wireless Communications and Networking Conference Workshops*, 173–178. IEEE, 2018.

51. Bhat, F. A., and S. Parvez. "Emerging Challenges in the Sustainable Manufacturing System: From Industry 4.0 to Industry 5.0." *Journal of the Institution of Engineers (India): Series C* (2024). https://doi.org/10.1007/s40032-024-01046-y
52. Javaid, M., et al. "Industry 4.0 Technologies and Their Applications in Fighting COVID-19 Pandemic." *Diabetes & Metabolic Syndrome: Clinical Research & Reviews* 14 (2020): 419–422. https://doi.org/10.1016/j.dsx.2020.04.032

10 The Integration of Augmented and Virtual Reality in Modern Manufacturing

Suresh Subramanian, Boopathi Sampath, Dhivya Priya Erode Loganathan, Elango Natarajan, and Elayaraja Radhakrishnan

10.1 INTRODUCTION

In today's manufacturing environment, characterized by complex supply chains, rapid product iterations, and increasing customization demands, traditional methods often fall short in addressing the evolving challenges. The need for agile, adaptable solutions that can streamline processes, enhance productivity, and minimize errors has never been more pressing. Modern manufacturing stands at the precipice of a technological revolution, propelled by the convergence of digital innovation and industrial processes. As industries strive for greater efficiency, flexibility, and competitiveness, the adoption of cutting-edge technologies has become imperative (Suresh, Natarajan, et al. 2024; Suresh, Kumar, et al. 2024; Kumar, Kumar et al. 2024; Kumar, Kumar et al. 2023; Burande et al. 2024). In this landscape, the manufacturing industry stands at the precipice of a technological revolution, driven by the integration of progressive technologies such as augmented reality (AR)/virtual reality (VR) (Ardiny and Khanmirza 2018; Yin et al. 2020). These immersive technologies are reshaping traditional manufacturing processes, offering unprecedented opportunities for efficiency gains, cost savings, and improved quality. As the boundaries between the physical and digital worlds blur, manufacturers are increasingly turning to AR and VR to unlock new capabilities, enhance workforce skills, and revolutionize product design and assembly. Over the past few decades, VR and AR have quietly evolved within research labs, often without garnering significant public attention. However, recent years have witnessed a dramatic shift as these technologies have entered the consumer market, propelled by advancements in hardware affordability and software algorithms. Now, VR and AR are not just novelties; they are transformative tools that find applications across diverse fields (Huang et al. 2018: Kumar, Rani et al. 2023).

Drawing upon real-world case studies from leading manufacturing companies such as Volkswagen Group, Boeing, Ford Motor Company, General Electric, and Tesla, Inc., we highlight successful implementations of AR and VR and

DOI: 10.1201/9781032725086-12

examine the tangible benefits they have realized (Ong and Nee 2004; Eswaran and Bahubalendruni 2022). These case studies serve as exemplars of how AR and VR can drive operational excellence, boost productivity, and enhance competitiveness in today's fast-paced manufacturing environment. Looking to the future, it explores emerging trends and prospects for AR and VR in manufacturing, including advances in technology, integration with the artificial intelligence (AI) and Internet of Things (IoT), the rise of Industry 4.0 and smart manufacturing, and potential disruptions and opportunities on the horizon (Kumar, Kumar Shrivastava et al. 2024; Rajendra et al. 2022; Rani et al. 2021; Sharmila et al. 2022; Suresh, Velmurugan et al. 2024; Batista et al. 2024). By staying abreast of these trends and harnessing the full potential of AR and VR, manufacturers can position themselves for long-term success and innovation leadership in the global marketplace. Furthermore, VR and AR technologies facilitate remote collaboration and support, allowing experts to provide guidance and assistance from anywhere in the world. This capability is particularly valuable in globalized manufacturing settings where teams may be distributed across different locations.

AR, an increasingly prominent technology in various industries, including advanced manufacturing, complements the narrative of technological innovation alongside VR. While VR has a long history spanning over 60 years, AR has emerged as a cutting-edge concept in recent years, drawing on advancements in computing and sensor technologies.

VR stands as a cornerstone of technological advancement with a rich history spanning over six decades. Its inception dates back to the 1950s, marked notably by the development of the Sensorama simulator, regarded as one of the earliest functional VR machines. Since then, dedicated researchers have tirelessly explored this domain, resulting in the creation of numerous prototypes and groundbreaking innovations. An ideal VR system strives to replicate the entirety of human sensory perception, catering to sight, hearing, touch, smell, and taste. This ambitious goal is realized through a diverse array of VR setups, ranging from expansive CAVE VR systems to desktop configurations and immersive head-mounted displays (Wu et al. 2023). Within the realm of advanced manufacturing and beyond, VR holds immense promise for revolutionizing processes and workflows. By transporting users to virtual environments where they can manipulate digital representations of physical objects, VR facilitates enhanced visualization, training, and prototyping.

Similar to VR, AR has roots in early experiments such as the Sensorama simulator developed in the 1950s. However, AR distinguishes itself by blending virtual elements seamlessly with the real world, offering users an augmented perception of their environment. This integration permits a client to interrelate with virtual objects while maintaining awareness of their physical surroundings (Reljić et al. 2021). AR permits users to visualize, manipulate, and interrelate with digital substance in real time, enhancing their understanding and engagement with the environment. Unlike VR, which sinks users in completely virtual environments, AR covers digital information onto the physical world, enriching awareness and communication without completely replacing reality (Azuma 1997; Katkuri et al. 2019). AR systems leverage various technologies, including computer vision, spatial mapping, and wearable displays, to deliver immersive experiences. These systems can range from handheld

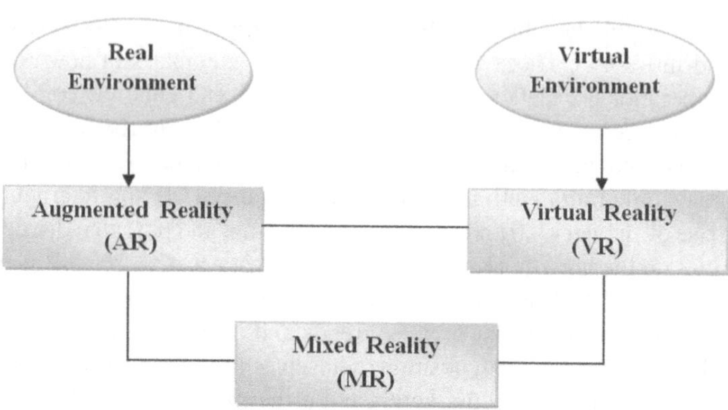

FIGURE 10.1 Difference between AR and VR.

devices, such as smart phones and tablets, to wearable devices like smart glasses or headsets. Figure 10.1 represents the relation between AR and VR.

This chapter embarks on a complete investigation of the implementation of AR and VR in modern manufacturing. It begins by defining AR and VR, elucidating their distinctions, and exploring the key components and underlying technologies that enable their functionality. Subsequently, it delves into the myriad applications of AR and VR in manufacturing, ranging from enhanced training and simulation to improved design and prototyping, streamlined assembly processes, remote assistance, and enhanced safety measures.

10.2 INTEGRATING AR IN TRAINING AND INDUSTRIAL APPLICATIONS

AR and VR technologies in training and industrial applications characterize a transformative move toward enhancing learning and operational efficiency. By immersing users in simulated environments, AR and VR offer unparalleled opportunities for hands-on training, skill development, and operational optimization (Damiani et al. 2018).

In training scenarios, AR and VR enable immersive simulations that replicate real-world environments, allowing trainees to practice tasks in a safe and controlled setting. For instance, in industrial settings, workers can undergo training in complex procedures, such as equipment maintenance or emergency response, without risking injury or damage to machinery (Naranjo et al. 2020). Through interactive simulations, trainees can gain practical experience, refine their skills, and build confidence in performing tasks before executing them in actual working conditions. Moreover, AR and VR technologies facilitate experiential learning by providing immediate feedback and personalized guidance to trainees. Virtual environments can adapt to entity study and skill levels, offering tailored challenges and assessments to maximize learning outcomes. This adaptive approach ensures that trainees receive targeted instruction and support, leading to accelerated skill acquisition and proficiency (Vidal-Balea et al. 2020).

In industrial applications, AR and VR solutions offer significant benefits beyond training. These technologies streamline operations by optimizing workflows, enhancing productivity, and minimizing errors. For example, AR-enabled maintenance platforms can overlay digital instructions and schematics onto physical equipment, guiding technicians through repair procedures with precision and efficiency. Similarly, VR-based design reviews allow engineers to visualize and refine product concepts collaboratively, leading to faster iterations and improved outcomes.

Furthermore, AR and VR play a crucial part in remote collaboration and knowledge sharing, particularly in decentralized work environments. Through virtual conferencing and remote assistance tools, distributed teams can collaborate seamlessly across geographical boundaries, accessing shared data and expertise in real time. This fosters cross-functional collaboration, accelerates decision-making, and drives innovation across the organization.

Here are some specific products and applications where AR/VR is implemented in the industrial training and education sector.

10.2.1 VR Welding Simulators

Welding is a fundamental process in modern manufacturing, with spot welding being especially crucial due to its efficiency and reliability in joining metal sheets, particularly in the automotive and electronics industries. The importance of spot welding lies in its ability to create strong, durable joints with minimal thermal distortion, making it ideal for mass production (Suresh, Natarajan et al. 2023; Suresh, Velmurugan et al. 2023; Suresh et al. 2019). VR welding simulators offer a transformative approach to training aspiring welders by providing an authentic welding experience within a controlled virtual environment. These simulators immerse trainees in lifelike scenarios, replicating the sights, sounds, and sensations of real-world welding operations.

Through VR technology, trainees can manipulate welding equipment and perform welding techniques with remarkable realism. Visual cues such as sparks, arc flashes, and molten metal behave realistically, creating an immersive experience that closely mirrors actual welding conditions. This level of realism not only improves trainee engagement but also fosters a deeper understanding of welding processes and techniques (Ipsita et al. 2022). A study offers insights into the factors affecting the effectiveness of AR welding training and its impact on user choices. Utilizing a modified technology acceptance model and involving 200 trainees, the research examines variables such as perceived enjoyment and system quality (Papakostas et al. 2022). Results indicate that these external variables predict perceived utility and ease of use. Specifically, the purpose to use AR simulators is completely influenced by system excellence and perceived ease of use. These consequences suggest impending for AR developers to progress in the excellence of AR-simulation teaching method, ultimately enhancing user skill and their willingness to use them.

One of the standout features of VR welding simulators is their ability to provide haptic feedback, allowing trainees to feel the resistance and vibrations associated with welding. This tactile feedback adds an extra layer of immersion, enabling trainees to develop a sense of muscle memory and refine their welding technique in a highly realistic manner (Wang et al. 2006). The study aimed to create a safe and efficient training method for novice welders. They developed a visual-haptic extended reality (VHXR) system, allowing trainees to practice manual arc welding without the risks connected with high temperatures and intense ultraviolet radiation. This hands-on training system supplies a realistic experience while ensuring the safety of novice welders during their learning process (Shankhwar et al. 2022).

Moreover, VR welding simulators incorporate real-time performance analysis tools that monitor trainees' actions and provide immediate feedback on their welding proficiency. Trainees can receive detailed metrics on parameters such as arc length, travel speed, and bead placement, enabling them to identify areas for improvement and make adjustments on the fly. By offering a combination of realistic immersion, haptic feedback, and performance analysis, VR welding simulators revolutionize the learning process for aspiring welders (Ni et al. 2017). Trainees can hone their skills in a risk-free environment, gaining confidence and proficiency before transitioning to real-world welding tasks. Ultimately, these simulators play a critical position in preparing the next generation of welders for accomplishment in the welding industry.

10.2.2 AR Equipment Maintenance Guides

AR Equipment Maintenance Guides revolutionize traditional maintenance procedures by leveraging AR technology to enhance efficiency and accuracy in equipment upkeep. These guides employ AR overlays to seamlessly integrate digital information onto physical tools, providing maintenance technicians with invaluable resources at their fingertips (Wu et al. 2020).

Through AR glasses or mobile devices, technicians can access dynamic step-by-step guides tailored to specific maintenance tasks. These guides offer interactive three-dimensional (3D) models of equipment components, allowing technicians to visualize intricate details and understand complex mechanisms with clarity. By presenting information directly within the context of the equipment, AR guides streamline the troubleshooting process, enabling technicians to identify and address issues promptly (Zhao et al. 2019). Moreover, AR Equipment Maintenance Guides facilitate real-time collaboration and knowledge sharing among maintenance teams. Technicians can overlay annotations or notes onto equipment components, highlighting key areas for inspection or repair. This collaborative feature enhances communication and coordination, particularly in large-scale maintenance operations or across distributed teams (Gao et al. 2020).

Figure 10.2 represents a screen-based video see-through display approach (Fiorentino et al. 2014), which can provide a less immersive augmentation without drawbacks. Additionally, AR guides enhance training initiatives by providing hands-on learning experiences in a simulated environment. New technicians can learn maintenance procedures efficiently by following guided instructions overlaid onto equipment, gaining practical experience without the risk of errors or damage (Scheffer et al. 2023). Overall, AR Equipment Maintenance Guides represent a paradigm shift in the maintenance industry, offering a user-friendly and intuitive solution for improving efficiency, accuracy, and collaboration in equipment upkeep. By

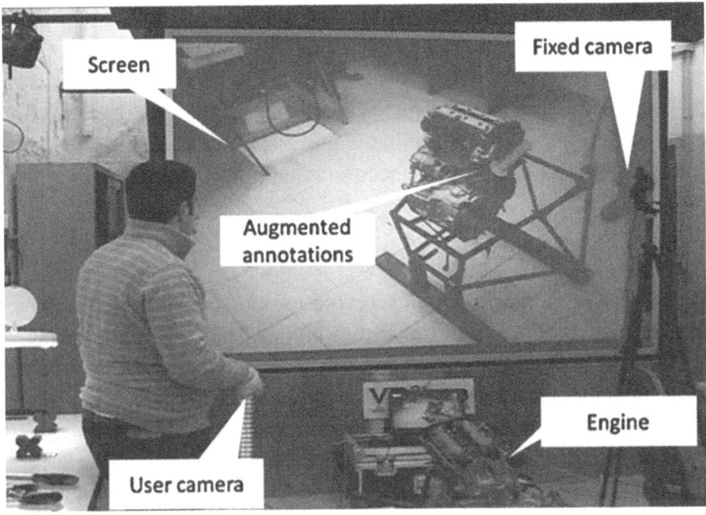

FIGURE 10.2 Interactive AR instructions on-screen. (Source: Fiorentino et al. 2014.)

harnessing the power of AR, maintenance technicians can execute their responsibilities with confidence, ensuring optimal performance and longevity of critical assets.

10.2.3 AR Industrial Safety Training

AR is a cutting-edge technology utilized to develop interactive safety training modules tailored to industrial settings. These modules immerse learners in realistic simulations where they can actively identify potential hazards, adhere to safety protocols, and practice emergency responses. For instance, within the context of manufacturing plants or construction sites, workers equipped with AR headsets can seamlessly integrate digital overlays into their physical surroundings (Tatić and Tešić 2017). As they navigate hazardous environments, these AR overlays provide real-time visualizations of safety warnings, precautionary measures, and step-by-step instructions. Workers can efficiently identify safety hazards, such as high-voltage areas or restricted zones, highlighted by AR markers overlaid onto machinery or structural elements.

Moreover, AR enables dynamic interaction with safety procedures by incorporating simulated scenarios that prompt users to respond appropriately. For example, during a fire drill simulation, workers may encounter virtual flames or smoke within their AR headset display, requiring them to execute evacuation procedures in real time. Through this hands-on approach, trainees gain valuable experience in navigating emergencies and reinforcing safety best practices (Gavish et al. 2015). Furthermore, AR enhances the effectiveness of safety training by offering personalized learning experiences. Trainees can access AR modules tailored to their specific job roles and responsibilities, ensuring that training content remains relevant and applicable to their daily tasks. This targeted approach increases engagement and knowledge retention, empowering workers to make knowledgeable decisions concerning safety protocols in their work environment (Wang et al. 2018). In summary, AR industrial safety training revolutionizes traditional safety education by providing immersive, interactive, and personalized learning experiences. By leveraging AR technology, organizations can effectively prepare their workforce to identify hazards, adhere to safety guidelines, and respond confidently to emergencies, ultimately fostering a safer and more secure working environment.

10.2.4 AR Automotive Maintenance Guides

AR Automotive Maintenance Guides revolutionize the way technicians conduct repairs and maintenance tasks in the automotive industry. By integrating AR technology into the workflow, these guides offer technicians digital overlays of repair instructions, component identification, and diagnostic information directly within their field of view. This hands-free approach removes the need for technicians to constantly refer to traditional manuals or digital screens, enhancing efficiency and reducing errors during maintenance and repairs (Arifitama et al. 2022; Singhal et al. 2022).

With AR maintenance guides, technicians can access step-by-step instructions overlaid onto the vehicle or specific components they are working on. This real-time guidance ensures that technicians follow the correct procedures, leading to faster turnaround times and higher quality repairs. Additionally, AR technology enables technicians to visualize complex systems and identify components more accurately, even in challenging or confined spaces. Furthermore, AR maintenance guides can provide diagnostic information overlaid onto the vehicle's components, helping technicians quickly pinpoint issues and streamline troubleshooting processes (Cachada et al. 2019). By presenting relevant data such as sensor readings, error codes, or system status indicators in real time, AR guides empower technicians to make informed decisions and expedite repairs. The hands-free nature of AR maintenance guides allows technicians to maintain focus on the task at hand while accessing crucial information seamlessly. This not only increases competence but also enhances safety in the workshop environment, as technicians can keep their hands and attention on the vehicle without distractions (Chouchene et al. 2022). Overall, AR Automotive Maintenance Guides represent a game-changing innovation in the automotive industry, offering technicians a more intuitive and efficient way to perform maintenance and repairs. By leveraging AR technology, these guides optimize workflow processes, reduce errors, and ultimately contribute to improved customer satisfaction and vehicle reliability.

10.2.5 AR Industrial Design and Prototyping

In the realm of industrial design and prototyping, AR emerges as a groundbreaking tool, revolutionizing traditional approaches to product development. By seamlessly blending virtual elements with physical environments, AR empowers designers to envision and refine product designs with unprecedented precision and efficiency (Ryead 2023). Through AR, designers gain the ability to visualize virtual models within real-world contexts, offering invaluable insights into how products will interact with their surroundings. This integration allows for the evaluation of factors such as ergonomics, spatial relationships, and aesthetic harmony, all before a single physical prototype is constructed. Designers can manipulate virtual models in real time, experimenting with different configurations and iterations to optimize functionality and user experience (Marner et al. 2011). Moreover, AR makes possible mutual design processes by facilitating stakeholders to interact with virtual prototypes at the same time, regardless of their geographical location. This fosters cross-functional collaboration and accelerates decision-making, ultimately leading to faster time-to-market and enhanced product quality.

In the part of prototyping, AR offers a cost-effective alternative to conventional methods by reducing the need for physical prototypes. Designers can conduct comprehensive testing and validation of virtual prototypes, recognizing and addressing potential issues early in the design process. This iterative progress minimizes the time and resources required for physical prototyping, streamlining the overall product development cycle (Murauer 2018; Fiorentino et al. 2002).

Furthermore, AR enhances the presentation and communication of design concepts to clients and stakeholders. By overlaying virtual models onto physical spaces or products, designers can effectively demonstrate design features and functionalities in a compelling and immersive manner. This aids in garnering feedback and buy-in from stakeholders, ensuring alignment with project objectives and requirements. Overall, AR's integration into industrial design and prototyping represents a paradigm shift in how products are conceptualized, developed, and brought to market. As technology continues to evolve, its potential to drive innovation and efficiency within the design process will only continue to grow, shaping the future of industrial design and product development.

10.3 MANUFACTURING ASSEMBLY SIMULATIONS IN AR AND VR

Assembly processes in automotive manufacturing are highly complex, involving the precise integration of numerous components to build a functioning vehicle. Traditional methods of assembly training often rely on static manuals or classroom-based instruction, which may not adequately prepare workers for the dynamic challenges of the assembly line. By leveraging these technologies, manufacturers can optimize assembly workflows, reduce errors, and enhance productivity. This is where AR and VR technologies step in, offering immersive and interactive solutions to simulate assembly processes effectively. This section explores how AR and VR technologies are revolutionizing assembly simulations, providing workers with immersive, interactive experiences. Case studies and examples demonstrate successful implementations and their impact on efficiency and quality. In the dynamic landscape of Industry 4.0, manufacturers are increasingly reevaluating their assembly systems to meet evolving customer demands and the rapid pace of technological innovation. To maintain competitiveness and ensure product success in a market characterized by shortened product lifecycles, flexible and efficient assembly methods and strategies are imperative. In this context, upgrading current assembly systems is essential, and assembly simulation in virtual environments emerges as a pivotal strategy.

By harnessing the capabilities of VR and AR technologies, assembly simulation can revolutionize the design and training processes, ultimately enhancing product quality and reducing time-to-market. These immersive technologies prioritize human experience, offering a more intuitive and interactive approach to simulating assembly tasks, particularly those involving workers. A fundamental aspect of assembly simulation in VR/AR is the creation of accurate digital models. These models, whether generated through computer-aided drafting (CAD) software or 3D reconstruction processes, must faithfully replicate physical parts in terms of dimensions, physics, and functionality. Unlike gaming VR/AR applications, the focus here lies in ensuring precision and realism to enable effective simulation of assembly processes. In VR, tracking the user's pose is crucial for dynamically adjusting digital content to simulate real-world interactions convincingly. In AR, the stakes are higher, as real-time environment sensing and pose estimation are essential for the seamless addition of virtual and physical elements. Optimal mapping between the virtual and real worlds is achieved through precise tracking, enhancing the immersive experience.

Visual interfaces take precedence in VR/AR assembly simulations, but supplementary auditory, haptic, and even olfactory interfaces can further enrich the user experience. Interactivity is key, requiring seamless manipulation of virtual objects akin to physical counterparts. Developing intuitive interfaces and interaction mechanisms is paramount to ensure user engagement and the effectiveness of the simulation.

Assembly training conducted in VR/AR environments offers unparalleled opportunities for workforce development. Workers can immerse themselves in lifelike simulations of assembly tasks, gaining practical experience and honing their skills in a risk-free virtual setting. By replicating real-world scenarios with fidelity, VR/AR assembly simulations facilitate more effective and efficient training programs.

10.3.1 Advantages of AR/VR Simulation in Assembly

1. ***Immersive Training Environments:*** AR and VR technologies make a realistic virtual environment where assembly workers can practice assembly tasks in a safe and prohibited setting. These simulations replicate the actual assembly line conditions, including the layout of the factory floor, the placement of tools and equipment, and the sequence of assembly steps (Checa and Bustillo 2020). By immersing workers in these virtual environments, they can gain hands-on experience and develop muscle memory for assembly tasks before working on the actual production line.
2. ***Interactive Learning Experiences:*** Unlike traditional training methods, AR and VR simulations enable interactive learning knowledge. Employees can interrelate with virtual components, manipulate those using virtual tools, and observe the consequences of their actions in real time (Dangelmaier et al. 2005). This interactivity enhances engagement and knowledge retention, as learners actively participate in the learning process rather than inactively absorbing information.
3. ***Error Identification and Correction:*** AR and VR simulations allow for real-time feedback on assembly performance. Workers can receive instant

feedback on their actions, identifying errors and correcting them before they occur on the actual assembly line. This proactive approach to error identification helps prevent costly mistakes and ensures that workers are sufficiently prepared to meet quality standards (Goel et al. 2000).

4. ***Adaptability and Customization:*** AR and VR simulations can be tailored to meet the specific needs of different assembly tasks and skill levels. Training modules can range from basic assembly procedures for novice workers to advanced troubleshooting scenarios for experienced technicians (Prasolova-Førland et al. 2017). Additionally, simulations can be updated and modified to reflect changes in assembly processes or introduce new product configurations, ensuring that workers remain up-to-date with the latest procedures.

10.3.2 AR/VR Assembly Simulation Methods

Employing VR and AR in manufacturing assembly simulation (MAS) opens doors to an experiential realm, granting manufacturers a seamless blend of intuition, immersion, and interactivity. These innovative solutions transcend traditional boundaries, elevating assembly processes by enhancing design precision, ensuring meticulous process verification, and refining workforce training methodologies. Creating digital representations for virtual entities stands as a pivotal endeavor within the domain of MAS. Traditional methods for crafting 3D models involve employing commercial CAD software, such as SolidWorks, and CATIA to generate shape corresponding to virtual objects. An alternative approach entails reconstructing 3D models from physical objects, beginning with the acquisition of data from real-world entities and culminating in the creation of digital shape (Remondino and El-Hakim 2006). This approach offers the advantage of automating the often time-intensive modeling process typically performed by human designers, thus significantly reducing both development time and costs. Illustrated in Figure 10.3, the process of 3D model reconstruction comprises some stages, including data acquisition, processing, spatial registration, modeling, and rendering (Heipke et al. 2016; El-Hakim 2000).

In the initial stages of transforming a tangible object into a digital 3D model, the primary task involves capturing the intricate details of its surface in digital form. Various methodologies have emerged to accomplish this, broadly classified into mechanical and optical approaches. Beyond these, alternative techniques like ultrasonography, radiography, and magnetic resonance imaging (MRI) also serve the reason for acquiring requisite data (Nayak et al. 2022). Figure 10.3 represents the various 3D data acquisition techniques.

Following the completion of the data acquisition process, the next step involves crafting a 3D representation of the object through surface reconstruction. This intricate process aims to extrapolate the surface geometry of the parts from a finite set of gathered measurements. Typically, the attain information exhibit disarray and noise, presenting a challenge to the reconstruction endeavor. Moreover, the object's surface might lack a defined topological structure, allowing for a wide range of shapes. The sequential stages involved in transforming the acquired digital data into comprehensive 3D models encompass the following: first, preprocessing, wherein invalid data

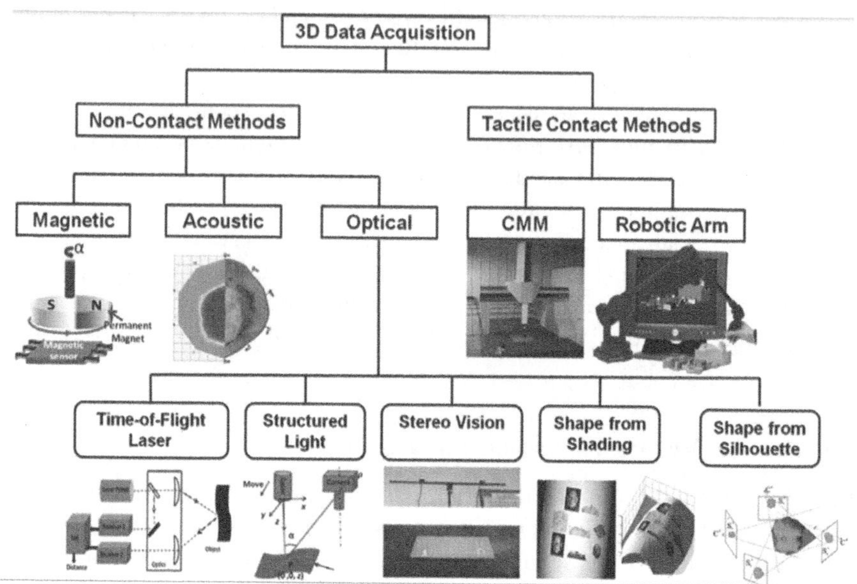

FIGURE 10.3 3D data acquisition techniques. (Source: Leu et al. 2013.)

points are eliminated, and the noise is mitigated; second, the determination of the overall topology of the surface; third, the actual creation of the polygonal surface; and finally, post-processing tasks such as edge refinement and triangle introduction are carried out to ensure the fidelity and precision (Osti et al. 2017; Kamble et al. 2024; Thakare et al. 2024; Mahatme et al. 2024).

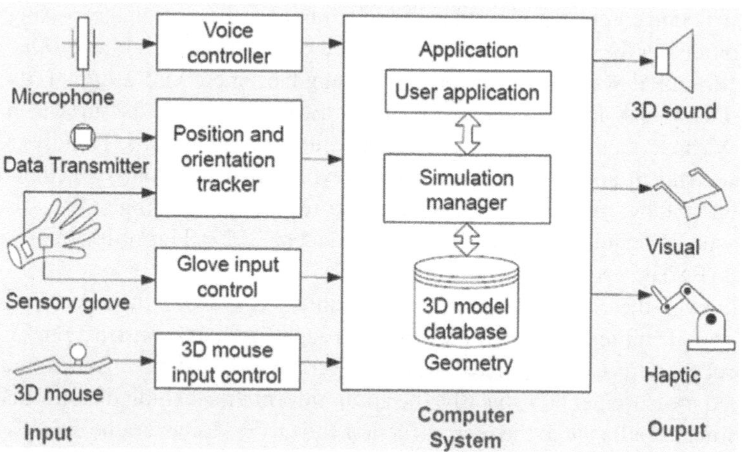

FIGURE 10.4 Typical VR/AR configuration with human-computer interfaces. (Source: Leu et al. 2013.)

In the realm of human-computer interaction, various sensory modalities intertwine to enhance immersion within VR/AR settings. Illustrated in Figure 10.4, the conventional setup of a VR/AR system integrates tangible input and output apparatuses, facilitating seamless interaction between the user and the digital environment. The efficacy of these sensory technologies profoundly influences the authenticity of the VR/AR experience, encompassing visual representation, auditory synthesis, and tactile feedback (Leu et al. 2013).

Ong et al. (2007) pioneered the formation of an assembly system utilizing AR, offering engineers an immersive and intuitive platform for designing and planning assemblies at the early stages of development. Through the integration of AR techniques, users are presented with blended surroundings where real and virtual objects coexist, facilitating a comprehensive consideration of the assembly environment as shown in Figure 10.5. This novel approach enables engineers to interact with and assess virtual prototypes within the actual assembly setting, empowering them to iteratively refine product designs for improved assembly processes.

Numerous corporations have incorporated MAS alongside VR/AR into their manufacturing processes, reaping substantial advantages. For instance, within the automotive sector, Toyota has pioneered an immersive virtual training system utilizing VR technology to educate assembly line workers. Participants are fully immersed in a virtual assembly line setting through the use of an HTC Vive headset. Throughout the training, individuals follow detailed assembly procedures and engage with virtual objects using controllers. The Volkswagen Group, in collaboration with VR studio Innoactive, has initiated a comprehensive program to introduce VR training to 10,000 employees. This initiative encompasses over 30 VR training modules, spanning tasks from vehicle assembly and onboarding of new team members to customer service training. Similarly, corporations such as Tesla, Mercedes-Benz, BMW, Volvo, Ford, and Bentley have all embraced VR/AR technologies within their manufacturing facilities to streamline assembly processes.

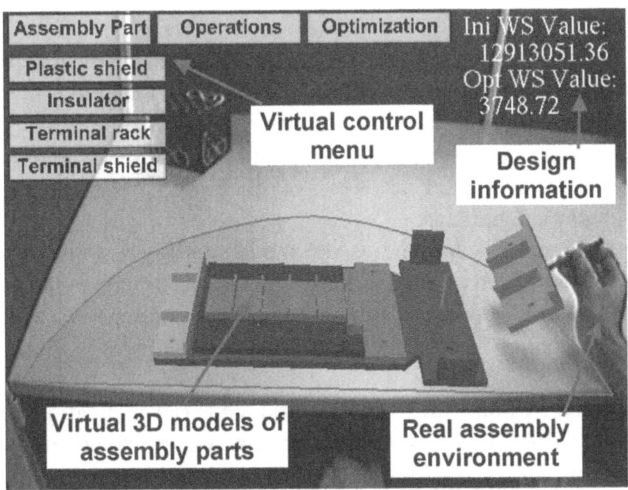

FIGURE 10.5 AR assembly environment. (Source: Ong et al. 2007.)

The integration of AR and VR technologies in assembly processes is poised for further advancement. Future developments may include the following:

1. ***Enhanced Realism:*** Advancements in hardware and software technologies will enable even more realistic simulations, with improved graphics, haptic feedback, and spatial audio to create immersive training experiences.
2. ***Remote Collaboration:*** AR and VR technologies will facilitate remote collaboration between assembly workers located in different geographical locations. Workers can connect virtually to troubleshoot issues, share best practices, and collaborate on complex assembly tasks in real time.
3. ***Data Analytics:*** The addition of data analytics tools will enable the compilation and study of performance metrics from AR and VR simulations. Manufacturers can influence this data to recognize areas for development, optimize assembly processes, and enhance workforce productivity.

In summary, manufacturing assembly simulations in AR/VR represent a transformative advance in enhancing assembly processes. By leveraging AR/VR technologies, manufacturers can streamline design iterations, improve worker training, and ultimately optimize assembly operations to meet the demands of Industry 4.0. Embracing these immersive technologies places human experience at the forefront, paving the way for more agile, efficient, and competitive manufacturing systems (Giri et al. 2024; Tufail et al. 2024).

10.4 CONCLUSION

- The integration of AR and VR in modern manufacturing represents a paradigm shift toward smarter, more agile production systems. By harnessing the power of these technologies, manufacturers can unlock new levels of efficiency, agility, and competitiveness, positioning themselves for success in the digital age.
- The implementation of AR and VR in manufacturing holds immense promise for driving operational excellence, enhancing workforce productivity, and delivering superior customer value.
- By embracing AR and VR and adopting best practices for their execution, manufacturers can unlock new levels of efficiency, agility, and competitiveness in today's digital age.
- VR and AR are revolutionizing the manufacturing assembly process by offering cost-effective simulations, interactive training modules, and seamless remote support. As these techniques continue to develop, their potential to optimize efficiency and drive innovation in manufacturing is boundless.
- Integration of AR and VR in training and industrial applications represents a standard shift in how businesses approach skill development, operational optimization, and collaboration. By harnessing the immersive power of these technologies, organizations can release new opportunities for efficiency, safety, and innovation in the digital age.

ACKNOWLEDGMENTS

The authors would like to acknowledge the financial support received from the Science and Engineering Research Board (SERB) under grant SSY/2023/001464. This support was instrumental in facilitating discussions and knowledge exchange that significantly contributed to the development of this chapter.

REFERENCES

Ardiny, Hadi, and Esmaeel Khanmirza. 2018. "The Role of AR and VR Technologies in Education Developments: Opportunities and Challenges." In *2018 6th RSI International Conference on Robotics and Mechatronics (IcRoM)*, 482–87. doi:10.1109/ICRoM.2018.8657615.

Arifitama, Budi, Zihan Noor Abdillah, Ade Syahputra, Silvester Dian Handy Permana, and Yaddarabullah. 2022. "A Mobile Augmented Reality Application for Automotive Sparepart Installation." *International Journal of Advanced Science Computing and Engineering* 4 (3): 193–202. doi:10.30630/ijasce.4.3.97.

Azuma, Ronald T. 1997. "A Survey of Augmented Reality." *Presence: Teleoperators and Virtual Environments* 6 (4): 355–85. doi:10.1162/pres.1997.6.4.355.

Batista, Rafael Cavicchioli, Abhishek Agarwal, Adash Gurung, Ajay Kumar, Faisal Altarazi, Namrata Dogra, Vishwanatha HM, Dundesh S. Chiniwar, and Ashish Agrawal. "Topological and lattice-based AM optimization for improving the structural efficiency of robotic arms." *Frontiers in Mechanical Engineering* 10 (2024): 1422539.

Burande, D.V., K. Kalita, R. Gupta, et al. 2024. Machine learning metamodels for thermo-mechanical analysis of friction stir welding. *International Journal on Interactive Design and Manufacturing (IJIDeM)*. doi:10.1007/s12008-024-01871-6.

Cachada, Ana, Luis Romero, David Costa, Hasmik Badikyan, Jose Barbosa, Paulo Leitao, Osmano Morais, Carlos Teixeira, Joao Azevedo, and Pedro Miguel Moreira. 2019. "Using AR Interfaces to Support Industrial Maintenance Procedures." In *IECON 2019 – 45th Annual Conference of the IEEE Industrial Electronics Society*, 1: 3795–800. IEEE. doi:10.1109/IECON.2019.8927815.

Checa, David, and Andres Bustillo. 2020. "A Review of Immersive Virtual Reality Serious Games to Enhance Learning and Training." *Multimedia Tools and Applications* 79 (9): 5501–27. doi:10.1007/s11042-019-08348-9.

Chouchene, Amal, Adriana Ventura Carvalho, Fernando Charrua-Santos, and Walid Barhoumi. 2022. "Augmented Reality-Based Framework Supporting Visual Inspection for Automotive Industry." *Applied System Innovation* 5 (3): 48. doi:10.3390/asi5030048.

Damiani, Lorenzo, Melissa Demartini, Guido Guizzi, Roberto Revetria, and Flavio Tonelli. 2018. "Augmented and Virtual Reality Applications in Industrial Systems: A Qualitative Review towards the Industry 4.0 Era." *IFAC-PapersOnLine* 51 (11): 624–30. doi:10.1016/j.ifacol.2018.08.388.

Dangelmaier, Wilhelm, Matthias Fischer, Jürgen Gausemeier, Michael Grafe, Carsten Matysczok, and Bengt Mueck. 2005. "Virtual and Augmented Reality Support for Discrete Manufacturing System Simulation." *Computers in Industry* 56 (4): 371–83. doi:10.1016/j.compind.2005.01.007.

El-Hakim, Sabry. 2000. "A Practical Approach to Creating Precise and Detailed 3D Models from Single and Multiple Views." *International Archives of Photogrammetry and Remote Sensing* 33.

Eswaran, M., and MVA Raju Bahubalendruni. "Challenges and opportunities on AR/VR technologies for manufacturing systems in the context of industry 4.0: A state of the art review." *Journal of Manufacturing Systems* 65 (2022): 260–278.

Fiorentino, Michele, Raffaele de Amicis, Giuseppe Monno, and Andre Stork. "Spacedesign: A mixed reality workspace for aesthetic industrial design." In *Proceedings. International symposium on mixed and augmented reality*, pp. 86–318. IEEE, 2002.

Fiorentino, Michele, Antonio E. Uva, Michele Gattullo, Saverio Debernardis, and Giuseppe Monno. 2014. "Augmented Reality on Large Screen for Interactive Maintenance Instructions." *Computers in Industry* 65 (2): 270–78. doi:10.1016/j.compind.2013.11.004.

Gao, Lingyan, Fang Wu, Lilan Liu, and Xiang Wan. 2020. "Construction of Equipment Maintenance Guiding System and Research on Key Technologies Based on Augmented Reality." In *Advanced Manufacturing and Automation IX*, edited by Yi Wang, Kristian Martinsen, Tao Yu, and Kesheng Wang, 275–82. Singapore: Springer Singapore.

Gavish, Nirit, Teresa Gutiérrez, Sabine Webel, Jorge Rodríguez, Matteo Peveri, Uli Bockholt, and Franco Tecchia. 2015. "Evaluating Virtual Reality and Augmented Reality Training for Industrial Maintenance and Assembly Tasks." *Interactive Learning Environments* 23 (6): 778–98, Routledge. doi:10.1080/10494820.2013.815221.

Giri, Jayant, T. Sathish, Taukeer Sheikh, Neeraj Sunehriya, Pallavi Giri, Rajkumar Chadge, Chetan Mahatme, and A. Parthiban. "Automatic liver segmentation using U-Net deep learning architecture for additive manufacturing." *Interactions* 245, no. 1 (2024): 90.

Goel, Puneet, Göksel Dedeoglu, Stergios I. Roumeliotis, and Gaurav S. Sukhatme. "Fault detection and identification in a mobile robot using multiple model estimation and neural network." In *Proceedings 2000 ICRA. Millennium Conference. IEEE International Conference on Robotics and Automation. Symposia Proceedings (Cat. No. 00CH37065)*, vol. 3, pp. 2302–2309. IEEE, 2000.

Heipke, Christian, Marguerite Madden, Zhilin Li, and Ian Dowman. 2016. "Theme Issue 'State-of-the-Art in Photogrammetry, Remote Sensing and Spatial Information Science.'" *ISPRS Journal of Photogrammetry and Remote Sensing* 115: 1–2. doi:10.1016/j.isprsjprs.2016.03.006.

Huang, Ta-Ko, Chi-Hsun Yang, Yu-Hsin Hsieh, Jen-Chyan Wang, and Chun-Cheng Hung. 2018. "Augmented Reality (AR) and Virtual Reality (VR) Applied in Dentistry." *The Kaohsiung Journal of Medical Sciences* 34 (4): 243–48. doi:10.1016/j.kjms.2018.01.009.

Ipsita, Ananya, Levi Erickson, Yangzi Dong, Joey Huang, Alexa K. Bushinski, Sraven Saradhi, Ana M. Villanueva, Kylie A. Peppler, Thomas S. Redick, and Karthik Ramani. 2022. "Towards Modeling of Virtual Reality Welding Simulators to Promote Accessible and Scalable Training." In *CHI Conference on Human Factors in Computing Systems*, 1–21. New York, NY, USA: ACM. doi:10.1145/3491102.3517696.

Kamble, Prashant D., Jayant Giri, Emad Makki, Neeraj Sunheriya, Shilpa B. Sahare, Rajkumar Chadge, Chetan Mahatme, Pallavi Giri, and Hitesh Panchal. 2024. "An application of hybrid Taguchi-ANN to predict tool wear for turning EN24 material." *AIP Advances* 14, no. 1.

Katkuri, Pavan Kumar, Archana Mantri, and Srilakshmi Anireddy. 2019. "Innovations in Tourism Industry & Development Using Augmented Reality (AR), Virtual Reality (VR)." In *TENCON 2019 – 2019 IEEE Region 10 Conference (TENCON)*, 2578–81. doi:10.1109/TENCON.2019.8929478.

Kumar, Ajay, Parveen Kumar, Naveen Sharma, and Ashish Kumar Srivastava (Eds.). 2024. *3D Printing Technologies*. Berlin, Boston: De Gruyter. doi:10.1515/9783111215112.

Kumar, Ajay, Sangeeta Rani, Sarita Rathee, and Surbhi Bhatia, eds. *Security and Risk Analysis for Intelligent Cloud Computing: Methods, Applications, and Preventions*. CRC Press, 2023.

Kumar, Ajay, Virendra Kumar Shrivastava, Parveen Kumar, Ashwini Kumar, and Vishal Gulati. 2024. "Predictive and Experimental Analysis of Forces in Die-Less Forming Using Artificial Intelligence Techniques." *Proceedings of the Institution of Mechanical Engineers, Part E: Journal of Process Mechanical Engineering*. doi:10.1177/09544089241235473.

Kumar, Love, Ajay Kumar, Rajiv Kumar Sharma, and Parveen Kumar. 2023. "Smart Manufacturing and Industry 4.0." In *Handbook of Smart Manufacturing*, 1–28. Boca Raton: CRC Press. doi:10.1201/9781003333760-1.

Leu, Ming C., Hoda A. ElMaraghy, Andrew Y.C. Nee, Soh Khim Ong, Michele Lanzetta, Matthias Putz, Wenjuan Zhu, and Alain Bernard. 2013. "CAD Model Based Virtual Assembly Simulation, Planning and Training." *CIRP Annals* 62 (2): 799–822. doi:10.1016/j.cirp.2013.05.005.

Mahatme, C., Mahatme, Chetan, Jayant Giri, Faruq Mohammad, Mohd Sajid Ali, Thanikodi Sathish, Neeraj Sunheriya, and Rajkumar Chadge. 2024. "Experimental and numerical investigation of PLA based different lattice topologies and unit cell configurations for additive manufacturing." *The International Journal of Advanced Manufacturing Technology*: 1–28.

Marner, Michael R., Ross T. Smith, Shane R. Porter, Markus M. Broecker, Benjamin Close, and Bruce H. Thomas. 2011. "Large Scale Spatial Augmented Reality for Design and Prototyping." In *Handbook of Augmented Reality*, edited by Borko Furht, 231–54. New York, NY: Springer New York. doi:10.1007/978-1-4614-0064-6_10.

Murauer, Nela. 2018. "Design Thinking: Using Photo Prototyping for a User-Centered Interface Design for Pick-by-Vision Systems." In *Proceedings of the 11th PErvasive Technologies Related to Assistive Environments Conference*, 126–32. PETRA '18. New York, NY, USA: Association for Computing Machinery. doi:10.1145/3197768.3201532.

Naranjo, Jose E., Diego G. Sanchez, Angel Robalino-Lopez, Paola Robalino-Lopez, Andrea Alarcon-Ortiz, and Marcelo V. Garcia. 2020. "A Scoping Review on Virtual Reality-Based Industrial Training." *Applied Sciences* 10 (22). doi:10.3390/app10228224.

Nayak, Krishna S., Yongwan Lim, Adrienne E. Campbell-Washburn, and Jennifer Steeden. 2022. "Real-Time Magnetic Resonance Imaging." *Journal of Magnetic Resonance Imaging* 55 (1): 81–99. doi:10.1002/jmri.27411.

Ni, D., A.W.W. Yew, S.K. Ong, and A.Y.C. Nee. 2017. "Haptic and Visual Augmented Reality Interface for Programming Welding Robots." *Advances in Manufacturing* 5 (3): 191–98. doi:10.1007/s40436-017-0184-7.

Ong, S.K., Y. Pang, and A.Y.C. Nee. 2007. "Augmented Reality Aided Assembly Design and Planning." *CIRP Annals* 56 (1): 49–52. doi:10.1016/j.cirp.2007.05.014.

Ong, S.K., and A.Y.C. Nee. 2004. "A Brief Introduction of VR and AR Applications in Manufacturing." In *Virtual and Augmented Reality Applications in Manufacturing*, edited by S. K. Ong and A. Y. C. Nee, 1–11. London: Springer London. doi:10.1007/978-1-4471-3873-0_1.

Osti, Francesco, Alessandro Ceruti, Alfredo Liverani, and Gianni Caligiana. 2017. "Semi-Automatic Design for Disassembly Strategy Planning: An Augmented Reality Approach." *Procedia Manufacturing* 11: 1481–88. doi:10.1016/j.promfg.2017.07.279.

Papakostas, Christos, Christos Troussas, Akrivi Krouska, and Cleo Sgouropoulou. 2022. "User Acceptance of Augmented Reality Welding Simulator in Engineering Training." *Education and Information Technologies* 27 (1): 791–817. doi:10.1007/s10639-020-10418-7.

Prasolova-Førland, Ekaterina, Judith Molka-Danielsen, Mikhail Fominykh, and Katherine Lamb. 2017. "Active Learning Modules for Multi-Professional Emergency Management Training in Virtual Reality." In *2017 IEEE 6th International Conference on Teaching, Assessment, and Learning for Engineering (TALE)*, 461–68. doi:10.1109/TALE.2017.8252380.

Rajendra, P., Mina Kumari, Sangeeta Rani, Namrata Dogra, Rahul Boadh, Ajay Kumar, and Mamta Dahiya. 2022. "Impact of Artificial Intelligence on Civilization: Future Perspectives." *Materials Today: Proceedings* 56: 252–56. doi:10.1016/j.matpr.2022.01.113.

Rani, Sangeeta, Ajay Kumar, Arko Bagchi, Snehlata Yadav, and Sachin Kumar. 2021. "RPL Based Routing Protocols for Load Balancing in IoT Network." *Journal of Physics: Conference Series* 1950 (1): 012073. doi:10.1088/1742-6596/1950/1/012073.

Reljić, Vule, Ivana Milenković, Slobodan Dudić, Jovan Šulc, and Brajan Bajči. 2021. "Augmented Reality Applications in Industry 4.0 Environment." *Applied Sciences* 11 (12). doi:10.3390/app11125592.

Remondino, Fabio, and Sabry El-Hakim. 2006. "Image-based 3D Modelling: A Review." *The Photogrammetric Record* 21: 269–91. doi:10.1111/j.1477-9730.2006.00383.x.

Ryead, Mohamed. 2023. "Utilization Evolving Prototype Technologies in Industrial Design in the Light of Industry 4.0." *International Design Journal* 13 (2): 227–43. doi:10.21608/idj.2023.288318.

Scheffer, S. E. (Sara), A. (Alberto) Martinetti, R. G. J. (Roy) Damgrave, and L. A. M. (Leo) van Dongen. 2023. "Supporting Maintenance Operators Using Augmented Reality Decision-Making: Visualize, Guide, Decide & Track." *Procedia CIRP* 119: 782–87. doi:10.1016/j.procir.2023.01.018.

Shankhwar, Kalpana, Tung-Jui Chuang, Yao-Yang Tsai, and Shana Smith. 2022. "A Visuo-Haptic Extended Reality–Based Training System for Hands-on Manual Metal Arc Welding Training." *The International Journal of Advanced Manufacturing Technology* 121 (1): 249–65. doi:10.1007/s00170-022-09328-4.

Sharmila, A., E.L. Dhivya Priya, and K.R. Gokul Anand. 2022. "A Survey of Societal Applications of IOT." In *IoT and Cloud Computing for Societal Good*, edited by Jitendra Kumar Verma, Deepak Saxena, and Vicente González-Prida, 73–99. Cham: Springer International Publishing. doi:10.1007/978-3-030-73885-3_6.

Singhal, Ayush, Manu Phogat, Deepak Kumar, Ajay Kumar, Mamta Dahiya, and Virendra Kumar Shrivastava. 2022. "Study of Deep Learning Techniques for Medical Image Analysis: A Review." *Materials Today: Proceedings* 56: 209–14. doi:10.1016/j.matpr.2022.01.071.

Suresh, S., D. Velmurugan, E.L. Dhivya Priya, and R. Selvapriya. 2024. "AI for Enhanced Drone Capabilities in Industry 5.0." In *Drone Applications for Industry 5.0*, 75–91. IGI Global. doi:10.4018/979-8-3693-2093-8.ch006.

Suresh, S., D. Velmurugan, J. Balaji, Elango Natarajan, P. Suresh, and S. Rajesh. 2023. "Influences of Nanoparticles in Friction Stir Welding Processes." In *Sustainable Utilization of Nanoparticles and Nanofluids in Engineering Applications*, 32–55. IGI Global. doi:10.4018/978-1-6684-9135-5.ch002.

Suresh, S., Elango Natarajan, P. Vinayagamurthi, K. Venkatesan, R. Viswanathan, and S. Rajesh. 2023. "Optimum Tool Traverse Speed Resulting Equiaxed Recrystallized Grains and High Mechanical Strength at Swept Friction Stir Spot Welded AA7075-T6 Lap Joints." In *Materials, Design and Manufacturing for Sustainable Environment*, edited by Elango Natarajan, S. Vinodh, and V. Rajkumar, 547–55. Singapore: Springer Nature Singapore. doi:10.1007/978-981-19-3053-9_41.

Suresh, S., Elango Natarajan, S. Boopathi, and Parveen Kumar. 2024. "8 Processing of Smart Materials by Additive Manufacturing and 4D Printing." In *3D Printing Technologies*, edited by Ajay Kumar, Parveen Kumar, Naveen Sharma, and Ashish Kumar Srivastava, 181–96. Berlin, Boston: De Gruyter. https://doi.org/10.1515/9783111215112-008.

Suresh, S., K. Venkatesan, and S. Rajesh. 2019. "Optimization of Process Parameters for Friction Stir Spot Welding of AA6061/Al2O3 by Taguchi Method." In *AIP Conference Proceedings*, 2128:030018. doi:10.1063/1.5117961.

Suresh, S., Parveen Kumar, T. Yuvaraj, D. Velmurugan, and Elango Natarajan. 2024. "4 Adoptability of Additive Manufacturing Process: Design Perceptive." In *3D Printing Technologies*, edited by Ajay Kumar, Parveen Kumar, Naveen Sharma, and Ashish Kumar Srivastava, 77–94. Berlin, Boston: De Gruyter. doi:10.1515/9783111215112-004.

Tatić, Dušan, and Bojan Tešić. 2017. "The Application of Augmented Reality Technologies for the Improvement of Occupational Safety in an Industrial Environment." *Computers in Industry* 85: 1–10. doi:10.1016/j.compind.2016.11.004.

Thakare, P.A., N. Kumar, V.B. Ugale, J. Giri, N. Sunheriya, and H.A. Al-Lohedan 2024. "Effect of Impact and Flexural Loading on Hybrid Composite Made of Kevlar and Natural Fibers. *AIP Advances* 14 (4). doi:10.1063/5.0195907.

Tufail, M.S., J. Giri, E. Makki, T. Sathish, R. Chadge, and N. Sunheriya. 2024. "Machinability of Different Cutting Tool Materials for Electric Discharge Machining: A Review and Future Prospects." *AIP Advances* 14 (4). doi:10.1063/5.0201614.

Vidal-Balea, Aida, Oscar Blanco-Novoa, Paula Fraga-Lamas, Miguel Vilar-Montesinos, and Tiago M. Fernández-Caramés. 2020. "Creating Collaborative Augmented Reality Experiences for Industry 4.0 Training and Assistance Applications: Performance Evaluation in the Shipyard of the Future." *Applied Sciences* 10 (24). doi:10.3390/app10249073.

Wang, Feifei, Enji Sun, Haoyu Wang, and Shuangyue Liu. 2018. "A Framework of Safety Training Based on Augmented Reality and Cloud Computing Platform in Mines." *International Journal of Georesources and Environment* 4 (3). doi:10.15273/ijge.2018.03.014.

Wang, Yizhong, Yonghua Chen, Zhongliang Nan, and Yong Hu. 2006. "Study on Welder Training by Means of Haptic Guidance and Virtual Reality for Arc Welding." In *2006 IEEE International Conference on Robotics and Biomimetics*, 954–58. doi:10.1109/ROBIO.2006.340349.

Wu, Dapeng, Zhigang Yang, Puning Zhang, Ruyan Wang, Boran Yang, and Xinqiang Ma. 2023. "Virtual-Reality Interpromotion Technology for Metaverse: A Survey." *IEEE Internet of Things Journal* 10 (18): 15788–809. doi:10.1109/JIOT.2023.3265848.

Wu, Yong, Jianguo Xu, lintao Li, Duode Ge, and Jun Zhang. 2020. "The Design of an Equipment Induction Maintenance System Based on AR Technology." *IOP Conference Series: Earth and Environmental Science* 440 (3). IOP Publishing:32041. doi:10.1088/1755-1315/440/3/032041.

Yin, Jun-Hao, Chin-Boon Chng, Pooi-Mun Wong, Nicholas Ho, Matthew Chua, and Chee-Kong Chui. 2020. "VR and AR in Human Performance Research—An NUS Experience." *Virtual Reality & Intelligent Hardware* 2 (5): 381–93. doi:10.1016/j.vrih.2020.07.009.

Zhao, Pinwang, Hongchao Wu, Xue Shi, Jing Li, Xiaojian Yi, and Shulin Liu. 2019. "Research on Maintenance Guiding System Based on Augmented Reality." In *2019 International Conference on Sensing, Diagnostics, Prognostics, and Control (SDPC)*, 944–49. doi:10.1109/SDPC.2019.00180.

11 CAD-CAM in Orthodontics

A Step Forward in Digitalization

Namrata Dogra, Ajay Kumar, Archana Jaglan, Kirti Sehrawat, Parveen Kumar, Mamta Dahiya, and Jahangir Khan

11.1 INTRODUCTION

CAD/CAM includes computer-aided design and computer-aided manufacturing. The objectives of CAD (computer-aided design) are to enable more rapid design repetitions and to facilitate a smooth transition to the CAM (computer-aided manufacturing) stage. Using computerized specifications, a computer uses computer-aided manufacturing to instruct machines to carry out tasks that would typically be managed by a machine operator. This requires initial expenditure but later it reduces the manpower and provides ease and comfort.

The field of dentistry has been successfully utilizing this technology for many years and orthodontics is also not spared from this significant technological change. Artificial intelligence (AI) has boosted the discoverability of digital data and the processing capacity [1].

Integration of machine-to-machine and Internet of Things (IoT) can improve automation, which can help in providing better communication. This can produce machines that can diagnose problems without the help of any human involvement [2].

The digital technology has improved and streamlined diagnosis, treatment planning and execution in the field of orthodontics. The benefits of CAD/CAM technology in orthodontics include clear aligner treatment, customized bracket system having patient-specific torque, titanium Herbst appliances, machine-milled indirect bonding jigs, robotic wire bendings, virtual articulators and facebows, digital impressions and digital models etc. [3–5].

The foundation that underpins CAD/CAM technology involves data collection unit, software and 3D printing as described in Figure 11.1.

3D printing, also called additive manufacturing, involves seven stages [6]. Different raw materials utilized by additive manufacturing are metals, polymers and resins [7].

A paradigm change has occurred with the use of intraoral scan in place of traditional impression making. The data collection unit gathers information from the

DOI: 10.1201/9781032725086-13

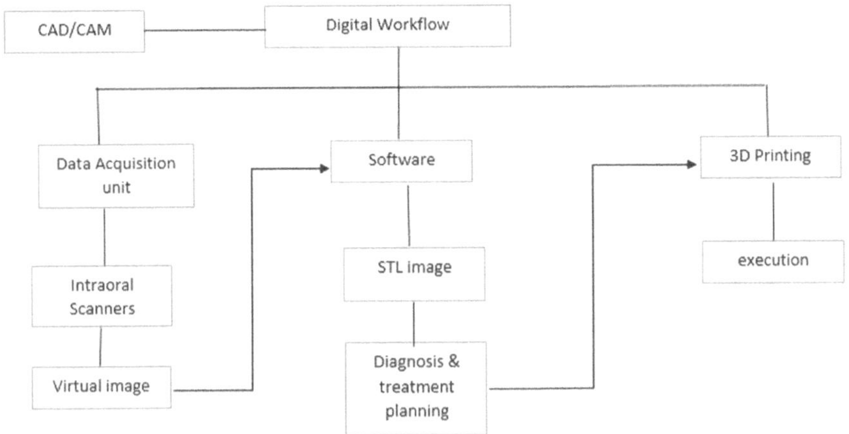

FIGURE 11.1 Digital workflow of CAD-CAM.

oral cavity, and then virtual impressions are created through the use of intraoral scanners. Digital intraoral scanners are the best to obtain digital models which are highly accurate [8]. Stone model can also be created through a traditional impression method. The conventional impression making is operator-dependent procedure which may be discomfortable for the patients, require more time as well as have less accuracy than virtual impression [9, 10]. The manipulation of scanned virtual impression is done using specific software and generated virtual image is then saved on computer in STL (standard triangulation language) format. These virtual cast aid the orthodontists to quickly get diagnostic information. 3D (three-dimensional) printing of acquired or edited image is the following step in the digital workflow. Orthodontics more frequently use the additive manufacturing printing techniques, including stereolithography (SLA), which uses ultraviolet light to solidify liquid resins, and fused deposition modeling (FDM) [9–11]. Additive manufacturing technique can help in printing of biocompatible materials [12–16].

Additive manufacturing involves production of a CAD through reverse engineering techniques [17]. CAD/CAM technology makes patient treatments quicker, more predictable, aesthetically pleasing and comfortable for the patients [18]. Despite all these advantages, orthodontists are still not using this technology often in their practice; the reason might be high cost and lack the technical expertise to implement it [5].

11.2 USES OF CAD/CAM IN ORTHODONTICS

CAD-CAM has a diversity of applications in the field of orthodontics as shown in Figure 11.2.

11.2.1 Diagnosis and Treatment Planning

CAD/CAM can be utilized for accurate diagnosis and treatment planning of impacted canines which are very difficult to diagnose. Figures 11.3 and 11.4 depict

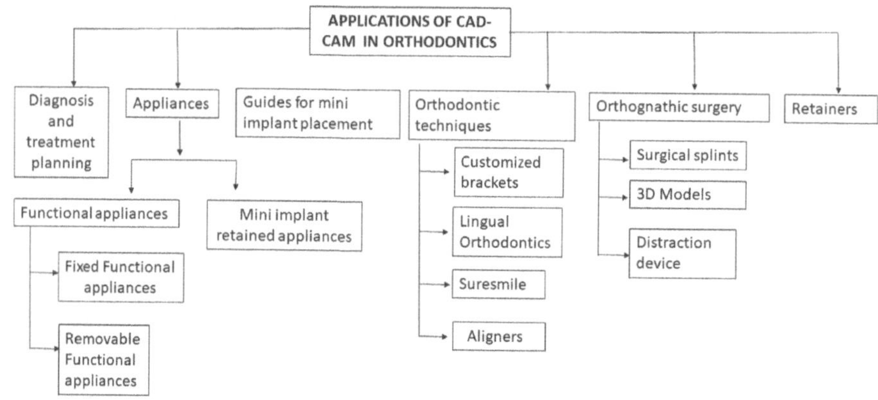

FIGURE 11.2 CAD-CAM applications in orthodontics.

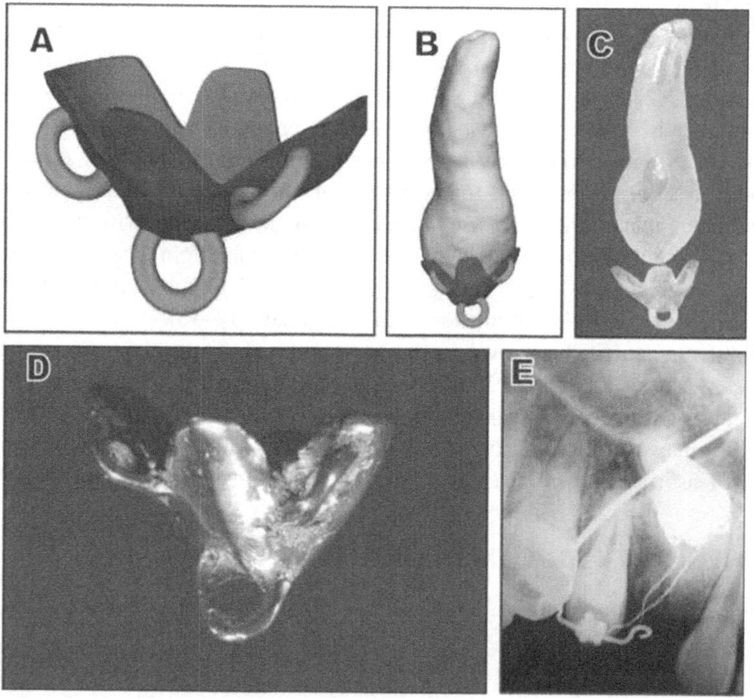

FIGURE 11.3 (A) Fabrication of attachment, (B) attachment placed over impacted canine, (C) attachment over prototype of impacted canine, (D) casted attachment, and (E) attachment on the canine seen in radiograph [15].

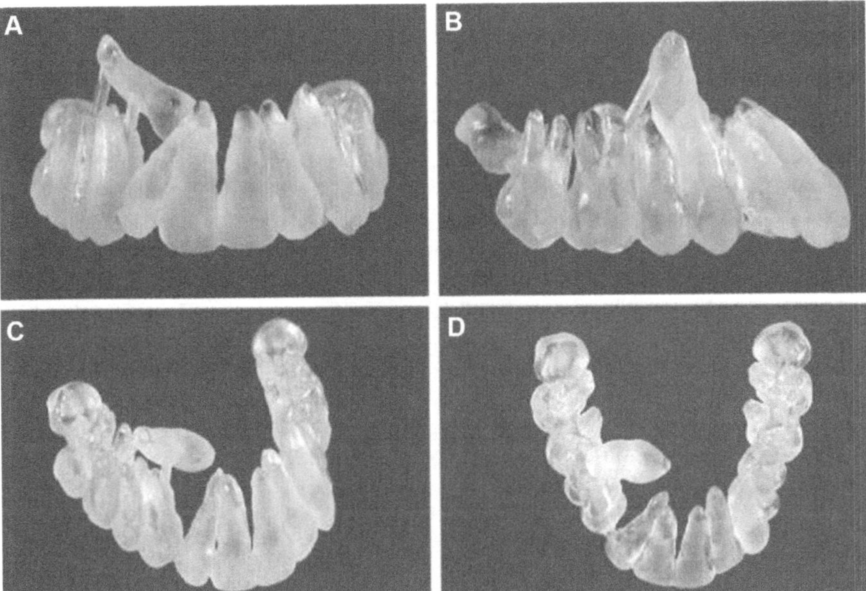

FIGURE 11.4 Prototype seen from different views: (A) front view, (B) side view, (C) oblique view, and (D) occlusal view [15].

rapid prototyping process that can efficiently generate 3D dental prototype and a custom-made attachment that can be bonded to the canine for its forced eruption. Also, the model generated can be utilized for patient education [19].

11.2.2 CAD/CAM Appliances

CAD/CAM technology can effectively construct fixed as well as removable functional and mini-implant retained appliances. Various applications of CAD-CAM technology are described in Table 11.1.

11.2.3 Functional Appliances

11.2.3.1 Removable Functional and Sleep-Apnea Devices

Andresen appliance and sleep apnea device: The latest CAD/CAM technique is capable of constructing a single build with wire components as presented in Figure 11.5. 3D images of a Class II Division 1 model were scanned and Phantom arm and Freeform software designed the appliance. Next, the additive manufacturing machine was used to build the design. The surfaces of the appliances developed were smooth without any sharp edges. CAD/CAM technology has the ability to insert wire and produce hinges as there is layer-by-layer fabrication process. Also, the thickness of the material can be accurately controlled by CAD [20].

Milled Twin block using CAD/CAM: Abier H. Sofrata et al designed twin block appliance with expansion screw for a Class II Division 1 female patient who was

TABLE 11.1
Applications of CAD-CAM

Sr. No.	Designing Software	Scanner	Material	Printer
Functional regulator 3	Software OnyxCeph	TRIOS 4	Resin (Dental LT Clear; Formlabs)	Form 3B; Formlabs, MA, USA
Hybrid hyrax expander	CAMbridge software (3Shape)	3Shape, Copenhagen, Denmark	Star metal alloy (Dentaurum, Ispringen, Germany)	Concept Laser, Lichtenfels, Germany
3D digital MARPE	Meshmixer, Blender for dental	TRIOS 3 scanner	Co–Cr alloy remanium star (R)	M100 EOS GmbH System
Miniimplant insertion guide	Driver Easy software		Acrylic MED620 material (Stratasys)	Stratasys OrthoDesk (Eden Prairie, MN)
Andresen appliance	(FreeForm Modeling Plus, version 11; Geo Magics SensAble Group, Wilmington, MA), in conjunction with a Phantom Desktop (Geo Magics SensAble Group)		VisiJet EX200, 3D Systems	
Occlusal splint	Initial splint-Dolphin Imaging® software Final splint – 3-matic (CAD software, Materialise, Belgium)	3Shape dental scanner	PolyJet photopolymer (MED610) medical materials	Objet30 OrthoDesk 3D Printer

nine years old. For the fabrication of appliance, the intraoral scanning and digital bite registration were done instead of conventional impression or construction bite process. A virtual articulator was used to digitally approximate the required mandibular advancement. The material utilized is biocompatible, clinical outcome is successful, and the precise fitting of the computer-aided designed and manufactured twin block all suggest that conventional orthodontic functional appliance therapy has been revolutionized successfully [21].

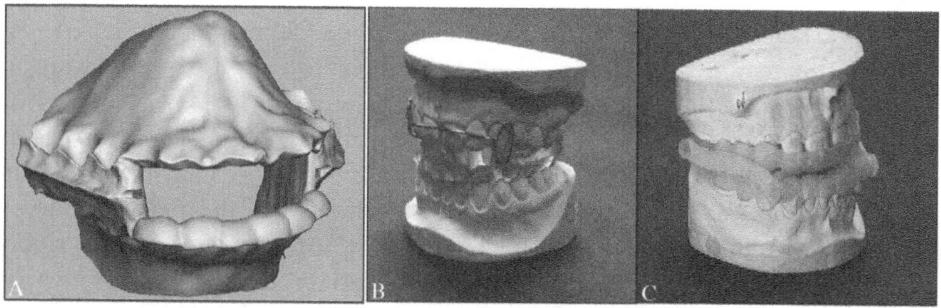

FIGURE 11.5 (A) Andresen appliance designed by CAD, (B) prototype of Andresen appliance on the casts, and (C) prototype of sleep-apnea on the cast [16].

Christoph Roser et al compared the efficiency of conventionally made functional regulator (FR3) and that made up of using CAD/CAM technique. No wire elements were used in the construction. The wire bending in functional regulator appliance is very complicated and hence this functional appliance is not very popular. But the possibility of this appliance to be manufactured using CAD/CAM has opened up new venues for its use. Other advantages of this CAD/CAM printed appliance are less chairside time, easy storage of digital model and easy reprinting of the appliance in cases where the patient loses the appliance. They concluded that if specific design criteria are followed, the CAD-FR3s have better mechanical properties than conventional FR3s [22].

11.2.3.2 Fixed Functional Appliances

CAD/CAM can design accurately the interface between the lingual appliance and the telescopic appliance as shown in Figure 11.6. It can be used to design the pivot base and can be attached to the molar in upper arch and canine in lower arch through custom-made bands. The precise 3D tube-and-plunger positioning required for proper and efficient operation of the telescopic mechanism is ensured by the specific CAD depiction of the interface. This customized appliance facilitates patient comfort as there is small telescope to tooth distance [23].

11.2.4 Mini-Implant Retained Appliances

CAD-CAM can be successfully used to fabricate hybrid hyrax expander. In this technique, following mini-implant insertion in the palate, intraoral scan is done and then the expander is digitally designed and laser printed as depicted in Figure 11.7.

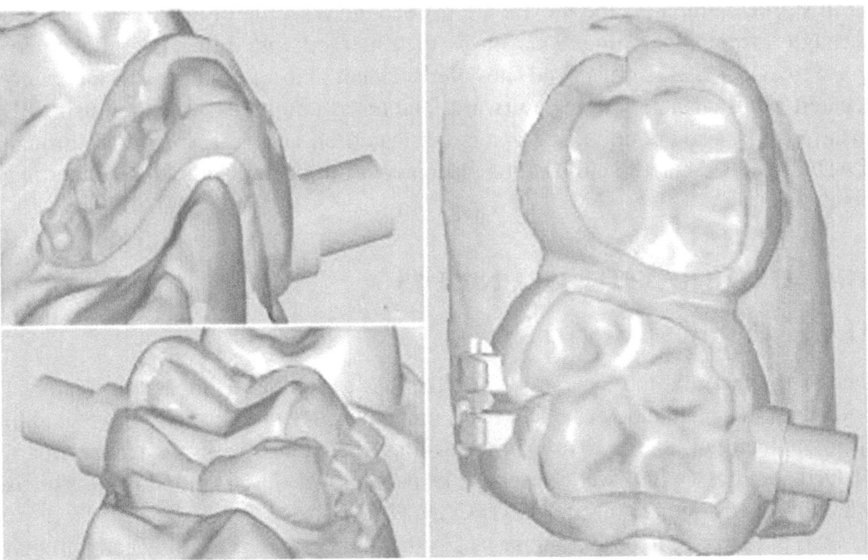

FIGURE 11.6 Bracket base extension to provide support to telescopes of Herbst [19].

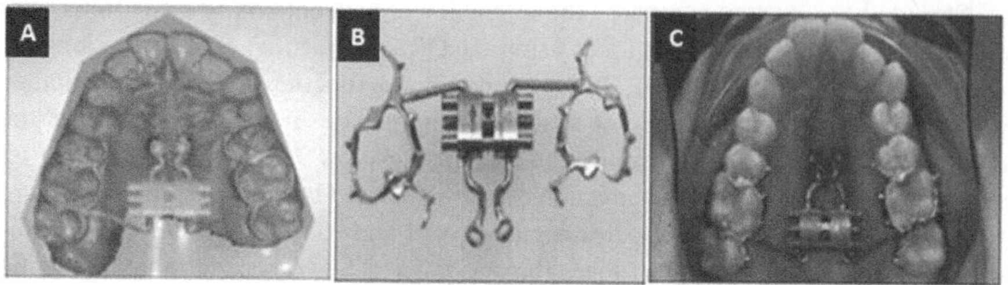

FIGURE 11.7 (A) Virtual planned hybrid hyrax expander, (B) hybrid hyrax expander prepared through laser-printing, and (C) hybrid hyrax expander fixed intraorally [20].

This CAD-CAM-supported technique is very patient friendly as it eliminates the need of frequent impression taking and involves less chair side time [24].

Other appliances printed through metal can be Hyrax-Halterman that can distalize and expand the molars [25].

11.3 GUIDE FOR MINI-IMPLANT PLACEMENT

Mini-implant stability and success depends upon many factors such as accurate placement in the attached gingiva, accurate angle and sufficiently away from the roots of the teeth. Root or sinus can be perforated by wrong mini-implant placement method. Thus, it is important to have a guide before placing the mini-implant. CAD/CAM can fabricate a surgical guide made of biocompatible materials through 3D printing [26]. Wilmes B. et al [27] described the use of CAD/CAM for fabrication of insertion guide for mini-implant placement and distalization appliance. In this procedure, after the molars are banded, an STL file of maxilla is obtained through intraoral scanning. This file is then merged with either lateral cephalogram image or CBCT image and sites of placement of mini-implants are accurately located using virtual planning software. The insertion guide is prepared using 3D printing. The orthodontic distalizer is fabricated on acrylic cast printed through CAD/CAM. This method allows the placement of mini-implant and distalizer in a single visit.

11.4 ORTHODONTIC TECHNIQUES

11.4.1 Customized Bracket System

Due to the fact that the location of brackets has a significant impact on treatment outcomes, this technology can help in accurate placement of brackets with regard to fixed appliances [28]. Fewer arch wire sessions and a low score in American Board of Orthodontics (ABO) were observed in patients who used orthodontic systems which were customized through CAD/CAM [29].

Insignia is one of best examples of orthodontic appliance fabricated through CAD/CAM. It is offered in conventional and self-ligating bracket system with the option to use aesthetic brackets. This system's primary benefit is its ability to customize the bracket slot. Another system is Sure Smile which is utilizing CAD/CAM technology as well. Sure smile differs from other personalized appliances in that robotic arch wire manufacturing follows afterward.

However, other research indicates that both the direct and indirect bonding methods have equal bracket placement precision [30]. A customized bracket system's clinical effectiveness was assessed by Hegele J. et al. They showed that brackets designed through CAD/CAM in comparison to a standard self-ligating bracket had a negligible impact on the effectiveness and results of therapy [31].

11.4.2 Lingual Orthodontics

The use of lingual orthodontic appliances is getting simpler because of advanced ideas like indirect bonding systems that use virtual setup models and virtual bracket positioning. The elimination of the soldering procedure by CAD/CAM technology has eased fabrication. In order to strengthen the bonding of the appliance, the lingual pads are provided with mesh. This is made possible by rapid prototyping technology. Soon-Yong Kwon et al describe a method for creating lingual orthodontic appliances using CAD/CAM and compare the cemented appliance's final position to its intended position on the lateral cephalogram. They concluded that using CAD technology, which combines 3D model images with cephalograms or CBCT scans, improves the precision of designing orthodontic appliances [32].

11.4.3 Sure Smile

The advancement in 3D imaging and techniques has made orthodontic diagnosis and treatment planning much easier. An example of such advanced diagnostic system is "Sure smile" which involves the use of CAD-CAM and 3D imaging. It was introduced by Dr. Rohit Sachdeva [33]. This system helps in production of customized fixed orthodontic appliances using robots. Sure smile can also fabricate other customized orthodontic devices accurately such as arch wires, aligners and indirect bonding trays. It helps to simulate the treatment well in advance and is able to visualize various treatment strategies. Shorter treatment time and better patient outcomes are noticed with the use of Sure smile technology.

11.4.4 Clear Aligner Therapy

With increasing esthetic demands of the patient, aligner therapy has gained much popularity. One of the most popular aligner system is Invisalign®, which involves a high-end CAD-CAM technology. This technology now allows treatment of even complex cases. Invisalign® is the leading brand in aligner market due to its advanced treatment planning and printing technology [34]. Thurzo A et al described a novel distalizer which was 3D designed and printed. This was used in conjunction with Invisalign® treatment [35] (Figure 11.8).

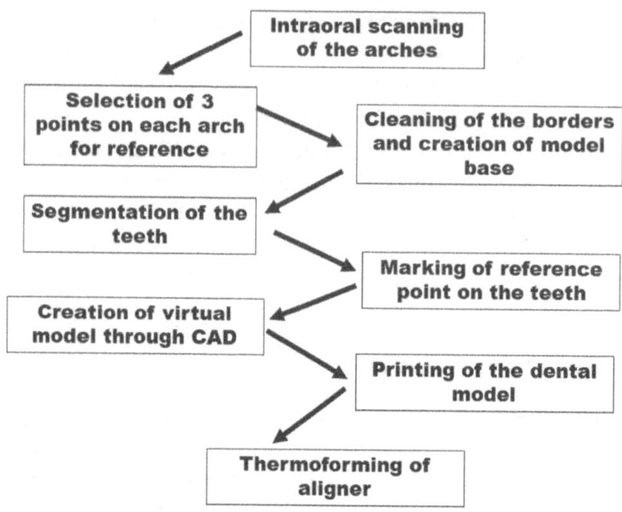

FIGURE 11.8 Steps followed in fabrication of aligners.

11.5 ORTHOGNATHIC SURGERY

11.5.1 Surgical Splint

The CAD/CAM surgical splint provides advantages to overcome potential flaws in the laboratory-based approach, such as the uncontrollable errors, inter-laboratory variations and time-consumption issue [36]. Lo LJ et al. described a novel occlusal splint fabricated through CAD/CAM for bi-jaw orthognathic surgery. The splint produced by this method was successful in reproducing the surgical plan during the surgery and it was very surgeon friendly [37].

Michele Cassetta et al. described surgical guide for corticotomy made by using CAD/CAM technology. This eliminates the requirement for flap elevation and overcomes the drawbacks of the corticotomy. No pain, early postoperative problems or unforeseen events were noticed [38].

11.5.2 3D Models for Orthognathic Surgery

3D models of maxilla, mandible and craniofacial bones generated through CAD/CAM can be helpful in surgical planning and simulation of surgical technique as presented in Figure 11.9. They can also be used for patient motivation and for discussion with fellow surgeons [39].

11.5.3 Distraction Device

Distraction osteogenesis is the procedure done to lengthen the bone using biological process by gradual separation of bony ends following incremental traction. For this procedure, surgical simulation is important in order to estimate the position of

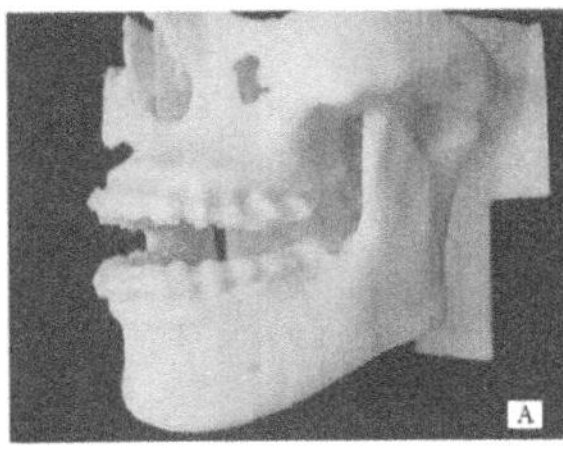

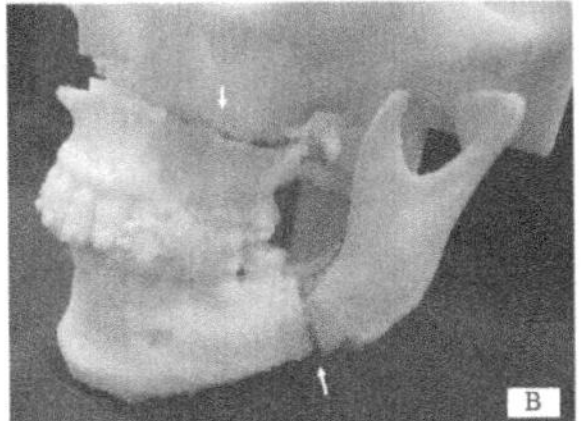

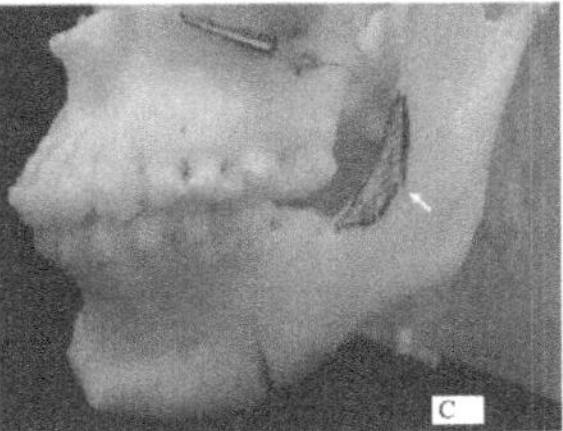

FIGURE 11.9 (A) 3-dimensional model with retrusion of maxilla and protrusion of mandible, (B) arrows indicate Le Fort I osteotomy and sagittal split ramus osteotomy, and (C) arrows indicate the amount of ramus trimming required [35].

osteotomy line as well as to choose the location of the distraction device. The factory-made distractor has flat surface and is difficult to bend according to the bone surface while keeping in mind the decided site for its placement [40, 41]. However, with the advent of CAD/CAM technology, guides can be made that can be used for precise positioning as well as distractor can be manufactured using the advanced technology [42].

Kang SH et al. introduced a CAD/CAM-fabricated distraction device as shown in Figure 11.10. This appliance has several advantages over conventional appliances such as better accuracy and fixation. The authors have evaluated the accuracy of this appliance for distraction osteogenesis in mandible [43].

11.6 RETAINERS

Multistranded and round wires are commonly used to provide fixed retention after active orthodontic treatment. But a lot of disadvantages are associated with this type of fixed retention such as frequent wire breakages, tongue irritation and inability to maintain proper oral hygiene [44–46].

A CAD/CAM-fabricated retainer has overcome these problems. In 2012, Pascal Schumacher introduced Memotain retainer. Figure 11.11 shows Memotain retainer which is made from nickel-titanium wire and is fabricated through CAD/CAM technology [47].

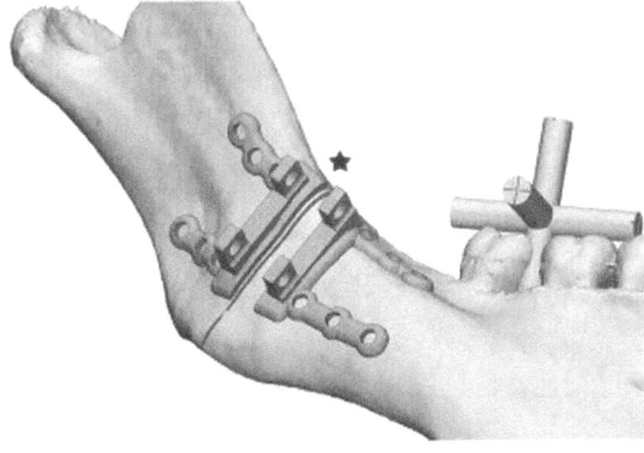

FIGURE 11.10 CAD/CAM-fabricated distraction device with two plates [39].

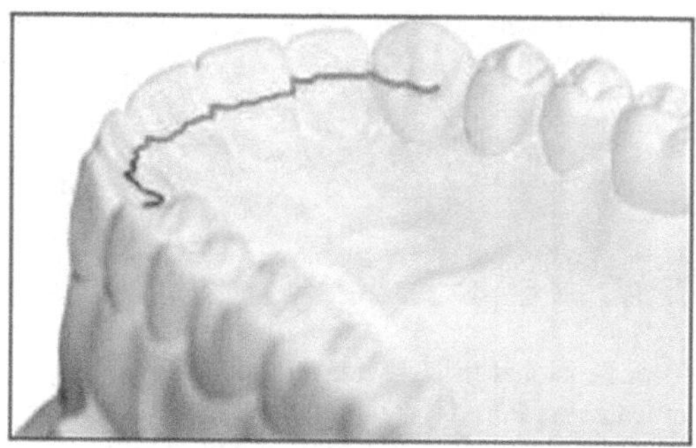

FIGURE 11.11 Memotain retainer [43].

In this, a Niti wire is custom cut from sheet and then electropolished. This results in a smooth and polished wire. This can be customized for every patient and its position can be digitally planned. This retainer has numerous advantages over conventional lingual retainers such as precision fit, less irritation to the tongue and less chance of breakage. Figure 11.12 depicts that Memotain can be easily adapted to the maxillary palatal surfaces without interfering with occlusion.

For patients who have nickel and chromium sensitivity, stainless steel retainers which contain nickel and chromium cannot be used. Hence a CAD/CAM-fabricated zirconium bar can be used [48].

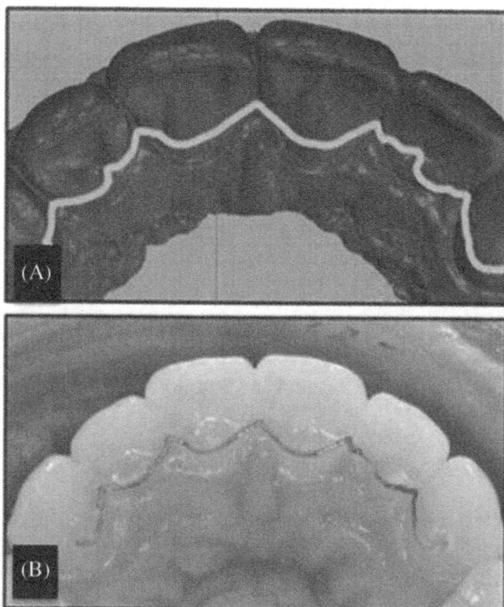

FIGURE 11.12 (A) Memotain designed in a way so that there is no occlusal interference and (B) Memotain bonded to the teeth [43].

TABLE 11.2
Different Types of Retainers

Author	Retainer	Material	Luting Agent
Kravitz ND	Memotain	Nickel-titanium	Flowable resin, such as Tetric EvoFlow (Ivoclar Vivadent, Amherst, NY)
Zreaqat M	Zirconium bar	Zirconium	Two dual-cured cement types (Variolink II and RelyX ARC)
Koller S	prime4me® RETAIN3R (Dentaurum, Ispringen, Germany)	Titanium Grade 5	A light-curing low-viscosity composite resin (Ortho Connect Flow, GC Orthodontics, Breckerfeld, Germany)

Zirconia is preferred as an alternative to stainless steel as it has good mechanical strength and good biocompatibility [49, 50].

Kang SH et al compared the stability among CAD/CAM-fabricated lingual retainer. Least deformation of flatness was seen in custom-cut group and there was no significant difference found between the groups in terms of stability [51].

Alwaras MB conducted a study to compare stability of teeth and health of periodontium among fixed retainer, removable retainer and nickel-titanium retainer fabricated through CAD/CAM. They concluded that the CAD/CAM-fabricated nickel-titanium retainer results in teeth stability comparable to other retainers and it has a greater fit. Also, it resulted in less accumulation of plaque and better gingival health [52].

Study done by Gera A et al concluded that when CAD/CAM nitinol and multistranded fixed retainers were compared for the rate of failure and satisfaction among the patients, there was no difference among the two evaluated over a period of six months after debonding [53]. Similar results were reported by Gelin E et al [54].

S Koller et al studied the accurate positioning of 3D CAD/CAM titanium retainers. prime4me® Scan was done intraorally to check the position. Comparison was made between this scan and virtual setup using software of 3D superimposition. They found out that there were only minor deviations between the two [55].

Various types of retainers have been summarized in Table 11.2.

11.7 CONCLUSION

The dental practice in today's era is continuously under renovation after the introduction of digital technologies such as CAD-CAM. The quality of treatment improves with the use of more precise and digital technology.

The future scope of CAD-CAM in orthodontics can be the fabrication of effective and more efficient In-House Aligners. AI has also been integrated with CAD-CAM in order to enhance its application scope in orthodontic practice. Integration of CAD/CAM in conjunction with reverse engineering (RE) and 3D printing has achieved new heights in terms of workflow in orthodontic clinics and has simplified the whole treatment process starting from diagnosis to the planning of treatment and finally orthodontic appliance fabrication. Integration of CAD/CAM in the field of orthodontics directly increases the operational cost of the treatment and requires experts to handle, operate and understand complexity of this process.

REFERENCES

1. P. Rajendra, M. Kumari, S. Rani, N. Dogra, R. Boadh, A. Kumar, and M. Dahiya, "Impact of artificial intelligence on civilization: Future perspectives," Mater Today, Vol. 56, pp 252–256, January 2022.
2. A. Kumar, H. Singh, P. Kumar, B. AlMangour, "Handbook of Smart Manufacturing: Forecasting the Future of Industry 4.0," CRC Press, July 2023.
3. R. Zandparsa, "Digital imaging and fabrication," Dent Clin North Am, Vol. 58, Issue 1, pp 135–158, January 2014.
4. G. Sannino, F. Germano, L. Arcuri, E. Bigelli, C. Arcuri, A. Barlattani, "CEREC CAD/CAM chairside system," Oral Implantol (Rome), Vol. 7, Issue 3, pp 57–70, July–September 2014.
5. T.M.A. da Cunha, I.D.S. Barbosa, K.K. Palma, "Orthodontic digital workflow: Devices and clinical applications," Dent Press J Orthod, Vol. 26 Issue 6, pp e21spe6, December 2021.
6. A. Kumar, R.K. Mittal, A. Haleem, "Advances in Additive Manufacturing: Artificial Intelligence, Nature-Inspired, and Biomanufacturing," Elsevier, November 2022.
7. A.K. Srivastava, A. Kumar, P. Kumar, P. Gautam, N. Dogra, "Research progress in metal additive manufacturing: Challenges and opportunities," Int J Interact Des Manuf, December 2023.
8. S. Sehrawat, A. Kumar, S. Grover, N. Dogra, J. Nindra, S. Rathee, M. Dahiya, A. Kumar, "Study of 3D scanning technologies and scanners in orthodontics," Mat Today: Proceedings, Vol. 56, Issue 1, pp 186–193, January 2022.
9. A. Jyosthna, D.L. Xavier, A.C. Evan, R. Piradhiba, N. Navaneetha, "A review–CAD/CAM in orthodontics," J Res Med Dent Sci, Vol. 10 Issue 2, pp 253–257, 2022.
10. N.D. Kravitz, C. Groth, P.E. Jones, J.W. Graham, W.R. Redmond, "Intraoral digital scanners," J Clin Orthod, Vol. 48 Issue 6, pp 337–347, June 2014.
11. A. Kumar, P. Kumar, N. Sharma, A.K. Srivastava, "3D Printing Technologies: Digital Manufacturing, Artificial Intelligence, Industry 4.0," Walter de Gruyter GmbH & Co KG; 29 Jan 2024.
12. A. Bhardwaj, A. Bhatnagar, A. Kumar, "Current trends of application of additive manufacturing in oral healthcare system," In Advances in Additive Manufacturing, pp 479–491, Jan 1 2023. Elsevier.
13. C. Mahatme, J. Giri, F. Mohammad, M. S. Ali, T. Sathish, N. Sunheriya, R. Chadge, "Experimental and numerical investigation of PLA based different lattice topologies and unit cell configurations for additive manufacturing," Int J Adv Manuf Technol, 2024 https://doi.org/10.1007/s00170-024-13882-4
14. J. Giri, T. Sathish, T. Sheikh, N. Sunehriya, P. Giri, R. Chadge, C. Mahatme, A. Parthiban, "Automatic liver segmentation using U-Net deep learning architecture for additive manufacturing," Interactions Vol. 245, Issue 1, 2024. https://doi.org/10.1007/s10751-024-01927-9
15. J. Giri, N. Sunheriya, T. Sathish, Y. Kadu, R. Chadge, P. Giri, A. Parthiban, C. Mahatme, "Optimization of process parameters to improve mechanical properties of fused deposition method using Taguchi method," Interactions, Vol. 245, Issue 1, 2024. https://doi.org/10.1007/s10751-024-01925-x
16. M.S. Tufail, J. Giri, E. Makki, T. Sathish, R. Chadge, N. Sunheriya, "Machinability of different cutting tool materials for electric discharge machining: A review and future prospects," AIP Adv 14, Issue 4, 2024. https://doi.org/10.1063/5.0201614

17. A. Kumar, P. Kumar, H. Singh, A. Haleem, R.K. Mittal, "Integration of reverse engineering with additive manufacturing," In Advances in Additive Manufacturing, pp 43–65, Jan 1 2023. Elsevier.
18. L.R. Christensen, "Digital workflows in contemporary orthodontics," APOS Trends Orthod, Vol. 7, Issue 1, pp 12–18, February 2017.
19. J. Faber, P.M. Berto, M. Quaresma, "Rapid prototyping as a tool for diagnosis and treatment planning for maxillary canine impaction," Am J Orthod Dentofacial Orthop, Vol. 129, Issue 4, pp 583–589, April 2006.
20. N. Mortadi, D. Eggbeer, J. Lewis, R.J. Williams, "CAD/CAM/AM applications in the manufacture of dental appliances," Am J Orthod Dentofacial Orthop, Vol. 142, Issue 5, pp 727–733, November 2012.
21. A.H. Sofrata, P.M. Cattaneo, M.A. Cornelis, "Computer-aided design and manufacture of a milled Twin-block: Workflow and use in a clinical patient. Have we entered the digital era," AJO-DO Clinical Companion, Vol. 2, Issue 5, pp 431–438, May 2022.
22. C. Roser, L.D. Hodecker, C. Koebel, C.J. Lux, D. Ruckes, S. Rues, A. Zenthöfer, "Mechanical properties of CAD/CAM-fabricated in comparison to conventionally fabricated functional regulator 3 appliances," Sci Rep, Vol. 11, Issue 1, pp 14719, July 2021.
23. D. Wiechmann, R. Schwestka-Polly, A. Hohoff, "Herbst appliance in lingual orthodontics," Am J Orthod Dentofacial Orthop, Vol. 134, Issue 3, pp 439–446, September 2008.
24. S. Graf, S. Vasudavan, B. Wilmes, "CAD-CAM design and 3-dimensional printing of mini-implant retained orthodontic appliances," Am J Orthod Dentofacial Orthop, Vol. 154, Issue 6, pp 877–882, December 2018.
25. S. Graf, N.E. Tarraf, N.D. Kravitz, "Three-dimensional metal printed orthodontic laboratory appliances," Semin Orthodont, Elsevier Vol. 27, Issue 3, pp 189–193, September 2021.
26. G. Vasoglou, I. Stefanidaki, K. Apostolopoulos, E. Fotakidou, M. Vasoglou, "Accuracy of mini-implant placement using a computer-aided designed surgical guide, with information of intraoral scan and the use of a cone-beam CT," Dent J (Basel), Vol. 10, Issue 6, pp 104, June 2022.
27. B. Wilmes, S. Vasudavan, D. Drescher, "CAD-CAM–fabricated mini-implant insertion guides for the delivery of a distalization appliance in a single appointment," Am J Orthod Dentofacial Orthop, Vol. 156, Issue 1, pp 148–156, July 2019.
28. B.C. Koo, C.H. Chung, R.L. Vanarsdall, "Comparison of the accuracy of bracket placement between direct and indirect bonding techniques," Am J Orthod Dentofacial Orthop, Vol. 116, Issue 3, pp 346–351, September 1999.
29. D.J. Weber, L.D. Koroluk, C. Phillips, T. Nguyen, W.R. Proffit, "Clinical effectiveness and efficiency of customized vs. conventional preadjusted bracket systems," J Clin Orthod, Vol. 47, Issue 4, pp 261–266, April 2013.
30. Y. Li, L. Mei, J. Wei, X. Yan, X. Zhang, W. Zheng, "Effectiveness, efficiency and adverse effects of using direct or indirect bonding technique in orthodontic patients: A systematic review and meta-analysis," BMC Oral Health, Vol. 19, Issue 1, pp 137, July 2019.
31. J. Hegele, L. Seitz, C. Claussen, U. Baumert, H. Sabbagh, A. Wichelhaus, "Clinical effects with customized brackets and CAD/CAM technology: A prospective controlled study," Prog Orthod, Vol. 22, Issue 1, pp 40, December 2021.
32. S.Y. Kwon, Y. Kim, H.W Ahn, K.B. Kim, K.R. Chung, "Computer-aided designing and manufacturing of lingual fixed orthodontic appliance using 2D/3D registration software and rapid prototyping," Int J Dent, Vol. 2014, Issue 2014, pp 164164, May 2014.
33. R. Sachdeva, "SureSmile technology in a patient–centered orthodontic practice," J Clin Orthod, Vol. 35, Issue 4, pp 245–253, April 2001.

34. I. Tamer, E. Oztaş, G. Masan, "Orthodontic treatment with clear aligners and the scientific reality behind their marketing: a literature review," Turk J Orthod, Vol. 32, Issue 4, pp 241–246, December 2019.
35. A. Thurzo, W. Urbanová, B. Novák, I. Waczulíková, I. Varga, "Utilization of a 3D printed orthodontic distalizer for tooth-borne hybrid treatment in class II unilateral malocclusions," Materials (Basel), Vol. 15, Issue 5, pp 1740, February 2022.
36. J. Gateno, J. Xia, J.F. Teichgraeber, A. Rosen, B. Hultgren, T. Vadnais, "The precision of computer-generated surgical splints," J Oral Maxillofac Surg, Vol. 61, Issue 7, pp 814–817, July 2003.
37. L.J. Lo, L.S Niu, C.H. Liao, H.H. Lin, "A novel CAD/CAM composite occlusal splint for intraoperative verification in single-splint two-jaw orthognathic surgery," Biomed J, Vol. 44, Issue 3, pp 353–362, June 2021.
38. M. Cassetta, M. Giansanti, "Accelerating orthodontic tooth movement: A new, minimally-invasive corticotomy technique using a 3D-printed surgical template," Med Oral Patol Oral Y Cir Bucal, Vol. 21, Issue 4, pp e483–e487, July 2016.
39. M. Narita, T. Takaki, T. Shibahara, M. Iwamoto, T. Yakushiji, T. Kamio, "Utilization of desktop 3D printer-fabricated cost-effective 3D models in orthognathic surgery," Maxillofac Plast Reconstr Surg, Vol. 42, Issue 1, pp 24, August 2020.
40. G. Badiali, F. Cutolo, A. Roncari, C. Marchetti, A. Bianchi, "Simulation-guided navigation for vector control in pediatric mandibular distraction osteogenesis," J Craniomaxillofac Surg, Vol. 45, Issue 6, pp 969–980, June 2017.
41. D.J. Singh, P.H. Glick, S.P. Bartlett, "Mandibular deformities: single-vector distraction techniques for a multivector problem," J Craniofac Surg, Vol. 20, Issue 5, pp 1468–1472, September 2009.
42. H. Yu, B. Wang, M. Wang, X. Wang, S.G. Shen, "Computer-assisted distraction osteogenesis in the treatment of hemifacial microsomia," J Craniofac Surg, Vol. 27, Issue 6, pp 1539–1542, September 2016.
43. S.H. Kang, H.J. Tak, H.W. Park, J.U. Kim, S.H. Lee, "Fully-customized distraction assembly for maxillofacial distraction osteogenesis: A novel device and its experimental accuracy verification," Head Face Med, Vol. 16, Issue 1, pp 31, November 2020.
44. E. Schneider, S. Ruf, "Upper bonded retainers," Angle Orthod, Vol. 81, Issue 6, pp 1050–1056, November 2011.
45. A.M. Renkema, A. Renkema, E. Bronkhorst, C. Katsaros, "Long-term effectiveness of canine-to-canine bonded flexible spiral wire lingual retainers," Am J Orthod Dentofacial Orthop, Vol. 139, Issue 5, pp 614–621, May 2011.
46. T. Taner, M. Aksu, "A prospective clinical evaluation of mandibular lingual retainer survival," Eur J Orthod, Vol. 34, Issue 4, pp 470–474, August 2012.
47. N.D. Kravitz, D. Grauer, P. Schumacher, Y.M. Jo, "Memotain: A CAD/CAM nickel-titanium lingual retainer," Am J Orthod Dentofacial Orthop, Vol. 151, Issue 4, pp 812–815, April 2017.
48. M. Zreaqat, R. Hassan, A.F. Hanoun, "A CAD/CAM zirconium bar as a bonded mandibular fixed retainer: A novel approach with two-year follow-up," Case Rep. Dent, Vol. 2017, Issue 2017, pp 1583403, July 2017.
49. T.F. Alghazzawi, "Advancements in CAD/CAM technology: Options for practical implementation," J Prosthodontic Res, Vol. 60, Issue 2, pp 72–84, April 2016.
50. Gautam C, Joyner J, Gautam A, Rao J, Vajtai R, "Zirconia based dental ceramics: Structure, mechanical properties, biocompatibility and applications," Dalton Trans, Vol. 45, Issue 48, pp 19194–19215, December 2016.
51. S.H. Kang, J.S. Kwon, C.J. Chung, J.Y. Cha, K.J. Lee, "Accuracy and stability of computer-aided customized lingual fixed retainer: A pilot study," Prog Orthod, Vol. 23 Issue 1, pp 39, November 2022.

52. M.B. Alrawas, Y. Kashoura, O. Tosun, U. Oz, "Comparing the effects of CAD/CAM nickel-titanium lingual retainers on teeth stability and periodontal health with conventional fixed and removable retainers: A randomized clinical trial," Orthod Craniofac Res, Vol. 24, Issue 2, pp 241–250, May 2021.
53. A. Gera, H. Pullisaar, P.M. Cattaneo, S. Gera, V. Vandevska-Radunovic, M.A. Cornelis, "Stability, survival, and patient satisfaction with CAD/CAM versus conventional multistranded fixed retainers in orthodontic patients: A 6-month follow-up of a two-centre randomized controlled clinical trial," Eur J Orthod, Vol. 45, Issue 1, pp 58–67, February 2023.
54. E. Gelin, L. Seidel, A. Bruwier, A. Albert, C. Charavet, "Innovative customized CAD/CAM nickel-titanium lingual retainer versus standard stainless-steel lingual retainer: A randomized controlled trial," Korean J Orthod, Vol. 50, Issue 6, pp 373–382, November 2020.
55. S. Koller, R.B. Craveiro, C. Niederau, T.L. Pollak, I. Knaup, M. Wolf, "Evaluation of digital construction, production and intraoral position accuracy of novel 3D CAD/CAM titanium retainers," J Orofac Orthop, Vol. 84, Issue 6, pp 384–391, November 2023.

12 Machine Learning/ Deep Learning Aspects of Manufacturing Techniques

Khalid Bouiti, Najoua Labjar, Mohamed Dalimi, Houda Labjar, Kaoutar Kara, Loubna Bouhachlaf, Hamid Nasrellah, and Souad El Hajjaji

12.1 INTRODUCTION

In contemporary manufacturing, technology and industry are fundamentally reshaping traditional paradigms. Machine learning (ML) and deep learning (DL), formidable branches of artificial intelligence (AI), are revolutionizing our conceptualization, design, and optimization of manufacturing processes [1].

In an era of abundant data, manufacturing processes generate vast amounts of information. ML with its ability to identify patterns, trends, and anomalies within this deluge of data is proving pivotal as a catalyst for informed decision-making [2]. ML algorithms demonstrate their ability to extract valuable insights from complex manufacturing data, whether applied to predictive maintenance to reduce downtime, quality control to improve product reliability, or supply chain optimization for efficient operations [3–5].

Beyond the realms of conventional ML lies the intricate domain of DL, drawing inspiration from the intricate neural networks of the human brain. DL models, propelled by layered neural networks, excel in tasks demanding nuanced understanding, including natural language processing (NLP), image recognition as well as intricate pattern recognition [6–8]. In the manufacturing domain, these capabilities open avenues to revolutionary applications, encompassing advanced defect detection, autonomous robotic systems, and adaptive manufacturing processes that continuously evolve based on real-time feedback [9, 10].

The combination of ML and DL with manufacturing techniques marks a new era of precision and efficiency. These technologies enable manufacturers to optimize their processes, minimize waste, and maximize output, from predictive modeling for resource allocation to real-time monitoring of production lines [3, 11]. Furthermore, algorithms are adaptable and have a learning capacity, which ensures continuous improvement. This creates a self-optimizing ecosystem that thrives on data-driven insights [12].

The integration of ML and DL provides significant advantages to manufacturing. However, it also presents challenges, including security concerns, ethical

DOI: 10.1201/9781032725086-14

considerations, and the need for a skilled workforce capable of navigating this technological terrain. These hurdles require careful consideration [13, 14].

The first section introduces ML, emphasizing learning from labeled data and the application of knowledge to new situations. DL, a subset, uses neural networks to autonomously learn complex patterns, demanding more resources but excelling in diverse applications. The distinction lies in autonomous feature extraction and hierarchical learning. Key networks, such as stacked autoencoders (SAEs) and convolutional neural networks (CNNs), lay the foundation for DL techniques.

The second section discusses the applications of ML/DL in manufacturing. ML optimizes processes, quality control, supply chain, and energy consumption, providing a competitive advantage. DL, with its advanced neural network capabilities, enhances manufacturing through anomaly detection, computer vision, predictive maintenance, process optimization, NLP, robotics, and generative design. These applications contribute to improving efficiency and reducing costs.

The integration of ML and DL with computer numerical control (CNC) machining is positioned to revolutionize manufacturing. The applications include predictive maintenance, optimization of cutting parameters, quality control, generative design processes, energy efficiency improvements, supply chain optimization, and enhanced human-machine collaboration. These technologies contribute to precision, speed, and efficiency, leading to a comprehensive optimization of the entire manufacturing cycle. The ongoing integration of cutting-edge algorithms and hardware advancements is expected to further evolve smart and adaptive manufacturing systems.

This chapter treats the role of ML and DL in control quality and defect detection in optimizing manufacturing processes. Predictive maintenance, fault detection, real-time monitoring, and advanced defect detection techniques enhance product quality, reduce operational costs, and contribute to overall efficiency. The continuous learning capabilities of these models, along with their integration with internet of things (IoT), enable adaptive responses to changing conditions, ensuring sustained improvements in defect detection accuracy. Automated defect detection not only saves costs and time but also leads to improved product quality and streamlined production workflows. The implementation of preventive maintenance based on defect detection data further contributes to the optimization of equipment performance and reduction in maintenance costs.

Also, predictive maintenance in smart manufacturing relies on a proactive strategy driven by data analytics, ML, and IoT technologies. It goes beyond preventing unplanned downtime by offering cost savings, risk mitigation, and workplace safety improvement. Integration with cyber-physical systems (CPS) ensures precise maintenance execution, aligning seamlessly with the principles of Industry 4.0. In the realm of ML models, functions such as anomaly detection, failure prediction, remaining useful life (RUL) estimation, and real-time monitoring spearhead a data-driven and proactive approach to predictive maintenance, enhancing equipment reliability and operational efficiency in the dynamic landscape of smart manufacturing.

Human-machine collaboration in the Industry 4.0 era involves critical elements like physical human-robot interaction (pHRI), safety measures, and real-time decision-making within CPS. Safety in CPS is ensured through various components, and AI/ML techniques play a crucial role in predicting human activity patterns.

ML and DL are pivotal in Industry 4.0 human-machine collaboration, contributing to equipment failure prediction, quality control enhancement, process optimization, and improved human-robot interaction. The concept of human-automation collaboration (HAC) is gaining traction, emphasizing the importance of human factors for successful collaboration, and leveraging ML/DL technologies significantly improves collaboration efficiency and quality control.

In the realm of manufacturing, ML and DL are catalysts for transformative change, addressing existing challenges and shaping future trends. From predicting equipment failures in predictive maintenance to refining precision in quality control through advanced computer vision, these technologies are enhancing efficiency and productivity. Reinforcement learning is anticipated to streamline complex processes, while supply chain optimization benefits from ML's predictive analytics. Human-machine collaboration aims for seamless integration, overcoming challenges through collaborative robots and AI-driven assistance. Addressing data quality, interoperability issues, and ensuring model explainability and security remain pivotal. As the industry anticipates future trends like edge computing, generative models, and ethical considerations, the integration of ML and DL promises increased automation, efficiency, and innovative solutions, ushering in a new era in manufacturing.

12.1.1 Fundamentals of Machine Learning and Deep Learning

12.1.1.1 Machine Learning Concept

ML, a subfield of AI, focuses on developing algorithms and statistical models for computers to execute tasks without explicit programming. It hinges on learning from data, where input features and output labels define datasets [15].

In the training phase, an ML model delves into a labeled dataset, adapting its internal parameters to grasp patterns. Algorithms, such as supervised (using labeled data), unsupervised (finding patterns without labels), and reinforcement learning (trial and error), characterize this phase [16, 17]. Success relies on the quality and quantity of data, leading to model evaluation with a separate, unseen dataset, utilizing metrics like accuracy, precision, recall, and F1 score.

Post-evaluation, the model applies its learned knowledge to make decisions on fresh, unseen data. A feedback loop often redefines models with new data, enhancing performance or adapting to evolving patterns [18–20]. The primary ML types include supervised (training on labeled data), unsupervised (patterns without labels), and reinforcement learning (interaction-based learning).

Features in ML act as descriptors for data samples, enabling the representation of intricate raw data. For instance, when handling images, storage approaches like grayscale matrices or Red, Green and Blue (RGB) tensors exist. Traditional models, requiring vector input, mandate feature extraction, especially when convolutional layers aren't employed [21–23].

A pivotal aspect is the label, signifying the output tied to a given data sample. In supervised learning, data samples are paired as (feature, label). A substantial amount of training data, forming the training set (strain), is crucial for the model's comprehension of underlying patterns [24].

Post-training set assembly, the focus shifts to selecting a suitable ML model, particularly a mapping function (hypothesis) capturing features-label relationships. The hypothesis set, chosen based on factors like family, complexity, and regularization, influences the model. This set encompasses multiple hypotheses characterized by parameters. Central to the journey is the learning process, involving optimal model parameter selection for a well-performing hypothesis and employing algorithms like stochastic gradient descent for iteration.

Hyperparameters, predefined before learning, significantly impact hypothesis selection and overall performance. Examples include regularization term-associated hyperparameters. These non-updated parameters shape the ML model's overall performance [25].

The intricate interplay between features, labels, training data, hypotheses, model parameters, learning processes, hyperparameters, and validation data forms the foundational landscape of ML concepts.

12.1.1.2 Deep Learning Concept

DL, a dynamic subset of ML, has transformed the landscape of AI, exhibiting remarkable success across diverse domains. Rooted in the emulation of human brain functions, DL leverages artificial neural networks to autonomously learn and make decisions [26]. Its rise to prominence can be attributed to its proficiency in tackling intricate challenges, leading to breakthroughs in fields like cancer detection, image recognition, and game development [27–29].

In recent years, deep learning has burgeoned, captivating researchers with its profound concepts and methodologies. Key operations like feedforward and backpropagation, coupled with the utilization of convolution to extract features, form the bedrock of DL. The significance of loss functions and optimization processes underscores the intricate nature of training artificial neural networks [6, 26].

DL distinguishes itself from traditional ML through several pivotal characteristics. Traditional ML relies on manual feature extraction, whereas DL algorithms autonomously glean features from raw data, eliminating the need for manual engineering [28, 30]. DL models excel in learning hierarchical representations, with each level being more abstract than the preceding one. This hierarchical learning empowers the models to comprehend complex patterns in the data.

The inherent complexity of DL models demands more computational resources and data compared to traditional ML algorithms. This is a trade-off for the sophisticated architectures and numerous parameters in DL models. Traditional models often offer greater interpretability, while DL models, particularly deep neural networks, are sometimes perceived as "black boxes" due to their intricate structures [30, 31].

DL models consistently outshine traditional methods in tasks such as image and speech recognition, NLP, and game playing. Their ability to capture intricate patterns in data underscores their superior performance [27, 32].

DL's prowess extends beyond theoretical frameworks, finding practical applications across diverse industries. Some noteworthy applications include healthcare (enhanced diagnostics and disease detection), finance (improved fraud detection and risk assessment), automotive (advancements in autonomous vehicle technologies),

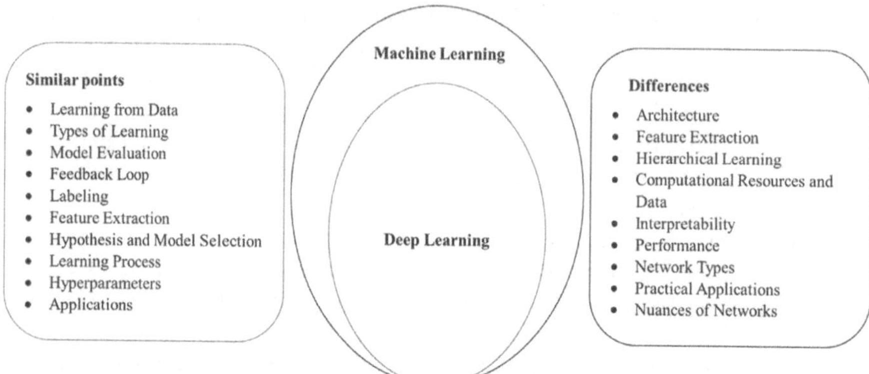

FIGURE 12.1 Machine learning and deep learning concept.

retail (enhanced customer experience through recommendation systems), and entertainment (innovations in content creation and personalization) [33, 34].

Understanding the basics is crucial for implementing DL techniques. Several foundational DL networks serve as the cornerstone. SAEs are unsupervised learning networks capturing higher level features through layer-wise input reconstruction [35, 36]. Deep belief networks (DBNs) are generative models blending directed and undirected graphical models, trained layer-wise with stochastic latent variables. CNNs are specialized networks designed for processing structured grid data, particularly images, using convolutional layers for hierarchical pattern learning. Recurrent neural networks (RNNs) are tailored for sequential data, maintaining a state to capture contextual information, applied in tasks like speech recognition and language modeling. Generative adversarial nets (GANs) are networks engaged in adversarial training, with one generating data and the other discerning between real and generated data [35, 37].

These foundational networks lay the groundwork for implementing DL techniques in various applications, including object detection and recognition. As we delve into the intricate details of these networks, we uncover the nuances of neural network architectures, activation functions, and the optimization algorithms that propel the evolution of DL. Figure 12.1 illustrates the concept of ML and DL.

12.2 APPLICATIONS OF ML AND DL IN MANUFACTURING

12.2.1 Machine Learning in Manufacturing

ML has gained importance in the manufacturing industry, offering various benefits and applications. ML algorithms analyze data from equipment sensors to predict maintenance needs, reduce downtime, and prevent failures. Real-time defect detection during the manufacturing process ensures high product quality [14]. Additionally, ML optimizes manufacturing processes, improves efficiency, reduces waste, and increases productivity. Supply chain operations benefit from ML algorithms optimizing inventory management, demand forecasting, and logistics, resulting in more efficient processes [38, 39].

Energy consumption in manufacturing processes can be optimized using ML techniques, leading to savings and environmental benefits. ML models excel in defect detection, identifying anomalies early to prevent defects. Customized manufacturing is facilitated by analyzing customer data and preferences and tailoring products to individual needs [40, 41]. Process monitoring with ML algorithms ensures real-time data analysis for consistency and quality.

By integrating ML into manufacturing, companies can gain a competitive advantage, improve operational efficiency, and drive innovation in their production processes. ML has gradually found applications in manufacturing, enhancing efficiency, productivity, and decision-making [3, 42, 43]. Predictive maintenance through ML algorithms analyzing sensor data allows proactive equipment maintenance, reducing downtime. Quality control benefits from real-time defect detection, ensuring high product quality and minimizing waste. Supply chain optimization, process optimization, and resource allocation are other areas where ML contributes by streamlining operations, improving efficiency, and reducing costs [44, 45].

The impact of ML extends to product design, customer feedback analysis, failure prediction, and feature optimization. The cause analysis becomes more robust as ML identifies complex data patterns that human analyst may overlook [46, 47]. Revolutionizing the manufacturing industry, ML improves efficiency, reduces downtime, and optimizes processes, significantly contributing to overall progress.

In the manufacturing sector, ML algorithms analyze large datasets, identify patterns, make predictions, and optimize processes. Predictive maintenance forecasts equipment failures, enabling proactive maintenance, minimizing downtime, and reducing costs [48, 49]. Quality control is strengthened by ML models that detect defects in real time, assuring only the highest quality products reach the market. Supply chain management benefits from ML demand forecasting, inventory optimization, and improved logistics operations [50].

Process optimization using ML techniques involves analyzing data from various sources, improving resource utilization, and reducing waste. Product design is refined by analyzing customer feedback, market trends, and historical data [51, 52]. Robotics and automation systems are also enhanced through ML, increasing productivity and flexibility. The versatility of ML in manufacturing offers benefits such as process optimization, predictive maintenance, quality control, and supply chain management, revolutionizing decision-making processes and stimulating innovation [14, 52].

12.2.2 Deep Learning in Manufacturing

DL, a subset of ML characterized by the utilization of neural networks with multiple layers, has witnessed significant integration into the manufacturing industry. This integration is primarily attributed to its adeptness in handling complex data and extracting valuable insights [27, 32]. The impact of DL in manufacturing spans various critical applications.

Anomaly detection stands out as a pivotal application, where DL models meticulously analyze sensor data to identify deviations from standard operations. This capability not only aids in the prevention of equipment failures but also elevates the standards of quality control within manufacturing processes [53, 54].

Computer vision, facilitated by advanced DL algorithms, assumes a crucial role in visual inspection tasks within manufacturing. Its ability to detect defects on production lines, ensure product quality, and automate quality control processes showcases the transformative potential of DL in enhancing visual inspection capabilities [55–57].

Predictive maintenance experiences a notable boost through the application of DL techniques. By analyzing intricate sensor data, these techniques predict equipment failures with heightened accuracy and earlier detection, fostering a proactive approach and minimizing both downtime and maintenance costs.

Manufacturing process optimization stands as another significant application, with DL models analyzing extensive datasets to identify patterns. This process leads to tangible improvements in efficiency and cost reduction, highlighting the potential for optimization through DL [58–60].

NLP is employed in manufacturing to analyze text data from diverse sources, such as maintenance logs and customer feedback [61]. This application contributes to the continuous improvement of manufacturing processes by extracting valuable insights from textual information [62, 63].

In the domain of robotics, DL has a pivotal role in applications such as robot control, object recognition, and autonomous navigation. These advancements not only enhance automation but also significantly improve overall efficiency on the factory floor [64].

Generative design processes benefit from the incorporation of DL techniques, allowing for the automatic generation and optimization of product designs based on specified constraints and objectives [65]. This application demonstrates the potential for innovation and efficiency improvement in the design phase of manufacturing (Figure 12.2).

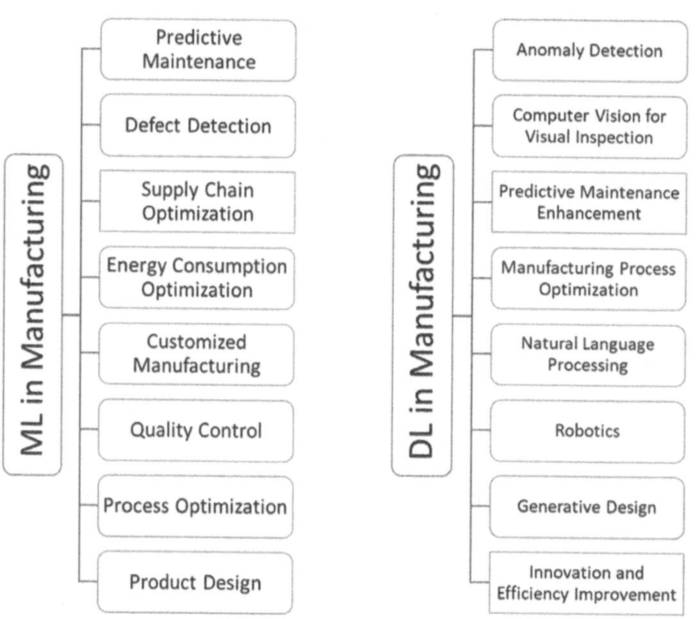

FIGURE 12.2 ML and DL applications.

By strategically leveraging the capabilities of DL in manufacturing, companies can drive innovation, improve operational efficiency, and gain a distinct competitive advantage in the ever-evolving industrial landscape. The ongoing integration of DL technologies underscores their pivotal role in shaping the future trajectory of manufacturing processes.

12.3 INTEGRATION OF ML/DL WITH CNC MACHINING

The integration of ML algorithms into CNC machining holds the potential to revolutionize the manufacturing industry, enhancing efficiency, accuracy, and overall productivity. Predictive maintenance emerges as a critical application, where algorithms analyze CNC machine data, including temperature, vibration, and tool wear [66, 67]. This analysis enables accurate predictions of maintenance needs, minimizing downtime, and preventing unexpected breakdowns.

Optimizing cutting parameters is another vital aspect where ML has a central role. Real-time data and historical performance are scrutinized to refine parameters like speed, feed rate, and depth of cut. This optimization leads to improved tool life, reduced cycle times, and enhanced surface finish, significantly contributing to overall efficiency [13, 68, 69].

Quality control experiences substantial improvements through ML algorithms, particularly support neural networks and vector machines. These algorithms analyze sensor data to detect anomalies and deviations in the machining process. Early identification of potential quality issues allows for prompt corrective actions, ensuring consistent product quality [50]. Additionally, ML facilitates adaptive control, enabling CNC machines to adjust cutting strategies in real time based on changing conditions, thereby optimizing performance and accuracy dynamically [70].

Generative design processes, powered by ML algorithms, contribute to the creation of innovative and optimized part geometries. This exploration of design possibilities based on performance criteria results in lightweight and structurally efficient components. Energy efficiency experiences advancements as well, with ML optimizing energy consumption in CNC machining operations through the analysis of usage patterns and the suggestion of energy-saving strategies.

The integration of ML and DL technologies with CNC machining promises significant enhancements in efficiency, precision, and overall performance. Predictive maintenance, a key application, involves ML algorithms analyzing sensor data to predict potential equipment failures, allowing for timely interventions and reduced downtime [67, 71–73].

ML also contributes to the optimization of cutting parameters by analyzing historical machining data and recommending the most efficient settings based on specific materials, tooling, and machining conditions [72, 73]. Similarly, DL algorithms optimize tool paths by analyzing the geometry of the part being machined, leading to more efficient movements, reduced cycle times, and minimized tool wear [74, 75].

Quality control and anomaly detection benefit from ML/DL models analyzing sensor data and images during machining. By comparing real-time data to acceptable outcomes, the system identifies defects or deviations, triggering alerts for immediate corrective action [76, 77]. Energy consumption optimization is another area where ML can contribute, leading to cost savings and environmental sustainability [78].

ML/DL enables adaptive control systems for CNC machines, adjusting machining parameters in real time based on dynamic conditions. Continuous monitoring provides real-time feedback regarding tool conditions, part quality, and potential issues [79, 80]. Generative design powered by DL assists in creating optimal tool paths, considering material properties, part geometry, and manufacturing constraints.

Furthermore, ML contributes to supply chain optimization by predicting material requirements, managing inventory levels, and streamlining procurement processes. This ensures continuous operation for CNC machines. Human-machine collaboration is facilitated, as ML/DL technologies provide insights and recommendations to operators, enhancing decision-making processes [81, 82].

The integration of ML/DL with CNC machining represents a paradigm shift in manufacturing, offering the potential for increased automation, improved quality, and enhanced overall efficiency in the production process [83, 84]. As technology advances, the continuous integration of cutting-edge algorithms and advancements in hardware is anticipated, contributing to the evolution of smart and adaptive manufacturing systems.

ML/DL technologies play a crucial role in improving precision, speed, and efficiency in CNC machining processes. By analyzing historical machining data, ML algorithms optimize cutting parameters, tailoring them to specific materials and tooling. This not only improves efficiency but also contributes to the precision of the machining process.

DL algorithms optimize tool paths based on the unique geometry of the part being machined, leading to more efficient movements, reduced cycle times, and minimized tool wear. The ability to adapt tool paths dynamically contributes to the overall efficiency of the machining process [85–87].

ML models provide a predictive maintenance capability by analyzing sensor data from CNC machines. Timely interventions based on these predictions reduce downtime, ensuring precision and efficiency. Real-time monitoring and feedback, enabled by ML/DL technologies, further enhance precision by addressing issues promptly during the machining process [71, 88].

Quality control and anomaly detection benefit from ML/DL models analyzing sensor data and images in real time. Defects or anomalies in machined parts are promptly identified, contributing to the production of high-quality components without extensive rework [14, 50].

Adaptive control systems, facilitated by ML/DL algorithms, adjust machining parameters based on dynamic process conditions. This adaptability ensures optimal performance, precision, and efficiency throughout the machining process.

DL-powered generative design assists in creating optimal tool paths by considering material properties and part geometry. This innovative approach not only enhances precision but also speeds up the design-to-production cycle, allowing for more efficient and streamlined manufacturing processes.

ML algorithms contribute to energy consumption optimization during machining operations. By analyzing energy patterns, the machining process can be optimized to reduce energy consumption without compromising precision, aligning with sustainability goals [89, 90].

Human-machine collaboration is improved as ML/DL technologies provide real-time insights and recommendations to operators. This collaboration ensures that operators can make informed decisions, contributing to the precision and efficiency of the CNC machining process.

Additionally, ML contributes to supply chain optimization by predicting material requirements and streamlining procurement processes. This optimization ensures that CNC machines have the necessary resources, minimizing delays and contributing to the overall efficiency of the manufacturing workflow. In conclusion, the integration of ML/DL with CNC machining processes brings advancements that positively impact precision, speed, and efficiency, optimizing the entire manufacturing cycle.

12.4 QUALITY CONTROL AND DEFECT DETECTION

12.4.1 ML in Quality Control and Defect Detection

ML is integral to quality control and defect detection in manufacturing, enhancing systems and capabilities. Through algorithms, anomalies in manufacturing processes are identified by analyzing sensor data, production metrics, and historical records. Predictive maintenance models prevent defects by foreseeing equipment failures [50, 91].

Fault detection and classification algorithms analyze data from sensors and production systems to identify patterns associated with specific defects. Feature selection algorithms identify key variables impacting product quality, optimizing quality control processes. ML models also optimize process parameters by analyzing data from various sensors and production stages [92].

Real-time monitoring of production processes is enabled by ML algorithms, allowing swift identification and addressing of quality issues. Pattern recognition algorithms identify complex patterns in manufacturing data that may indicate potential defects, prompting corrective actions [93].

In manufacturing, advanced ML algorithms and techniques contribute to automated and efficient defect detection. CNNs and DL models analyze images, sensor data, and other information for automated and accurate defect detection [8, 94]. This proactive approach enhances product quality, prevents faulty products from reaching the market, and reduces operational costs.

Automated defect detection powered by ML minimizes the need for manual inspection and rework, optimizing production efficiency and reducing waste. Real-time monitoring allows immediate defect detection, enabling prompt corrective actions [95, 96]. ML is also utilized for predictive maintenance, minimizing downtime and ensuring continuous production [97].

Continuous improvement is achieved through ML algorithms that adapt to new data, enhancing defect detection accuracy and efficiency. Incremental learning in the proposed clothing manufacturing system allows the model. The system uses mobile devices to capture images for defect detection, and quality control officers can provide feedback through a mobile application [98, 99].

In defect detection, DL models show potential in analyzing large volumes of data generated during manufacturing processes. The integration of embedded sensor equipment and real-time defect detection systems enhances efficiency and accuracy. ML contributes to improving product consistency, reducing waste, and overall enhancement of product quality.

Development trends in defect detection involve integrating various non-destructive methods for multi-modal defect detection and a growing focus on 3D defect detection methods. Challenges related to the accuracy of online detection techniques are addressed through ongoing research, aiming to reduce production costs and enhance product quality, paving the way for the intelligent transformation of the manufacturing industry.

12.4.2 ML/DL for Sensor Data Analysis

ML/DL algorithms have a pivotal role in extracting pertinent features from sensor data, encompassing variables like temperature, pressure, vibration, and image data [100, 101]. This extraction process is essential for identifying patterns associated with defects, with the utilization of advanced DL techniques, including CNNs, which prove particularly effective in handling intricate features for precise defect detection [102, 103].

ML algorithms, employing unsupervised learning techniques like clustering or autoencoders, demonstrate the capability to detect anomalies in sensor data, potentially indicating defects in the production line. By discerning outliers or irregular patterns, these algorithms can promptly flag potential quality issues, enabling real-time intervention and ensuring the maintenance of product quality standards [50, 104].

DL algorithms, specifically CNNs and RNNs, exhibit proficiency in classifying sensor data into distinct categories based on predefined defect types or quality levels [8, 105]. Training these models on labeled data empowers them to differentiate between normal and defective patterns, thereby facilitating accurate and efficient classification.

ML/DL algorithms analyze sensor data to predict equipment failures or maintenance needs before they manifest. By monitoring sensor readings and historical data, these algorithms can forecast potential defects in machinery, enabling proactive maintenance actions to prevent quality issues and minimize downtime, ultimately enhancing operational efficiency [81].

It provides real-time monitoring of sensor data streams, enabling the immediate detection of anomalies or deviations from expected values. This continuous monitoring feature ensures swift intervention to address quality issues and uphold product quality standards throughout the entirety of the production process.

The models exhibit the capacity for continuous learning from new sensor data, leading to improved defect detection accuracy over time. Regular updates incorporating fresh data and feedback from production processes enable these models to adapt to changing conditions, ensuring a sustained enhancement in their ability to maintain high product quality standards [14, 50].

ML/DL algorithms seamlessly integrate with IoT devices and sensors, establishing a connected manufacturing environment. Leveraging IoT data streams, these algorithms can analyze a diverse range of sensor data sources, providing comprehensive insights into product quality and defect detection [106, 107].

In the context of quality control and defect detection in manufacturing, ML and DL algorithms serve as instrumental tools for analyzing sensor data, offering versatile applications such as anomaly detection, pattern recognition, predictive maintenance, process optimization, fault detection and classification, real-time monitoring, and feature engineering [108, 109, 110]. These applications collectively contribute to the continuous improvement of product quality and the efficiency of manufacturing processes.

12.4.3 Automated Defect Detection Performance

Automated defect detection using ML and DL algorithms presents a valuable opportunity for cost savings and increased efficiency in manufacturing processes. One of the primary advantages lies in early detection, as these systems can identify defects at the initial stages of production [55]. This preemptive identification mitigates the need for expensive rework or scrapping of defective products. Furthermore, addressing issues promptly during early stages reduces the likelihood of defects permeating through the production line, thereby minimizing their impact on overall product quality.

Another significant benefit is the reduction in inspection costs. Automated defect detection eliminates the necessity for manual product inspections, saving both time and labor costs associated with conventional inspection methods [100]. This automation not only expedites the inspection process but also enhances throughput, ultimately leading to increased efficiency and reduced operational costs.

The implementation of automated defect detection systems contributes to improved product quality. These systems consistently and accurately identify defects, resulting in higher overall product quality and reduced defect rates; manufacturers can bolster customer satisfaction, enhance brand reputation, and improve their competitiveness in the market [111, 112].

Optimized production processes are another noteworthy advantage of automated defect detection. These systems offer real-time feedback on production processes, empowering manufacturers to make timely adjustments and optimizations [111]. Through the analysis of data provided by automated defect detection systems, manufacturers can identify root causes of defects, implement corrective actions, and streamline production workflows, leading to enhanced efficiency and reduced defects [55, 71].

Additionally, automated defect detection systems facilitate preventive maintenance by predicting equipment failures or defects before they occur. This capability enables proactive maintenance measures to prevent costly downtime and production disruptions. By implementing predictive maintenance strategies based on data from automated defect detection, manufacturers can optimize equipment performance, extend asset lifespan, and effectively reduce maintenance costs.

12.5 PREDICTIVE MAINTENANCE IN SMART MANUFACTURING

12.5.1 Predictive Maintenance Concept

Predictive maintenance, as a proactive strategy, is integral to smart manufacturing, where its role extends beyond mere prevention of unplanned downtime and costly repairs. It represents a sophisticated approach that leverages data analytics, ML, and IoT technologies to forecast potential equipment failures [113, 114]. By scheduling timely interventions during planned downtime, manufacturers can minimize disruptions, optimize production processes, and enhance overall equipment effectiveness [113].

This strategic approach not only lowers maintenance costs by avoiding unnecessary tasks but also mitigates the risk of catastrophic failures, contributing to substantial cost savings. Real-time monitoring and early identification of potential issues further ensure that equipment operates at optimal performance levels, leading to heightened productivity and reduced cycle times in manufacturing processes [115, 116]. Moreover, the improvement in workplace safety by reducing the likelihood of equipment failures adds an additional layer of value to the implementation of predictive maintenance.

In the smart manufacturing landscape, the integration of predictive maintenance into CPS architecture ensures continuous monitoring of equipment health. This intelligent integration enables maintenance activities to be executed precisely when necessary, optimizing resources, reducing downtime, and fostering overall operational excellence [117, 118]. Moreover, the implementation of predictive maintenance systems, driven by data-driven AI models, empowers manufacturers to predict equipment faults, optimize maintenance schedules, and enhance overall reliability and performance [119].

This proactive maintenance approach is particularly vital in the Industry 4.0 era, where efficiency, productivity, and cost-effectiveness take precedence. Predictive maintenance, with its ability to anticipate and address issues before they escalate, aligns seamlessly with the principles of Industry 4.0, contributing to a competitive advantage for manufacturers. As the manufacturing landscape evolves, the role of predictive maintenance becomes increasingly crucial for ensuring the reliability, efficiency, and sustainability of production processes in the dynamic and data-centric environment of smart manufacturing [76, 113, 120].

12.5.2 Prediction Equipment Failures and Maintenance Optimization by ML Models

ML models perform a critical role in predictive maintenance in smart manufacturing environments, contributing significantly to predicting equipment failures and optimizing maintenance schedules. Using data, sensor readings, and equipment performance metrics, these models provide a comprehensive solution for maintenance efficiency [121, 122].

An essential aspect of this role lies in anomaly detection, where ML models analyze sensor data to identify deviations from normal operating conditions. These anomalies serve as early indicators, allowing maintenance teams to proactively intervene and prevent potential equipment failures or degradation [123–125].

Failure prediction, another critical function, involves the examination of historical data on equipment failures and maintenance records. ML models discern patterns and trends, predicting when specific components or systems are likely to fail. This foresight empowers maintenance teams to schedule preventive activities strategically, mitigating the risk of unplanned downtime [123, 125].

RUL estimation is facilitated by evaluating the current condition and usage patterns of equipment. ML models estimate how long a component or system will operate before failure, optimizing maintenance schedules to replace or repair equipment before critical failures occur [103, 126, 127].

The optimization of maintenance schedules encompasses a thorough analysis of parameters, including environmental conditions, equipment usage, and historical maintenance data. ML models ensure maintenance activities are scheduled at the most cost-effective times, minimizing downtime and maximizing equipment availability [55, 128].

Prescriptive maintenance, an advanced feature, goes beyond predictive maintenance by offering recommendations for effective maintenance actions. ML models equipped with prescriptive analytics suggest actions such as adjusting operating parameters, replacing components, or scheduling maintenance tasks, contributing to optimal maintenance strategies [129].

Real-time monitoring is a dynamic aspect, with ML models continuously analyzing streaming sensor data. This real-time capability allows for the immediate detection of anomalies and prediction of failures, enabling swift maintenance interventions to prevent downtime [130–132].

In essence, ML models spearhead a data-driven and proactive approach to predictive maintenance. This transition enhances equipment reliability, reduces downtime, and improves overall operational efficiency within the dynamic landscape of smart manufacturing environments.

12.6 HUMAN-MACHINE COLLABORATION IN MANUFACTURING

In the era of Industry 4.0, human-machine collaboration stands as a fundamental aspect of manufacturing. It brings together the strengths of humans and machines, enhancing productivity, flexibility, and overall production performance. Various critical elements contribute to this collaboration, including pHRI on the factory floor. Designing collaborative workspaces, implementing safety measures, and creating intuitive interfaces are crucial for ensuring seamless and secure collaboration [133, 134].

Real-time decision-making, particularly within CPS, becomes imperative for smart interactions with humans and other machines. Ensuring safety in CPS involves components such as safety monitors, reflex blocks, reactive recovery units, and normal operations modules, guaranteeing efficient and secure operation in dynamic environments [135–137]. AI and ML techniques have a significant role in predicting human activity patterns, inferring actions, monitoring scenes, task modeling, planning, and real-time decision-making.

Emphasized by Industry 4.0, flexibility and adaptability underscore the need for automated adaptation in manufacturing systems and collaborative robots. This ensures smooth transitions between different product variants without disrupting production, aligning with the overarching goal of flexibilizing production systems [133, 138].

Human-machine collaboration is a multidisciplinary field, integrating robotics, AI, ML, safety engineering, and human factors. The aim is to create efficient and safe working environments where humans and machines collaborate harmoniously to achieve common production goals.

ML and DL technologies play a pivotal role in enabling human-machine collaboration in the Industry 4.0 era. These technologies predict potential equipment failures through ML algorithms analyzing sensor data. Quality control is enhanced through ML/DL models trained to detect defects, automating the inspection process and ensuring only high-quality products reach the market [139, 140]. Moreover, ML algorithms optimize manufacturing processes, analyzing data to identify bottlenecks, inefficiencies, and opportunities for improvement. The integration of ML/DL techniques in human-robot interaction ensures a more natural and intuitive collaboration on the factory floor, enabling adaptive control for machines based on changing conditions or production requirements [133, 138, 141–145].

Human-machine collaboration, or HAC, is a concept gaining traction in manufacturing. It involves interactive cooperation between human operators and intelligent automation systems within shared workspaces. Factors influencing successful collaboration include stress, anxiety, and safety concerns due to close proximity between humans and robots [133]. Automation reliability, perceived attentional control (PAC), and attitudes toward robots/automation significantly impact the effectiveness of collaboration.

Understanding and addressing these human factors are crucial for the successful implementation of human-machine collaboration in manufacturing. Organizations can optimize collaborative systems by identifying and evaluating these factors, thereby enhancing productivity in the manufacturing environment. Leveraging ML and DL technologies significantly contributes to human-machine collaboration in manufacturing, improving automation systems and decision-making processes for enhanced efficiency, quality control, and safety in Industry 4.0 operations.

12.7 CHALLENGES AND FUTURE TRENDS

ML and DL are pivotal in reshaping manufacturing, offering solutions to challenges while paving the way for future trends. Predictive maintenance, a crucial aspect, confronts the challenge of accurately forecasting equipment failures to minimize downtime. The trajectory leans toward improved algorithms that enhance predictive accuracy, enabling better maintenance planning and reduced operational disruptions.

Quality control, another significant domain, faces the ongoing challenge of ensuring product quality and swiftly identifying defects in real time. The future sees the integration of advanced computer vision systems and anomaly detection algorithms, refining precision in quality control processes. Process optimization, an intricate task in manufacturing, wrestles with the challenge of streamlining complex processes for optimal efficiency. The future trend involves the application of reinforcement learning, facilitating adaptive process optimization and energy efficiency.

Supply chain optimization is a multifaceted challenge encompassing forecasting, inventory management, and demand planning. ML's impact on this sector is set to intensify, with predictive analytics and advanced algorithms enhancing decision-making and overall efficiency. Addressing the consumer demand for customization presents a challenge, met by the evolving trend of employing DL-driven production systems. These systems cater to mass customization, fostering adaptive manufacturing to meet individualized consumer preferences.

Human-machine collaboration in manufacturing strives to seamlessly integrate ML/DL systems with human workers, fostering increased productivity and safety. Challenges in this realm prompt ongoing efforts to develop collaborative robots and AI-driven assistance that complement and augment human capabilities. Data quality and availability remain a persistent challenge, with solutions lying in substantial investments in data collection infrastructure, cleaning processes, and integration with IoT devices.

Interoperability issues in integrating ML/DL solutions with existing manufacturing systems necessitate standardization efforts and the development of compatible interfaces. Explainability and trust in complex ML/DL models present a challenge, addressed by the ongoing development of interpretable models and transparent decision-making processes.

The pervasive skill gap in personnel equipped to implement and maintain ML/DL systems is met with initiatives aimed at investing in training programs and educational endeavors. Looking toward the future, trends include the deployment of edge computing to bring ML/DL processing closer to manufacturing equipment for real-time decision-making. Generative models, particularly generative adversarial networks (GANs), find applications in creating synthetic data to optimize manufacturing processes.

Explainable AI gains prominence, with models providing transparent explanations for decisions to enhance trust and understanding. The trajectory also points toward fully autonomous manufacturing systems driven by AI, while AI-driven design transforms product development, emphasizing improved functionality, cost-efficiency, and sustainability. The future of manufacturing includes ethical considerations, with the establishment of guidelines and frameworks to ensure responsible AI use in the industry. In summary, the integration of ML and DL in manufacturing holds promise for increased automation, heightened efficiency, and innovative solutions to persisting challenges.

12.8 CONCLUSION

In the dynamic landscape of contemporary manufacturing, the integration of ML/DL stands as a catalyst for transformative change. This chapter has explored the profound impact of these intelligent technologies on reshaping traditional paradigms, ushering in an era of precision, efficiency, and innovation.

The applications of ML/DL in manufacturing are diverse and far-reaching. From optimizing processes, quality control, and supply chain efficiency to enhancing defect detection and predictive maintenance, these technologies offer a competitive advantage. The integration with CNC machining heralds a revolution in manufacturing, promising comprehensive optimization of the entire production cycle.

However, this technological integration presents challenges that demand careful consideration. Security concerns, ethical considerations, and the need for a skilled workforce capable of navigating this evolving terrain are critical aspects that require attention.

The chapter has provided a detailed exploration of their role in quality control, defect detection, and predictive maintenance. The continuous learning capabilities of these models, coupled with integration with the IoT, ensure adaptive responses to changing conditions, contributing to sustained improvements in defect detection accuracy.

Human-machine collaboration in the Industry 4.0 era is a pivotal aspect of this transformative journey. AI/ML techniques have a crucial role in predicting human activity patterns, enhancing equipment reliability, and improving collaboration efficiency, quality control, and safety.

As the manufacturing industry anticipates future trends such as edge computing, generative models, and ethical considerations, the ML/DL integration promises increased automation, efficiency, and innovative solutions. These technologies are set to usher in a new era in manufacturing, addressing existing challenges and shaping the industry's future landscape.

REFERENCES

1. Biswas, Aakanksha, Aditi Kumari, D. S. Gaikwad, and Dhananjay K. Pandey. 2023. Revolutionizing biological science: The synergy of genomics in health, bioinformatics, agriculture, and artificial intelligence. *OMICS: A Journal of Integrative Biology* 27. Mary Ann Liebert, Inc., publishers: 550–569. https://doi.org/10.1089/omi.2023.0197.
2. Jordan, Michael I., and Tom M. Mitchell. "Machine learning: Trends, perspectives, and prospects." *Science* 349, no. 6245 (2015): 255–260. https://doi.org/10.1126/science.aaa8415.
3. Rai, Rahul, Manoj Kumar Tiwari, Dmitry Ivanov, and Alexandre Dolgui. 2021. Machine learning in manufacturing and industry 4.0 applications. *International Journal of Production Research* 59: 4773–4778. https://doi.org/10.1080/00207543.2021.1956675.
4. Batista, Rafael Cavicchioli, Abhishek Agarwal, Adash Gurung, Ajay Kumar, Faisal Altarazi, Namrata Dogra, Vishwanatha HM, Dundesh S. Chiniwar, and Ashish Agrawal. "Topological and lattice-based AM optimization for improving the structural efficiency of robotic arms." *Frontiers in Mechanical Engineering* 10 (2024): 1422539.
5. Burande, Dinesh V., Kanak Kalita, Rohit Gupta, et al. 2024. Machine learning metamodels for thermo-mechanical analysis of friction stir welding. *International Journal on Interactive Design and Manufacturing*. https://doi.org/10.1007/s12008-024-01871-6.
6. Tadesse, Helen, Balkeshwar Singh, Habtamu Deresso, et al. 2024. Investigation of production bottlenecks and productivity analysis in soft drink industry: A case study of East Africa Bottling Share Company. *International Journal on Interactive Design and Manufacturing*. https://doi.org/10.1007/s12008-023-01715-9.
7. Pedrycz, Witold, and Shyi-Ming Chen, ed. 2020. *Deep Learning: Concepts and Architectures*, vol. 866. Studies in Computational Intelligence. Cham: Springer International Publishing. https://doi.org/10.1007/978-3-030-31756-0.
8. Kumar, Ajay, Virendra Kumar Shrivastava, Parveen Kumar, Ashwini Kumar, and Vishal Gulati. 2024. Predictive and experimental analysis of forces in die-less forming using artificial intelligence techniques. *Proceedings of the Institution of Mechanical Engineers, Part E: Journal of Process Mechanical Engineering*. https://doi.org/10.1177/09544089241235473.
9. Kumar, Ajay, Gopal Jee Mishra, Vishal Gulati, et al., 2023. A comprehensive review on heat-assisted incremental sheet forming. *International Journal on Interactive Design and Manufacturing*. https://doi.org/10.1007/s12008-023-01670-5.

10. Mineo, Carmelo, Momchil Vasilev, B. Cowan, C. N. MacLeod, S. Gareth Pierce, C. Wong, E. Yang, R. Fuentes, and E. J. Cross. "Enabling robotic adaptive behaviour capabilities for new industry 4.0 automated quality inspection paradigms." *Insight-Non-Destructive Testing and Condition Monitoring* 62, no. 6 (2020): 338–344.https://doi.org/10.1784/insi.2020.62.6.338.
11. Morariu, Cristina, Octavian Morariu, Silviu Răileanu, and Theodor Borangiu. 2020. Machine learning for predictive scheduling and resource allocation in large scale manufacturing systems. *Computers in Industry* 120: 103244. https://doi.org/10.1016/j.compind.2020.103244.
12. Ailon, Nir, Bernard Chazelle, Kenneth L. Clarkson, Ding Liu, Wolfgang Mulzer, and C. Seshadhri. 2011. Self-improving algorithms. *SIAM Journal on Computing* 40: 350–375. https://doi.org/10.1137/090766437.
13. Mohammed, Muzakkiruddin Ahmed. 2022. Machine learning approaches, technologies, recent applications, advantages and challenges on manufacturing and industry 4.0 applications. *International Journal for Research in Applied Science and Engineering Technology* 10: 1114–1121. https://doi.org/10.22214/ijraset.2022.46362.
14. Wuest, Thorsten, Daniel Weimer, Christopher Irgens, and Klaus-Dieter Thoben. 2016. Machine learning in manufacturing: Advantages, challenges, and applications. *Production & Manufacturing Research* 4: 23–45. https://doi.org/10.1080/21693277.2016.1192517.
15. Duarte, Denio, and Niclas Ståhl. 2019. Machine Learning: A Concise Overview. In *Data Science in Practice*, vol. 46, ed. Alan Said and Vicenç Torra, 27–58. Studies in Big Data. Cham: Springer International Publishing. https://doi.org/10.1007/978-3-319-97556-6_3.
16. Ratner, Alexander J., Christopher M. De Sa, Sen Wu, Daniel Selsam, and Christopher Ré. "Data programming: Creating large training sets, quickly." *Advances in Neural Information Processing Systems* 29 (2016).
17. Erdogmus, Deniz, Y. N. Rao, and José C. Principe. 2005. Supervised Training of Adaptive Systems with Partially Labeled Data. In *Proceedings. (ICASSP '05). IEEE International Conference on Acoustics, Speech, and Signal Processing, 2005*, 5, 321–324. Philadelphia, Pennsylvania, USA: IEEE. https://doi.org/10.1109/ICASSP.2005.1416305.
18. Abinaya, G., Gyan Ranjan, and P. Aswin Karthik. 2019. Continuous Learning Mechanism of NLU-ML Models Boosted by Human Feedback. In *2019 International Conference on Computational Intelligence in Data Science (ICCIDS)*, 1–6. Chennai, India: IEEE. https://doi.org/10.1109/ICCIDS.2019.8862102.
19. Taori, Rohan, and Tatsunori Hashimoto. "Data feedback loops: Model-driven amplification of dataset biases." In *International Conference on Machine Learning*, pp. 33883–33920. PMLR, 2023.
20. Jordan, Alexander H., and Pino G. Audia. 2012. Self-enhancement and learning from performance feedback. *Academy of Management Review* 37: 211–231. https://doi.org/10.5465/amr.2010.0108.
21. Hassan, Adel, and Muath Sabha. 2023. Feature extraction for image analysis and detection using machine learning techniques. *International Journal of Advanced Networking and Applications* 14: 5499–5508. https://doi.org/10.35444/IJANA.2023.14401.
22. Jogin, Manjunath, M. S. Madhulika, G. D. Divya, R. K. Meghana, and S. Apoorva. "Feature extraction using convolution neural networks (CNN) and deep learning." In 2018 3rd IEEE international conference on recent trends in electronics, information & communication technology (RTEICT), pp. 2319–2323. IEEE, 2018. https://doi.org/10.1109/RTEICT42901.2018.9012507.
23. Suhaidi, Mustazzihim, Rabiah Abdul Kadir, and Sabrina Tiun. 2021. A review of feature extraction methods on machine learning. *Journal of Information System and Technology Management* 6: 51–59.

24. Kaur, Amandeep, Kalpna Guleria, and Naresh Kumar Trivedi. 2021. Feature Selection in Machine Learning: Methods and Comparison. In *2021 International Conference on Advance Computing and Innovative Technologies in Engineering (ICACITE)*, 789–795. Greater Noida, India.
25. Tanyildizi, Erkan, and Fadime Demirtas. 2019. Hiper Parametre Optimizasyonu Hyper Parameter Optimization. İn *2019 1st International Informatics and Software Engineering Conference (UBMYK)*, 1–5. Ankara, Turkey: IEEE. https://doi.org/10.1109/UBMYK48245.2019.8965609.
26. Vogt, Michael. 2019. An Overview of Deep Learning and Its Applications. In *Fahrerassistenzsysteme 2018*, ed. Torsten Bertram, 178–202. Proceedings. Wiesbaden: Springer Fachmedien Wiesbaden. https://doi.org/10.1007/978-3-658-23751-6_17.
27. LeCun, Yann, Yoshua Bengio, and Geoffrey Hinton. 2015. Deep learning. *Nature* 521: 436–444. https://doi.org/10.1038/nature14539.
28. Shen, Dinggang, Guorong Wu, and Heung-Il Suk. 2017. Deep learning in medical image analysis. *Annual Review of Biomedical Engineering* 19: 221–248. https://doi.org/10.1146/annurev-bioeng-071516-044442.
29. Liang, Hong, Xiao Sun, Yunlei Sun, and Yuan Gao. 2017. Text feature extraction based on deep learning: A review. *EURASIP Journal on Wireless Communications and Networking* 2017: 211. https://doi.org/10.1186/s13638-017-0993-1.
30. Testolin, Alberto, Michele Piccolini, and Samir Suweis. 2019. Deep learning systems as complex networks. Edited by Thilo Gross. *Journal of Complex Networks*. https://doi.org/10.1093/comnet/cnz018.
31. Yuan, Yubai, Yujia Deng, Yanqing Zhang, and Annie Qu. 2020. Deep learning from a statistical perspective. *Stat* 9: e294. https://doi.org/10.1002/sta4.294.
32. Schmidhuber, Jürgen. 2015. Deep learning in neural networks: An overview. *Neural Networks* 61: 85–117. https://doi.org/10.1016/j.neunet.2014.09.003.
33. Siegismund, Daniel, Vasily Tolkachev, Stephan Heyse, Beate Sick, Oliver Duerr, and Stephan Steigele. 2018. Developing deep learning applications for life science and pharma industry. *Drug Research* 68: 305–310. https://doi.org/10.1055/s-0043-124761.
34. Mijwil, Maad M., Dhamyaa Salim Mutar, Enas Sh. Mahmood, Murat Gök, Süleyman Uzun, and Ruchi Doshi. 2022. Deep learning applications and their worth: A short review. *Asian Journal of Applied Sciences* 10. https://doi.org/10.24203/ajas.v10i5.7078.
35. Gao, Yiping, Xinyu Li, and Liang Gao. 2020. Discriminative stacked autoencoder for feature representation and classification. *Science China Information Sciences* 63: 120111. https://doi.org/10.1007/s11432-019-2722-3.
36. Navlakha, Saket. 2017. Learning the structural vocabulary of a network. *Neural Computation* 29: 287–312. https://doi.org/10.1162/NECO_a_00924.
37. Mao, Xudong, and Qing Li. 2021. *Generative Adversarial Networks for Image Generation*. Singapore: Springer Singapore. https://doi.org/10.1007/978-981-33-6048-8.
38. Eni Lima Nasrin, et al., 2023. From data to decisions leveraging machine learning in supply-chain management. *Tuijin Jishu/Journal of Propulsion Technology* 44: 4218–4225. https://doi.org/10.52783/tjjpt.v44.i4.1644.
39. Bousqaoui, Halima, Said Achchab, and Kawtar Tikito. 2017. Machine Learning Applications in Supply Chains: An Emphasis on Neural Network Applications. In *2017 3rd International Conference of Cloud Computing Technologies and Applications (CloudTech)*, 1–7. Rabat: IEEE. https://doi.org/10.1109/CloudTech.2017.8284722.
40. Schmidt, Christopher, Wen Li, Sebastian Thiede, Sami Kara, and Christoph Herrmann. 2015. A methodology for customized prediction of energy consumption in manufacturing industries. *International Journal of Precision Engineering and Manufacturing-Green Technology* 2: 163–172. https://doi.org/10.1007/s40684-015-0021-z.

41. Dietmair, Anton, Alexander Verl, and Philipp Eberspaecher. 2011. Model-based energy consumption optimisation in manufacturing system and machine control. *International Journal of Manufacturing Research* 6: 380. https://doi.org/10.1504/IJMR.2011.043238.
42. Purmala, Yulio Agefa. 2021. Implementation of machine learning to increase productivity in the manufacturing industry: A literature review. *Operations Excellence.*
43. Pham, Duc T., and Ashraf A. Afify. "Machine-learning techniques and their applications in manufacturing." *Proceedings of the Institution of Mechanical Engineers, Part B: Journal of Engineering Manufacture* 219, no. 5 (2005): 395–412.
44. Trần, Ngọc Trung, Hùng Trường Triệu, Vũ Tùng Trần, Hữu Hải Ngô, and Quang Khoa Đào. 2021. An overview of the application of machine learning in predictive maintenance. *Petrovietnam Journal* 10: 47–61. https://doi.org/10.47800/PVJ.2021.10-05.
45. Strauss, Patrick, Markus Schmitz, Rene Wostmann, and Jochen Deuse. 2018. Enabling of Predictive Maintenance in the Brownfield through Low-Cost Sensors, an IIoT-Architecture and Machine Learning. In *2018 IEEE International Conference on Big Data (Big Data)*, 1474–1483. Seattle, WA, USA: IEEE. https://doi.org/10.1109/BigData.2018.8622076.
46. Choudhury, Prithwiraj, Ryan T. Allen, and Michael G. Endres. 2021. Machine learning for pattern discovery in management research. *Strategic Management Journal* 42: 30–57. https://doi.org/10.1002/smj.3215.
47. Bakshi, Kapil, and Kiran Bakshi. 2018. Considerations for Artificial Intelligence and Machine Learning: Approaches and Use Cases. In *2018 IEEE Aerospace Conference*, 1–9. Big Sky, MT: IEEE. https://doi.org/10.1109/AERO.2018.8396488.
48. Phan, Thuy Linh Jenny, Ingolf Gehrhardt, David Heik, Fouad Bahrpeyma, and Dirk Reichelt. 2022. A systematic mapping study on machine learning techniques applied for condition monitoring and predictive maintenance in the manufacturing sector. *Logistics* 6: 35. https://doi.org/10.3390/logistics6020035.
49. Nacchia, Milena, Fabio Fruggiero, Alfredo Lambiase, and Ken Bruton. 2021. A systematic mapping of the advancing use of machine learning techniques for predictive maintenance in the manufacturing sector. *Applied Sciences* 11: 2546. https://doi.org/10.3390/app11062546.
50. Escobar, Carlos A., and Ruben Morales-Menendez. 2018. Machine learning techniques for quality control in high conformance manufacturing environment. *Advances in Mechanical Engineering* 10: 168781401875551. https://doi.org/10.1177/1687814018755519.
51. Suzuki, Yusuke, Shinya Iwashita, Toshiki Sato, Hitoshi Yonemichi, Hironori Moki, and Tsuyoshi Moriya. 2018. Machine Learning Approaches for Process Optimization. In *2018 International Symposium on Semiconductor Manufacturing (ISSM)*, 1–4. Tokyo, Japan: IEEE. https://doi.org/10.1109/ISSM.2018.8651142.
52. Weichert, Dorina, Patrick Link, Anke Stoll, Stefan Rüping, Steffen Ihlenfeldt, and Stefan Wrobel. 2019. A review of machine learning for the optimization of production processes. *The International Journal of Advanced Manufacturing Technology* 104: 1889–1902. https://doi.org/10.1007/s00170-019-03988-5.
53. Pang, Guansong, Chunhua Shen, Longbing Cao, and Anton van den Hengel. 2022. Deep learning for anomaly detection: A review. *ACM Computing Surveys* 54: 1–38. https://doi.org/10.1145/3439950.
54. Omar, Salima, Asri Ngadi, and Hamid H. Jebur. 2013. Machine learning techniques for anomaly detection: An overview. *International Journal of Computer Applications* 79: 33–41. https://doi.org/10.5120/13715-1478.
55. Yang, Jing, Shaobo Li, Zheng Wang, Hao Dong, Jun Wang, and Shihao Tang. 2020. Using deep learning to detect defects in manufacturing: A comprehensive survey and current challenges. *Materials* 13: 5755. https://doi.org/10.3390/ma13245755.

56. Voulodimos, Athanasios, Nikolaos Doulamis, Anastasios Doulamis, and Eftychios Protopapadakis. 2018. Deep learning for computer vision: A brief review. *Computational Intelligence and Neuroscience* 2018: 1–13. https://doi.org/10.1155/2018/7068349.
57. Guo, Yanming, Yu Liu, Ard Oerlemans, Songyang Lao, Song Wu, and Michael S. Lew. "Deep learning for visual understanding: A review." *Neurocomputing* 187 (2016): 27–48.
58. Sadati, Najibesadat, Ratna Babu Chinnam, and Milad Zafar Nezhad. 2018. Observational data-driven modeling and optimization of manufacturing processes. *Expert Systems with Applications* 93: 456–464. https://doi.org/10.1016/j.eswa.2017.10.028.
59. Zimmerling, Clemens, Patrick Schindler, Julian Seuffert, and Luise Kärger. 2021. Deep Neural Networks as Surrogate Models for Time-Efficient Manufacturing Process Optimisation. In *ESAFORM 2021*. https://doi.org/10.25518/esaform21.3882.
60. Luckow, Andre, Ken Kennedy, Marcin Ziolkowski, et al. 2018. Artificial Intelligence and Deep Learning Applications for Automotive Manufacturing. In *2018 IEEE International Conference on Big Data (Big Data)*, 3144–3152. Seattle, WA, USA: IEEE. https://doi.org/10.1109/BigData.2018.8622357.
61. Dallo, Khan Ali Marwani. 2023. Natural language processing for business analytics. *Advances in Engineering Innovation* 3: 37–40. https://doi.org/10.54254/2977-3903/3/2023038.
62. Chowdhury, Gobinda G. 2003. Natural language processing. *Annual Review of Information Science and Technology* 37: 51–89. https://doi.org/10.1002/aris.1440370103.
63. Kang, Yue, Zhao Cai, Chee-Wee Tan, Qian Huang, and Hefu Liu. 2020. Natural language processing (NLP) in management research: A literature review. *Journal of Management Analytics* 7: 139–172. https://doi.org/10.1080/23270012.2020.1756939.
64. Mihalca, Vlad Ovidiu, Flaviu Birouaş, Florin Avram, and Arnold Nilgesz. 1970. A review regarding deep learning technology in mobile robots. *Recent Innovations in Mechatronics* 5. https://doi.org/10.17667/riim.2018.1/8.
65. Regenwetter, Lyle, Amin Heyrani Nobari, and Faez Ahmed. "Deep generative models in engineering design: A review." *Journal of Mechanical Design* 144, no. 7 (2022): 071704.
66. Lee, Wo Jae, Haiyue Wu, Huitaek Yun, Hanjun Kim, Martin B. G. Jun, and John W. Sutherland. 2019. Predictive maintenance of machine tool systems using artificial intelligence techniques applied to machine condition data. *Procedia CIRP* 80: 506–511. https://doi.org/10.1016/j.procir.2018.12.019.
67. *Traini, Emiliano, Giulia Bruno, Gianluca D'Antonio, and Franco Lombardi.* 2019. Machine learning framework for predictive maintenance in milling. *IFAC-PapersOnLine* 52: 177–182. https://doi.org/10.1016/j.ifacol.2019.11.172.
68. Zhang, Yun, and Xiaojie Xu. 2021. Machine learning cutting force, surface roughness, and tool life in high speed turning processes. *Manufacturing Letters* 29: 84–89. https://doi.org/10.1016/j.mfglet.2021.07.005.
69. Denkena, Berend, Marc-André Dittrich, and Florian Uhlich. 2016. Self-optimizing cutting process using learning process models. *Procedia Technology* 26: 221–226. https://doi.org/10.1016/j.protcy.2016.08.030.
70. Preez, Anli Du, and Gert Adriaan Oosthuizen. 2019. Machine learning in cutting processes as enabler for smart sustainable manufacturing. *Procedia Manufacturing* 33: 810–817. https://doi.org/10.1016/j.promfg.2019.04.102.
71. Luo, Weichao, Tianliang Hu, Yingxin Ye, Chengrui Zhang, and Yongli Wei. 2020. A hybrid predictive maintenance approach for CNC machine tool driven by Digital Twin. *Robotics and Computer-Integrated Manufacturing* 65: 101974. https://doi.org/10.1016/j.rcim.2020.101974.
72. Nasir, Vahid, and Farrokh Sassani. 2021. A review on deep learning in machining and tool monitoring: Methods, opportunities, and challenges. *The International Journal*

of Advanced Manufacturing Technology 115: 2683–2709. https://doi.org/10.1007/s00170-021-07325-7.
73. Aggogeri, Francesco, Nicola Pellegrini, and Franco Luis Tagliani. 2021. Recent advances on machine learning applications in machining processes. *Applied Sciences* 11: 8764. https://doi.org/10.3390/app11188764.
74. Shojaeipour, Shahed. 2010. A novel method for automated tool path optimisation for CNC machining operations. *Solid State Phenomena* 166–167: 357–362. https://doi.org/10.4028/www.scientific.net/SSP.166-167.357.
75. Meyghani, Bahman, and Mokhtar Awang. 2019. A novel tool path strategy for modelling complicated perpendicular curved movements. *Key Engineering Materials* 796: 164–174. https://doi.org/10.4028/www.scientific.net/KEM.796.164.
76. Garcia Plaza, E., P. J. Nunez Lopez, and E. M. Beamud Gonzalez. "Multi-sensor data fusion for real-time surface quality control in automated machining systems." *Sensors* 18, no. 12 (2018): 4381. https://doi.org/10.3390/s18124381.
77. Liang, Y. C., Sheng Wang, W. D. Li, and Xin Lu. "Data-driven anomaly diagnosis for machining processes." *Engineering* 5, no. 4 (2019): 646–652. https://doi.org/10.1016/j.eng.2019.03.012.
78. Janošovský, Ján, Miroslav Variny, and Otto Mierka. 2020. Cost-saving opportunities in the energy management of papermaking processes. *Chemical Engineering & Technology* 43: 1194–1204. https://doi.org/10.1002/ceat.201900483.
79. Vasiloni, Anton Mircea, and Mircea Viorel Dragoi. 2014. Smart adaptive CNC machining – State of the art. *Applied Mechanics and Materials* 657: 859–863. https://doi.org/10.4028/www.scientific.net/AMM.657.859.
80. Srinivasa Prasad, B., D. Siva Prasad, A. Sandeep, and G. Veeraiah. "Condition monitoring of CNC machining using adaptive control." *International Journal of Automation and Computing* 10 (2013): 202–209. https://doi.org/10.1007/s11633-013-0713-1.
81. Carvalho, Thyago P., Fabrízzio A. A. M. N. Soares, Roberto Vita, Roberto Da P. Francisco, João P. Basto, and Symone G. S. Alcalá. 2019. A systematic literature review of machine learning methods applied to predictive maintenance. *Computers & Industrial Engineering* 137: 106024. https://doi.org/10.1016/j.cie.2019.106024.
82. Usuga Cadavid, Juan Pablo, Samir Lamouri, Bernard Grabot, Robert Pellerin, and Arnaud Fortin. 2020. Machine learning applied in production planning and control: A state-of-the-art in the era of industry 4.0. *Journal of Intelligent Manufacturing* 31: 1531–1558. https://doi.org/10.1007/s10845-019-01531-7.
83. Yamazaki, Taku. 2016. Development of A hybrid multi-tasking machine tool: Integration of additive manufacturing technology with CNC machining. *Procedia CIRP* 42: 81–86. https://doi.org/10.1016/j.procir.2016.02.193.
84. Bains, Preetkanwal Singh, Sarabjeet Singh Sidhu, and H. S. Payal. 2016. Fabrication and machining of metal matrix composites: A review. *Materials and Manufacturing Processes* 31: 553–573. https://doi.org/10.1080/10426914.2015.1025976.
85. Lazoglu, I., C. Manav, and Y. Murtezaoglu. 2009. Tool path optimization for free form surface machining. *CIRP Annals* 58: 101–104. https://doi.org/10.1016/j.cirp.2009.03.054.
86. Castelino, Kenneth, Roshan D'Souza, and Paul K. Wright. 2003. Toolpath optimization for minimizing airtime during machining. *Journal of Manufacturing Systems* 22.
87. Wah, Pang King, Katta G. Murty, Ajay Joneja, and Leung Chi Chiu. 2002. Tool path optimization in layered manufacturing. *IIE Transactions* 34: 335–347. https://doi.org/10.1080/07408170208928874.
88. Duro, João A., Julian A. Padget, Chris R. Bowen, H. Alicia Kim, and Aydin Nassehi. 2016. Multi-sensor data fusion framework for CNC machining monitoring. *Mechanical Systems and Signal Processing* 66–67: 505–520. https://doi.org/10.1016/j.ymssp.2015.04.019.

89. Kant, Girish, and Kuldip Singh Sangwan. 2014. Prediction and optimization of machining parameters for minimizing power consumption and surface roughness in machining. *Journal of Cleaner Production* 83: 151–164. https://doi.org/10.1016/j.jclepro.2014.07.073.
90. Jia, Shun, Qinghe Yuan, Wei Cai, Meiyan Li, and Zhaojun Li. 2018. Energy modeling method of machine-operator system for sustainable machining. *Energy Conversion and Management* 172: 265–276. https://doi.org/10.1016/j.enconman.2018.07.030.
91. Durakbasa, Numan M., and M. Güneş Gençyılmaz, ed. 2020. *Proceedings of the International Symposium for Production Research 2019*. Lecture Notes in Mechanical Engineering. Cham: Springer International Publishing. https://doi.org/10.1007/978-3-030-31343-2.
92. Ciaburro, Giuseppe. 2022. Machine fault detection methods based on machine learning algorithms: A review. *Mathematical Biosciences and Engineering* 19: 11453–11490. https://doi.org/10.3934/mbe.2022534.
93. Liu, Yumin, Haofei Zhou, Fugee Tsung, and Shuai Zhang. 2019. Real-time quality monitoring and diagnosis for manufacturing process profiles based on deep belief networks. *Computers & Industrial Engineering* 136: 494–503. https://doi.org/10.1016/j.cie.2019.07.042.
94. Ran, Xinyi. 2022. Research on the optimization of defect detection based on convolutional neural network architecture. Edited by A. Luqman, Q. Zhang, and W. Liu. In *SHS Web of Conferences* 144, 02018. https://doi.org/10.1051/shsconf/202214402018.
95. Essa, Ehab, M. Shamim Hossain, A. S. Tolba, Hazem M. Raafat, Samir Elmogy, and Ghulam Muahmmad. 2020. Toward cognitive support for automated defect detection. *Neural Computing and Applications* 32: 4325–4333. https://doi.org/10.1007/s00521-018-03969-x.
96. Cerezci, Feyza, Serap Kazan, Muhammed Ali Oz, Cemil Oz, Tugrul Tasci, Selman Hizal, and Caglayan Altay. 2020. Online metallic surface defect detection using deep learning. *Emerging Materials Research* 9: 1266–1273. https://doi.org/10.1680/jemmr.20.00197.
97. Rani, Sangeeta, Khushboo Tripathi, and Ajay Kumar. 2023. Machine learning aided malware detection for secure and smart manufacturing: A comprehensive analysis of the state of the art. *International Journal on Interactive Design and Manufacturing (IJIDeM)*. https://doi.org/10.1007/s12008-023-01578-0.
98. Shahrabadi, Somayeh, Yusbel Castilla, Miguel Guevara, Luís G. Magalhães, Dibet Gonzalez, and Telmo Adão. 2022. Defect detection in the textile industry using image-based machine learning methods: A brief review. *Journal of Physics: Conference Series* 2224: 012010. https://doi.org/10.1088/1742-6596/2224/1/012010.
99. San-Payo, Gonçalo, João Carlos Ferreira, Pedro Santos, and Ana Lúcia Martins. 2020. Machine learning for quality control system. *Journal of Ambient Intelligence and Humanized Computing* 11: 4491–4500. https://doi.org/10.1007/s12652-019-01640-4.
100. Cao, Yankai, Huaizhe Yu, Nicholas L. Abbott, and Victor M. Zavala. 2018. Machine learning algorithms for liquid crystal-based sensors. *ACS Sensors* 3: 2237–2245. https://doi.org/10.1021/acssensors.8b00100.
101. Toh, Gyungmin, and Junhong Park. 2020. Review of vibration-based structural health monitoring using deep learning. *Applied Sciences* 10: 1680. https://doi.org/10.3390/app10051680.
102. Weimer, Daniel, Bernd Scholz-Reiter, and Moshe Shpitalni. 2016. Design of deep convolutional neural network architectures for automated feature extraction in industrial inspection. *CIRP Annals* 65: 417–420. https://doi.org/10.1016/j.cirp.2016.04.072.
103. Wang, Tian, Yang Chen, Meina Qiao, and Hichem Snoussi. 2018. A fast and robust convolutional neural network-based defect detection model in product quality control. *The*

International Journal of Advanced Manufacturing Technology 94: 3465–3471. https://doi.org/10.1007/s00170-017-0882-0.

104. Jiang, Yu, Wei Wang, and Chunhui Zhao. 2019. A Machine Vision-Based Realtime Anomaly Detection Method for Industrial Products Using Deep Learning. In *2019 Chinese Automation Congress (CAC)*, 4842–4847. Hangzhou, China: IEEE. https://doi.org/10.1109/CAC48633.2019.8997079.
105. Swain, Debabala, Prasant Kumar Pattnaik, and Pradeep K. Gupta, ed. 2020. *Machine Learning and Information Processing: Proceedings of ICMLIP 2019*, vol. 1101. Advances in Intelligent Systems and Computing. Singapore: Springer Singapore. https://doi.org/10.1007/978-981-15-1884-3.
106. Shahbazi, Zeinab, and Yung-Cheol Byun. 2021. Smart manufacturing real-time analysis based on blockchain and machine learning approaches. *Applied Sciences* 11: 3535. https://doi.org/10.3390/app11083535.
107. Alexopoulos, Kosmas, Nikolaos Nikolakis, and George Chryssolouris. 2020. Digital twin-driven supervised machine learning for the development of artificial intelligence applications in manufacturing. *International Journal of Computer Integrated Manufacturing* 33: 429–439. https://doi.org/10.1080/0951192X.2020.1747642.
108. Kumar, Ajay, Parveen Kumar, Naveen Sharma, and Srivastava Ashish Kumar. 2024. *3D Printing Technologies: Digital Manufacturing, Artificial Intelligence, Industry 4.0.* De Gruyter. https://doi.org/10.1515/9783111215112.
109. Kumar, Ajay, Sangeeta Rani, Sarita Rathee, and Surbhi Bhatia, ed. 2023. *Security and Risk Analysis for Intelligent Cloud Computing: Methods, Applications, and Preventions.* Boca Raton: CRC Press. https://doi.org/10.1201/9781003329947.
110. Cinar, Goktug T., Jeffrey Thompson, and Soundar Srinivasan. 2015. Cost-Sensitive Optimization of Automated Inspection. In *2015 IEEE International Conference on Big Data (Big Data)*, 1211–1219. Santa Clara, CA, USA: IEEE. https://doi.org/10.1109/BigData.2015.7363875.
111. Trinks, Sebastian. "Real Time Quality Assurance and Defect Detection in Industry 4.0." Proceedings http://ceur-ws. org ISSN 1613 (2021): 0073.
112. Bandara, Prasanna, Thilan Bandara, Tharaka Ranatunga, Vibodha Vimarshana, Sulochana Sooriyaarachchi, and Chathura De Silva. 2018. Automated Fabric Defect Detection. In *2018 18th International Conference on Advances in ICT for Emerging Regions (ICTer)*, 119–125. Colombo, Sri Lanka: IEEE. https://doi.org/10.1109/ICTER.2018.8615491.
113. Nangia, Shikhil, Sandhya Makkar, and Rohail Hassan. 2020. IoT based predictive maintenance in manufacturing sector. *SSRN Electronic Journal.*
114. Him, Leong Chee, Yu Yong Poh, and Lee Wah Pheng. 2019. IoT-Based Predictive Maintenance for Smart Manufacturing Systems. In *2019 Asia-Pacific Signal and Information Processing Association Annual Summit and Conference (APSIPA ASC)*, 1942–1944. Lanzhou, China: IEEE. https://doi.org/10.1109/APSIPAASC47483.2019.9023106.
115. Nezami, Farnaz Ghazi, and Mehmet Bayram Yildirim. 2013. A sustainability approach for selecting maintenance strategy. *International Journal of Sustainable Engineering* 6: 332–343. https://doi.org/10.1080/19397038.2013.765928.
116. Chang, Qing, Jun Ni, Pulak Bandyopadhyay, Stephan Biller, and Guoxian Xiao. 2007. Maintenance opportunity planning system. *Journal of Manufacturing Science and Engineering* 129: 661–668. https://doi.org/10.1115/1.2716713.
117. Nikolakis, Nikolaos, Apostolos Papavasileiou, Konstantinos Dimoulas, Kiriakos Bourmpouchakis, and Sotirios Makris. 2018. On a versatile scheduling concept of maintenance activities for increased availability of production resources. *Procedia CIRP* 78: 172–177. https://doi.org/10.1016/j.procir.2018.09.065.

118. Perdpunya, Tawatchai, Anuntasup Sukpradit, Orasa Limpaporn, and Maleerat Maliyaem. 2021. CPS-Based Automation Model Prediction for Inspection. In *2021 13th International Conference on Knowledge and Smart Technology (KST)*, 108–112. Bangsaen, Chonburi, Thailand: IEEE. https://doi.org/10.1109/KST51265.2021.9415846.
119. Ajay, Hari Singh, Parveen, and Bandar AlMangour, ed. 2023. *Handbook of Smart Manufacturing: Forecasting the Future of Industry 4.0.* Boca Raton: CRC Press. https://doi.org/10.1201/9781003333760.
120. Sang, Go Muan, Lai Xu, and Paul de Vrieze. 2020. Predictive Maintenance in Industry 4.0.
121. Correia, Miguel Clemente. 2018. Machine Learning for Predictive Maintenance of Industrial Equipment. Thesis, Lisboa, Portugal: Universidade de Lisboa.
122. Cline, Brad, Radu Stefan Niculescu, Duane Huffman, and Bob Deckel. 2017. Predictive Maintenance Applications for Machine Learning. In *2017 Annual Reliability and Maintainability Symposium (RAMS)*, 1–7. Orlando, FL, USA: IEEE. https://doi.org/10.1109/RAM.2017.7889679.
123. Rabatel, Julien, Sandra Bringay, and Pascal Poncelet. 2011. Anomaly detection in monitoring sensor data for preventive maintenance. *Expert Systems with Applications* 38: 7003–7015. https://doi.org/10.1016/j.eswa.2010.12.014.
124. Nassif, Ali Bou, Manar Abu Talib, Qassim Nasir, and Fatima Mohamad Dakalbab. 2021. Machine learning for anomaly detection: A systematic review. *IEEE Access* 9: 78658–78700. https://doi.org/10.1109/ACCESS.2021.3083060.
125. Sodemann, Angela A., Matthew P. Ross, and Brett J. Borghetti. 2012. A review of anomaly detection in automated surveillance. *IEEE Transactions on Systems, Man, and Cybernetics, Part C (Applications and Reviews)* 42: 1257–1272. https://doi.org/10.1109/TSMCC.2012.2215319.
126. Ahmadzadeh, Farzaneh, and Jan Lundberg. 2014. Remaining useful life estimation: Review. *International Journal of System Assurance Engineering and Management* 5: 461–474. https://doi.org/10.1007/s13198-013-0195-0.
127. Lei, Yaguo, Naipeng Li, Szymon Gontarz, Jing Lin, Stanislaw Radkowski, and Jacek Dybala. 2016. A model-based method for remaining useful life prediction of machinery. *IEEE Transactions on Reliability* 65: 1314–1326. https://doi.org/10.1109/TR.2016.2570568.
128. Doganay, Kivanc, and Markus Bohlin. "Maintenance plan optimization for a train fleet." *WIT Transactions on The Built Environment* 114, no. 12 (2010): 349–358. https://doi.org/10.2495/CR100331.
129. Butte, Sujata, A. R. Prashanth, and Sainath Patil. 2018. Machine Learning Based Predictive Maintenance Strategy: A Super Learning Approach with Deep Neural Networks. In *2018 IEEE Workshop on Microelectronics and Electron Devices (WMED)*, 1–5. Boise, ID: IEEE. https://doi.org/10.1109/WMED.2018.8360836.
130. Tegen, Agnes, Paul Davidsson, and Jan A. Persson. "An Interactive Learning Scenario for Real-time Environmental State Estimation Based on Heterogeneous and Dynamic Sensor Systems." In European Conference on Machine Learning and Principles and Practice of Knowledge Discovery in Databases, Dublin, Ireland (10-14/9, 2018). ECML, 2018.
131. Park, James J., Doo-Soon Park, Young-Sik Jeong, and Yi Pan, ed. 2020. *Advances in Computer Science and Ubiquitous Computing: CSA-CUTE 2018*, vol. 536. Lecture Notes in Electrical Engineering. Singapore: Springer Singapore. https://doi.org/10.1007/978-981-13-9341-9.
132. Khan, Nida, Sayeed Ghani, and Sajjad Haider. 2018. Real-time analysis of a sensor's data for automated decision making in an IoT-based smart home. *Sensors* 18: 1711. https://doi.org/10.3390/s18061711.

133. Othman, Uqba, and Erfu Yang. 2023. Human–Robot collaborations in smart manufacturing environments: Review and outlook. *Sensors* 23: 5663. https://doi.org/10.3390/s23125663.
134. Dobra, Zoltan, and Krishna S. Dhir. 2020. Technology jump in the industry: Human–Robot cooperation in production. *Industrial Robot: The International Journal of Robotics Research and Application* 47: 757–775. https://doi.org/10.1108/IR-02-2020-0039.
135. Nikolakis, Nikolaos, Vasilis Maratos, and Sotiris Makris. 2019. A cyber physical system (CPS) approach for safe human-robot collaboration in a shared workplace. *Robotics and Computer-Integrated Manufacturing* 56: 233–243. https://doi.org/10.1016/j.rcim.2018.10.003.
136. Hamzah, Muzaffar, Md. Monirul Islam, Shahriar Hassan, Md. Nasim Akhtar, Most. Jannatul Ferdous, Muhammed Basheer Jasser, and Ali Wagdy Mohamed. 2023. Distributed control of cyber physical system on various domains: A critical review. *Systems* 11: 208. https://doi.org/10.3390/systems11040208.
137. Alzahrani, Ahmad, Mohammed Alshehri, Rayed AlGhamdi, and Sunil Kumar Sharma. 2023. Improved wireless medical cyber-physical system (IWMCPS) based on machine learning. *Healthcare* 11: 384. https://doi.org/10.3390/healthcare11030384.
138. Soori, Mohsen, Roza Dastres, Behrooz Arezoo, and Fooad Karimi Ghaleh Jough. 2024. Intelligent robotic systems in Industry 4.0: A review. *Journal of Advanced Manufacturing Science and Technology*. https://doi.org/10.51393/j.jamst.2024007.
139. Kotsiopoulos, Thanasis, Panagiotis Sarigiannidis, Dimosthenis Ioannidis, and Dimitrios Tzovaras. 2021. Machine Learning and Deep Learning in smart manufacturing: The smart grid paradigm. *Computer Science Review* 40: 100341. https://doi.org/10.1016/j.cosrev.2020.100341.
140. Mazzei, Daniele, and Reshawn Ramjattan. 2022. Machine learning for Industry 4.0: A systematic review using deep learning-based topic modelling. *Sensors* 22: 8641. https://doi.org/10.3390/s22228641.
141. Mahatme, C., J. Giri, F. Mohammad, M. S. Ali, T. Sathish, N. Sunheriya, and R. Chadge. 2024. Experimental and numerical investigation of PLA based different lattice topologies and unit cell configurations for additive manufacturing. *The International Journal of Advanced Manufacturing Technology/International Journal, Advanced Manufacturing Technology*. https://doi.org/10.1007/s00170-024-13882-4.
142. Giri, J., T. Sathish, T. Sheikh, N. Sunehriya, P. Giri, R. Chadge, C. Mahatme, and A. Parthiban. 2024. Automatic liver segmentation using U-Net deep learning architecture for additive manufacturing. *Interactions* 245(1). https://doi.org/10.1007/s10751-024-01927-9.
143. Giri, J., N. Sunheriya, T. Sathish, Y. Kadu, R. Chadge, P. Giri, A. Parthiban, and C. Mahatme. 2024. Optimization of process parameters to improve mechanical properties of fused deposition method using Taguchi method. *Interactions* 245(1). https://doi.org/10.1007/s10751-024-01925-x.
144. Tufail, M. S., J. Giri, E. Makki, T. Sathish, R. Chadge, and N. Sunheriya. 2024. Machinability of different cutting tool materials for electric discharge machining: A review and future prospects. *AIP Advances* 14(4). https://doi.org/10.1063/5.0201614.
145. Natarajan, M., T. Pasupuleti, J. Giri, H. A. Al-Lohedan, L. N. Katta, F. Mohammad, N. Sunheriya, R. Chadge, C. Mahatme, P. Giri, S. Mallik, and T. Sathish. 2024. Optimization of wire spark erosion machining of Grade 9 titanium alloy (Grade 9) using a hybrid learning algorithm. *AIP Advances* 14(1). https://doi.org/10.1063/5.0177658.

13 Revaluation of 3D Printing
Technology and Process Parameters

Sharda Kumari, Shalini Anand, Ashish Kumar Srivastava, and Vishwa Ratan Mishra

13.1 INTRODUCTION

The term "Additive Manufacturing" (AM) was first mentioned by a committee named ASTM F42. AM is different from subtracting manufacturing in that it uses the deposition of one layer over another to create an object. At the same time, the latter is concerned with removing extra stock from a raw material. AM can convert a suitable CAD file into a 3D-printed object. The machines can convert the 3D CAD design into process parameters for guiding the movement of the nozzle for the deposition of material in layers. Dr. Mldeo Kodama pioneered the AM process by first utilizing layer-by-layer photopolymerization in 1980. AM materials are categorized based on input materials, such as solid, liquid, and powder. In addition, AM techniques are classified as per the pattern of depositing and solidifying layers.

1. Point deposition:
   a. Laser sintering and melting
   b. Electron beam melting (EBM)
2. Line deposition:
   a. Fused deposition modeling (FDM)/fused filament fabrication (FFF)
   b. Material jetting
3. Area deposition:
   a. Stereolithography (SLA)
   b. Digital light processing (DLP)
   c. Sheet lamination

Net-shaped products like medical equipment and structural aircraft components are produced using the AM method [1–4]. It evolved AM as a disruptive technology for manufacturing; the designers now have a wider scope of re-designing their products. AM cannot be taken as mass manufacturing technology, as it is particularly best suited to the production of products involving complex geometries. AM has significantly contributed to the successful implementation of Industry 4.0 as remote

DOI: 10.1201/9781032725086-15

designing and printing have modulated the scope for wider ranges and complexity of products. The advantages of AM have derived evolution of the design for manufacture and assembly (DfMA) which emphasizes on productivity of assembly rather than any individual part. A better optimized design bolsters the faster, safer, and cost-effective design of an assembly.

The aforesaid advantages forced AM to develop its own set of regulations, and thus, it significantly diverged from DfMA regulations. The set of regulations is known as design for AM (DfMA) which refers to design for AM [5, 6]. The new set of regulations facilitated pathways for engineers to design and print parts by utilizing less-weight and high-strength materials [7].

Jiang et al.'s [8] production of necessary and replacement components, together with product customization, allowed for the study of AM and necessitated improved structural performance and reasonable design techniques. Three categories comprise the DfMA process: process design, system design, and parts design. These categories ought to be viewed as a series of steps, with objectives established at each stage [9].

The evaluation of 3D printing from 1984 to 2024 is shown in Figure 13.1. Klebe invented bioprinting in 1988 with the assistance of Hewlett-Packard (HP), which produced an inkjet printer that was used to print cells using the cyto-writing technique [10]. This resulted in the first 3D printer becoming available in 1988, the addition of an SLA type printer [11], the addition of Scott Crump's FDM technique in 1989 [12], and the successful printing of an organ—a bladder—for transplantation in 1999 [13]. The first extrusion-based 3D printer (3D Bioplotter) was developed in 2000, and its use in kidney production via micro-extrusion was made possible in 2002 [14]. In 2003, Wilson and Boland [15] created the first inkjet bioprinter by modifying a standard HP inkjet.

Replicating rapid prototyping, or RepRap, was founded in 2005 to develop a less costly 3D printer that could produce most of its parts [16]. The first prosthetic limb was printed in 2008 thanks to the development of selective laser sintering (SLS) [17, 18]. The first 3D blood vessel was also printed in 2009 [19]. Research on the first liver tissue 3D printing in 2014 was spurred by the development of the first 3D-printed jaw in 2012 [20].

Nonetheless, the first desktop 3D printer was formally introduced [21, 22]. In 2015, Beheshtizadeh et al. created a 3D-printed scaffold for periodontal therapy [23]. The first commercial 3D human skin was developed in 2028 [24], and it was improved and reprinted in 2019. In 2019, Noor et al. created a viable, decremented-dimensioned heart with success [25]. The personalized application of 3DP in the medical sector was made possible in 2020 [26].

Specifically, in the fields of periodontics and dentistry, 3DP has yielded numerous favorable outcomes. 3D printing technologies have been used to investigate the healing of gingival lesions (such as gingival recessions, gingivectomies, or smile restoration) as well as the regeneration of periodontal tissues (such as cement, periodontal ligaments, and alveolar bone) [27, 28]. Many developments in the fields of biological evaluation, materials, process parameters, 3D printing, and recycling polymers will be made between 2020 and 2024. By the end of 2030, the 3D printing market will probably surpass a valuation of USD 91.74 billion.

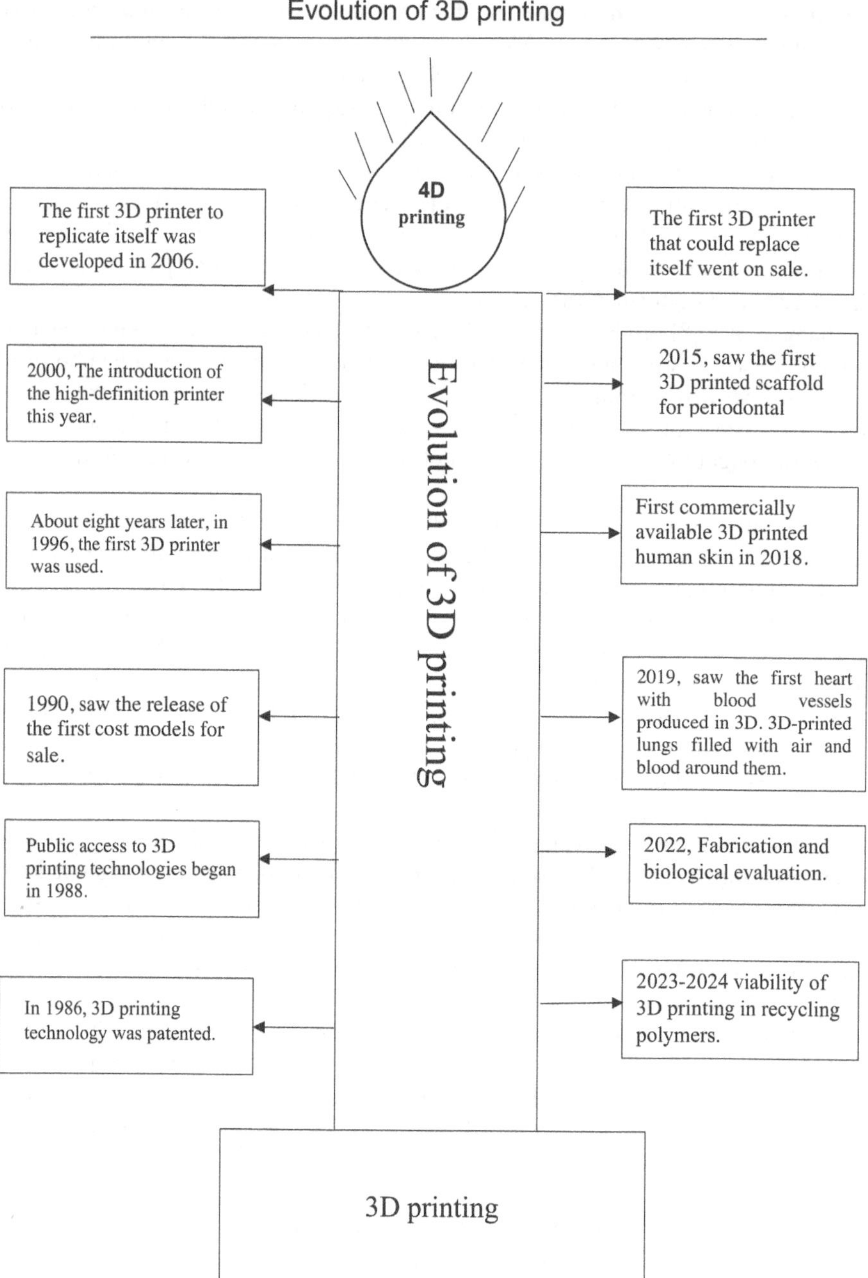

FIGURE 13.1 Evolution of 3D printing.

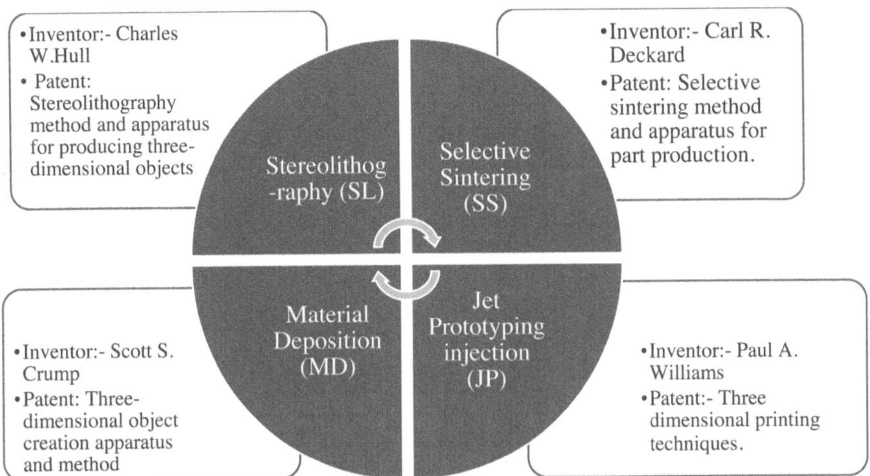

FIGURE 13.2 Invention in additive manufacturing.

13.1.1 Invention in 3D Printing

Material extrusion (ME), powder bed fusion (PBF), binder jetting (BJ), material jetting (MJ), SL, vat photopolymerization (VP), and direct energy deposition (DED) are the seven categories into which the American Society of Standards divided AM. There is no disagreement over the best machine or technology because each has its own unique set of uses and benefits. These days, 3D printing techniques are being applied to create a wide range of items, not only prototypes. AM process have various categories such as: Material extrusion (ME), powder bed fusion (PBF), SL, vat photopolymerization (VP), binder jetting (BJ), material jetting (MJ), directed energy deposition (DED) (Figure 13.2).

13.2 MATERIALS AND METHODS

Many different materials have been created and successfully employed in the 3D printing process. The mechanical and other properties of these materials vary. A new class of materials for usage in 3D printing is also constantly being developed by researchers. Much of the recent rise in 3D printing's use can be attributed to the technology's simplicity and versatility, which make it ideal for use in a variety of applications. An overview of the development, methods, components, and categorization of 3D printing technology is given in this section. Table 13.1 shows different processes, materials, and their properties.

13.2.1 Process Parameters

Researchers have been optimizing the process parameters of AM to obtain the best suited product at an optimum speed of printing. The optimization of 3D printing parameters has been dominated by experimental research which yielded better

TABLE 13.1
Different Processes, Materials, and Their Properties

S. No.	AM Technique	Advantage	Disadvantage	Materials	Process	Application
1.	Extrusion of material (ME)	• Wide variety of raw materials • Low initial cost • Adaptable • Able to construct fully functional pieces	• More time spent building with less accuracy • Difficulty printing acute angles • Vertical anisotropy	Composites Ceramics Polymers	The filament-shaped material is constantly forced into the nozzle.	Food applications, jigs and fixtures, aerospace parts, prototypes, and supporting parts
2.	Fusion of powder beds (PBF)	• Polymer powder does not require any kind of support structure. • Recyclable for metal and polymer powder • Capacity to print a range of materials with intricate geometric designs • Improved precision and sharpness for the metal powder-printed component	• Limited produced item size • High beginning cost (expensive machines) • Low printing rate • A polymer's rough surface finish	Metals Ceramics Composites and polymers Blended	Thermal energy causes the construction materials to fuse.	Small series types, supporting parts, and prototypes
3.	Binder jetting (BJ)	• Polymer powder does not require any kind of support structure. • Recyclable for metal and polymer powder	• Post-processing is necessary to improve strength and eliminate moisture from printed parts, which have rough surfaces and low strength.	Metals Ceramics Composites Polymers and hybrid materials	As the particles are linked together, layer after layer of the component is created.	Green parts, molds, cores, casting patterns, and prototypes

		• Capacity to print a range of materials with intricate geometric patterns • Improved metal powder resolution and accuracy				
4.	Material jetting (MJ)	• Create parts with excellent resolution and precision. • There is no model material waste; a range of hues and materials are possible	• Waste of non-recyclable support materials • To reduce damage, post-processing is not advised for minor features.	Biochemistry Composites Polymers Ceramics Hybrids	Droplets of construction materials deposited	Casting patterns and prototypes
5.	Lamination of sheets (SL)	• More color and material options • Faster printing speed • Minimal internal tension, warping, and strain. • No formal support is needed.	• The removal of the support structure can be challenging. • Stage of the materials • Warpage of the lamination could result from the laser's heat.	Metals, Papers Ceramics and Polymers Blended	Raw material sheets are fused.	Fast tooling patterns and fewer intricate pieces

(Continued)

TABLE 13.1 *(Continued)*

Different Processes, Materials, and Their Properties

S. No.	AM Technique	Advantage	Disadvantage	Materials	Process	Application
6.	Photopolymerization in vats (VP)	• Generate parts with excellent precision, resolution, and surface quality • Cover a broad spectrum of materials • Accelerate the printing process. • Low-specific energy imaging	• A support system is necessary. • It's necessary to post-process and post-cure	Ceramics, Polymers and resins containing photo-active monomers	Light cured the liquid polymers in a vat.	High heat applications, snap fittings, presentation models, and rapid tooling patterns
7.	Deposition of direct energy (DED)	• High material utilization; • Suitable for large metal components; • Effective for add-on and repair characteristics; • Extensive flexibility over grain configurations	• Many materials are not appropriate for this printing method; complex geometry is not achievable; printed objects have low precision and surface polish.	• Hybrid • Metals	During deposition, thermal energy is focused on melting materials.	Aerospace parts, medical implants, high heat applications, and quick tooling

product behaviors by optimizing the parameters. Table 13.2 depicts several parameters and their effect on the properties of the product. As a conclusion of the parameters shown in the table, it can be said that the lesser the layer thickness, the greater the surface finish and mechanical properties of the 3D-printed product. Along with the layer thickness, it is found that the speed of printing also plays a pivotal role in product quality. Slower printing speed ensures no discontinuation of the layer being deposited and better outcome as strength. Slower speed also ensures a better surface finish. However, slower speed led to a slow rate of production; thus, it needs to be optimized based on layer thickness, solidification rate, and property of the material being fused for deposition. One of the very important parameters is the temperature of the depositing nozzle; a higher nozzle temperature ensures proper melting of material and strong adherence between two layers, thus increasing the layer-to-layer bonding. The presented literature review reveals that a flat orientation and 0° raster angle are best suited for better product quality. A fishbone diagram is shown to understand the parameters and their effect on quality (Figure 13.3). Table 13.2 highlights results of process parameters optimization of different AM methods.

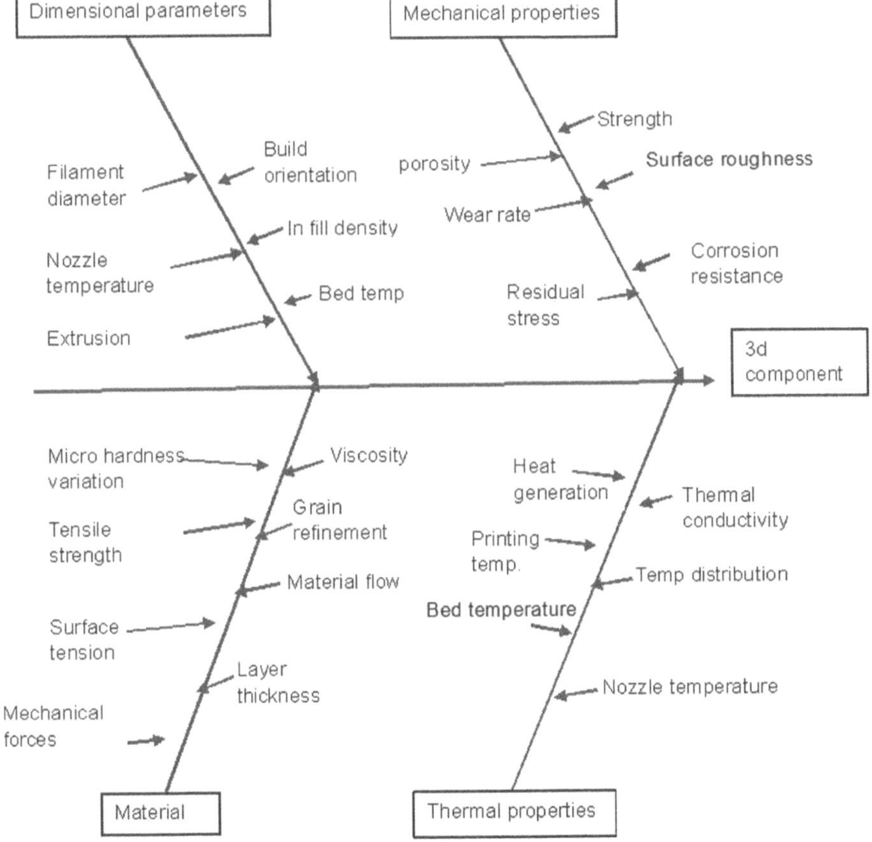

FIGURE 13.3 Fish bone diagram.

TABLE 13.2
Process Parameters Optimization of Various Methods

S. No.	AM Techniques	Parameters	Properties	Result	Reff.
1.	Fused filament fabrication	Printing orientation: • Vertical • Horizontal Layer thickness: 0 mm, 1 mm, 2 mm, 3 mm	• Electron transport kinetics • Conductive material resistivity	• The maximum conductivity was obtained with electrodes printed in a vertical position with a layer thickness of 0.1 mm. • More resistive electrodes were produced by printing in a horizontal orientation and with increased layer thickness.	Abdalla et al [29]
2.	Fused deposition modeling (FDM)	Nozzle temp. (in c): 400, 410, 420, 430, 440 Platform temp (in c): 240, 250, 260, 270, 280 Printing speed (mm/sec): 10, 15, 20, 25 Layla thickness (mm): 0.1, 0.15, 0.2, 0.25	• Tensile strength; • Flexural strength; • Impact strength	• Nozzle temperature: it was found that increasing the temperature increased the strength in both tensile and flexural tests. • Platform temperature: all strengths rise with temperature. • Thickness of layer: as layer thickness increases, strength decreases.	Wang et al [30]

3.	Material extrusion	Extruding rate (mm^3/s) 2703 29.2 32.3 35.4 39.7 44.1 50.2 52.4 54.7 60.5	Printing speed (mm/s) 18.3 20.6 22.4 25.7 37.9	Characteristics of food paste	• 22.4 mm/s and 39.7 mm^3/s are the most appropriate printing speeds and extruding rates; 22.1 mm/s and 20 mm^3/s are the least appropriate.	Liu and Ciftci [31]
4.	3D cementitious material printing	Printing speed mm/s 50 52 54 56 58 60		Three-dimensional printed cementitious material deformation	• As printing speed decreases, the printed filament's distortion rises.	Liu et al [32]

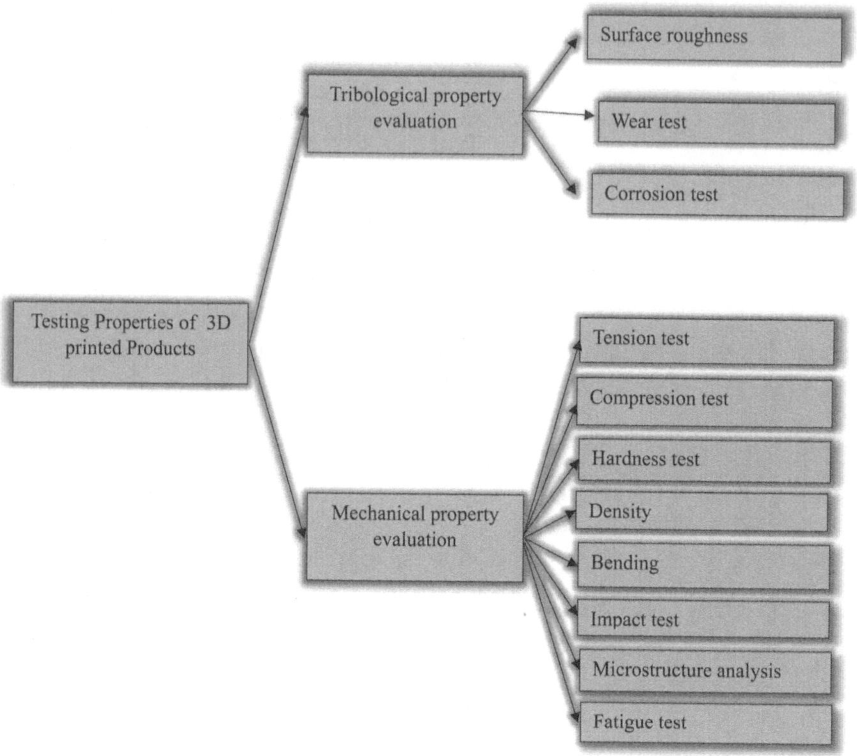

FIGURE 13.4 Mechanical and tribological property evaluation of AM part.

13.2.2 Properties of Additive Manufactured Components

13.2.2.1 Mechanical and Tribological Properties

Figure 13.4 lists the several tribological and mechanical tests that were performed on AM components [33, 34].

13.3 RANGE OF 3D PRINTING

A large part of the recent rise in 3D printing's use can be attributed to its simplicity, which makes it a versatile technology with applications in many different sectors. Initially, 3D printing was costly to use. 3D printer models and materials are expensive. The cost of 3D printing has significantly decreased as a result of technological breakthroughs [35]. Figure 13.5 shows a variety of 3D printing uses. Applications have been the subject of debate in [36, 37]. Figure 13.6 shows percentage of utilization of various AM methods. Figure 13.7 shows percentage applications of various 3D printing techniques.

Laser power fusion is regarded as the most versatile metal AM process, which has been proven to manufacture near-net shape up to 99.9% relative density, with geometrically complex and high performance metallic parts at reduced time. Various unusual behavior YS, UTS, and EL for LPF is described in Figure 13.8 [38–53].

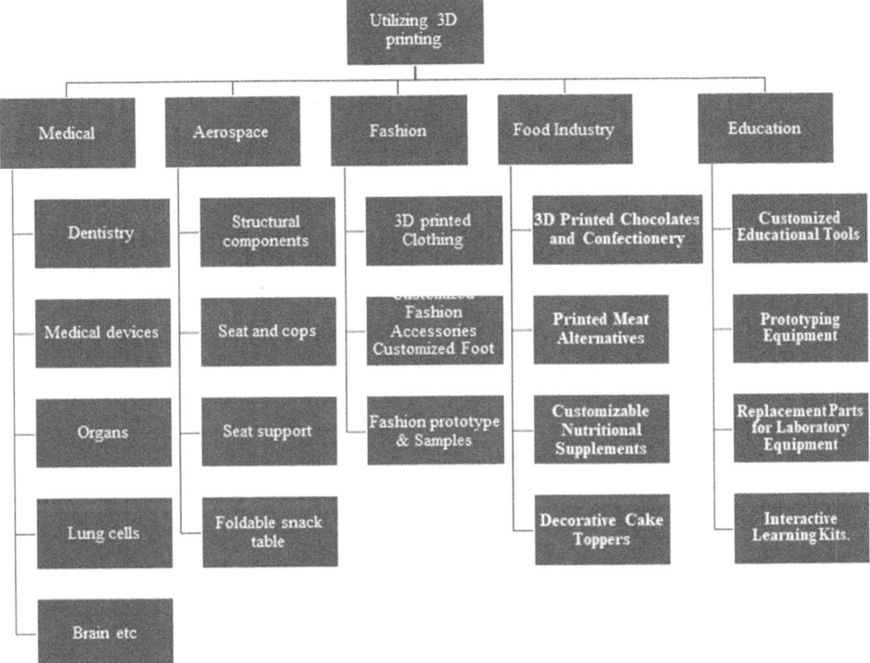

FIGURE 13.5 Scope of 3D-printed products.

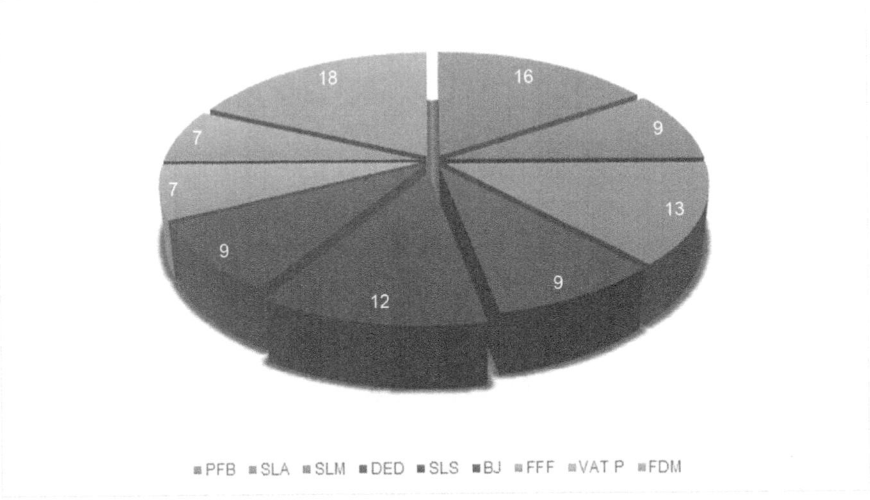

FIGURE 13.6 % Age of contribution.

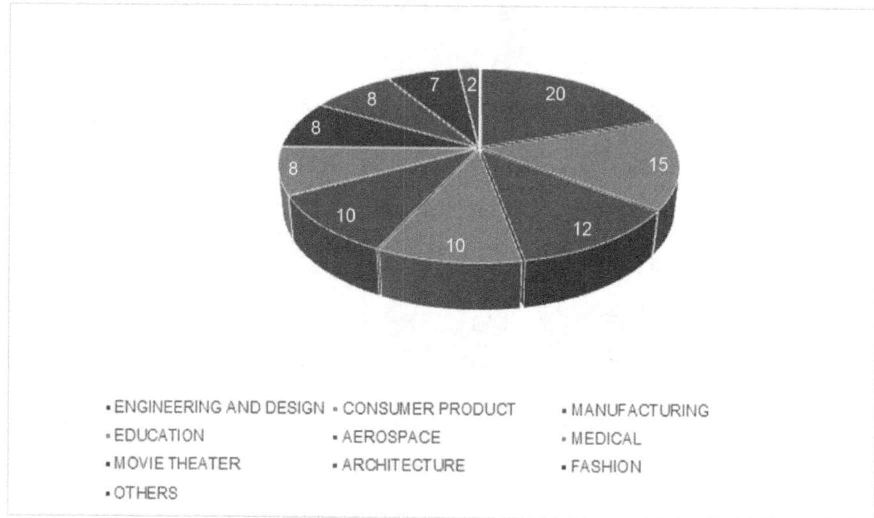

FIGURE 13.7 % Contribution of AM processes in manufacturing in various industries.

13.3.1 Future Perspective

Because 3D-printed parts have found their applications from home to the oil and gas industry, it can be assumed that the next era of manufacturing components (specifically complex designed) will be dominated by AM. This is due to ample benefits such as the ability to print complex-shaped jobs at higher speeds with almost no

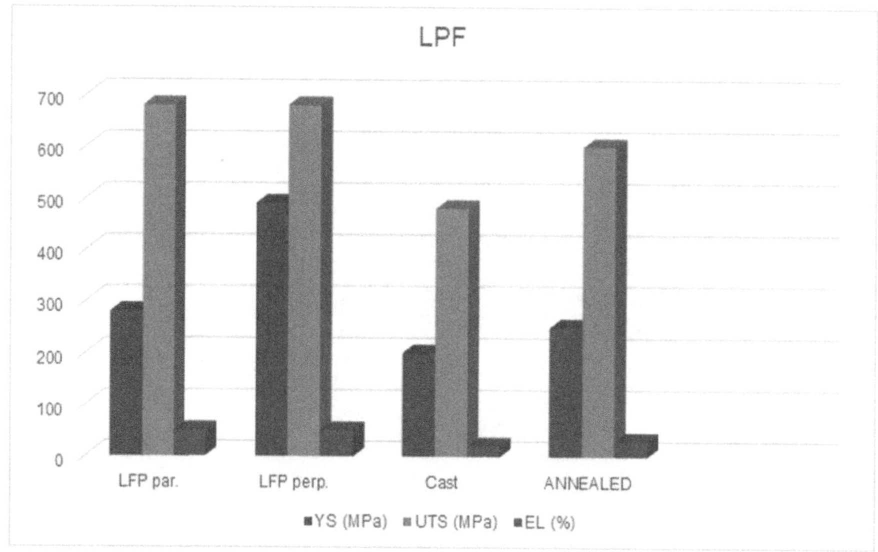

FIGURE 13.8 Unusual behavior of US, UTS, and EL for LPF.

production lead time. Though 3D printing is enough to make any shape and size of product (depending upon machine size), AM enthusiasts are looking ahead to 4D printing, which uses shape memory polymers for a high level of customization. Several drawbacks such as limitation of materials, directional properties, and surface finish have limited the scope of AM to rapid prototyping rather than end-use products. The disadvantages of the slow speed of printing can be overcome by providing adequate facilities for metal deposition and subsequent solidification. The use of multiple nozzles has been suggested by researchers as well. The following disadvantages of the AM process need to be addressed to produce objects that can be utilized in structural components without compromising strength:

1. Void formation leading to improperly distributed porosity.
2. Layer-to-layer adhesion depends upon fusion rate and solidification temperature.
3. Addition of metals.
4. Anisotropic behaviors of 3D-printed products.

The advent of new-age AM techniques like Continuous Liquid Interface Production (CLIP), Multi Jet Fusion (MJF), and High-Speed Sintering (HSS) has taken the utilization of AM products to the next level.

13.4 CONCLUSIONS

In recent years, 3D printing technology has become a flexible and effective advanced production method. Numerous industries, especially the manufacturing industry, have made extensive use of this technology. The three main advantages of 3D printing are reduced waste, increased design freedom, and the capacity to build intricate structures. This review includes a range of methods used in the process of 3D printing. We talked about the different materials that are used, their characteristics, their uses in different industries, and the factors that can be optimized to produce superior mechanical qualities. The findings show that the mechanical properties of FDM are relatively low when compared to the processes of selective laser melting (SLM) and SLS. Nevertheless, this method has great processing speed, low cost, and ease of usage. The three most important factors in 3D printing are build orientation, raster angle, and layer thickness. It will be beneficial to optimize the processing parameters for maximum utilization in a variety of industries, including electronics, medical, aerospace, and others.

REFERENCES

1. Sehrawat, S., Kumar, A., Prabhakar, M., & Nindra, J. (2022). The expanding domains of 3D printing are about the specialty of orthodontics. Materials Today: Proceedings, 50, 1611–1618. https://doi.org/10.1016/j.matpr.2021.09.124
2. Batista, R. C., Agarwal, A., Gurung, A., Kumar, A., Altarazi, F., Dogra, N., HM, V., Chiniwar, D. S., & Agrawal, A. (2024). Topological and lattice-based AM optimization for improving the structural efficiency of robotic arms. Frontiers in Mechanical Engineering, 10, 1422539.

3. Srivastava, A. K., Tiwari, S., Pachauri, P., Gupta, N., Sunil, B., & Kumar, A. (2024). Bonding strength and microstructural features of Al5083-AZ31B alloys laminated sheet through friction stir additive manufacturing. Journal of Adhesion Science and Technology, 38(4), 583–596.
4. Aggoune, S., Hamadi, F., Abid, C. et al. (2024). Instabilities in the formation of single tracks during selective laser melting process. International Journal on Interactive Design and Manufacturing. https://doi.org/10.1007/s12008-024-01887-y
5. Kumar, P., Hussain, S. S., Kumar, A., Srivastava, A. K., Hussain, M., & Singh, P. K. (2024). 10 Finite element method investigation on delamination of 3D printed hybrid composites during the drilling operation. In 3D Printing Technologies: Digital Manufacturing, Artificial Intelligence, Industry 4.0, 223. De Gruyter.
6. Goyal, G., Kumar, A., & Sharma, D. (2024). 12 Recent applications of rapid prototyping with 3D printing: A review. In 3D Printing Technologies: Digital Manufacturing, Artificial Intelligence, Industry 4.0, 245. De Gruyter
7. Goyal, G., Kumar, A., & Gupta, A. (2024). 16 Recent developments in 3D printing: A critical analysis and deep dive into innovative real-world applications. In 3D Printing Technologies: Digital Manufacturing, Artificial Intelligence, Industry 4.0, 335. De Gruyter.
8. Jiang, R., Kleer, R., & Piller, F. T. (Apr 2017). Predicting the future of additive manufacturing: A Delphi study on economic and societal implications of 3D printing for 2030. Technological Forecasting and Social Change, 117, 84–97. https://doi.org/10.1016/j.techfore.2017.01.006
9. Tuvayanond, W., & Prasittisopin, L. (Feb 2023). Design for manufacture and assembly of digital fabrication and additive manufacturing in construction: A review. Buildings, 13(2), 429. https://doi.org/10.3390/buildings13020429
10. Klebe, RJ. (1988). Cytoscribing: A method for microfossil research, and three-dimensional synthetic tissues. Experimental Cell Research, 179(2), 362–373. https://doi.org/10.1016/0014-4827(88)90275-3
11. Kumar, A., Kumar, P., Sharma, N., & Srivastava, A. K. (Eds.). (2024). 3D Printing Technologies: Digital Manufacturing, Artificial Intelligence, Industry 4.0. https://doi.org/10.1515/9783111215112
12. Apparatus and Method for Creating Three-Dimensional Objects (A System and a Method for Building Three-Dimensional Objects in a Layer-by-Layer Manner via Fused Deposition Modeling). U.S. Patent 5121329, 9 June 1989.
13. Odde, D. J., & Renn, M. J. (1999). Laser-guided direct writing for applications in biotechnology. Trends in Biotechnology, 17(10), 385–389. https://doi.org/10.1016/S0167-7799(99)01355-4
14. Landers, R., Hübner, U., Schmelzeisen, R., & Mülhaupt, R. (2002). Rapid prototyping of scaffolds derived from thermoreversible hydrogels and tailored for applications in tissue engineering. Blomaterials, 23(23), 4437–4447. https://doi.org/10.1016/S0142-9612(02)00139-4
15. Wilson, W. C. Jr, & Boland, T. (2003). Cell and organ printing 1: Protein and cell printers. The Anatomical Record Part A: Discoveries in Molecular, Cellular, and Evolutionary Biology, 272(2), 491–496. https://doi.org/10.1002/ar.a.10057
16. Jones, R., Haufe, P., Sells, E., Iravani, P., Olliver, V., Palmer, C., & Bowyer, A. (2011). RepRap-the replicating rapid prototyper. Robotica, 29(1), 177–191. https://doi.org/10.1017/S026357471000069X
17. Charoo, N. A., Barakh Ali, S. F., Mohamed, E. M., Kuttolamadom, M. A., Ozkan, T., Khan, M. A., & Rahman, Z. (2020). Selective laser sintering 3D printing–an overview of the technology and pharmaceutical applications. Drug Development and Industrial Pharmacy, 46(6), 869–877.
18. Sangani, K. (2013). Make it to fake it [3D printing technologies]. Engineering & Technology, 8(9), 38–41. https://doi.org/10.1049/et.2013.0901

19. Gu, Z., Fu, J., Lin, H., & He, Y. (2020). Development of 3D bioprinting: From printing methods to [biomedical applications. Asian Journal of Pharmaceutical Sciences, 15(5), 529–557. https://doi.org/10.1016/j.ajps.2019.11.003
20. Nickels, L. (2012). World's first patient-specific jaw implant. Metal Powder Report, 67(2), 12–14. https://doi.org/10.1016/S0026-0657(12)70128-5
21. Duan, B. (2017). State-of-the-art review of 3D bioprinting for cardiovascular tissue engineering. Annals of Biomedical Engineering, 45, 195–209.
22. Izdebska-Podsiadły, J. (2022). History of the development of additive polymer technologies. In Polymers for 3D Printing (pp. 3–11). William Andrew Publishing. https://doi.org/10.1016/B978-0-12-818311-3.00007-0
23. Beheshtizadeh, N., Lotfibakhshaiesh, N., Pazhouhnia, Z., Hoseinpour, M., & Nafari, M. (2020). A review of 3D bio-printing for bone and skin tissue engineering: A commercial approach. Journal of Materials Science, 55(9), 3729–3749.
24. Noor, N., Shapira, A., Edri, R., Gal, I., Wertheim, L., & Dvir, T. (2019). 3D printing of personalized thick and perfusable cardiac patches and hearts. Advanced Science, 6(11), 1900344. https://doi.org/10.1002/advs.201900344
25. Vaz, V. M., & Kumar, L. (2021). 3D printing is a promising tool in personalized medicine. AAPS Pharmscitech, 22, 1–20. https://doi.org/10.1208/s12249-020-01905-8
26. Qiao, S., Sheng, Q., Li, Z., Wu, D., Zhu, Y., Lai, H., & Gu, Y. (2020). 3D-printed Ti6Al4V scaffolds coated with freeze-dried platelet-rich plasma as a bioactive interface for enhancing osseointegration in osteoporosis. Materials & Design, 194, 108825. https://doi.org/10.1016/j.matdes.2020.108825
27. Reddy, M. S., Shetty, S. R., Shetty, R. M., Vannala, V., Sk, S., & Salem, R. S. (2020). Focus on periodontal engineering by 3D printing technology systematic review. Journal of Oral Research, 9(6), 522–531.
28. Kim, J. D., Choi, J. S., Kim, B. S., Choi, Y. C., & Cho, Y. W. (2010). Piezoelectric inkjet printing of polymers: Stem cell patterning on polymer substrates. Polymer, 51(10), 2147–2154. https://doi.org/10.1016/j.polymer.2010.03.038
29. Abdalla, A., Hamzah, H. H., Keattch, O., Covill, D., & Patel, B. A. (2020). "Augmentation of conductive pathways in carbon black/PLA 3D-printed electrodes achieved through varying printing parameters. Electrochimica Acta, 354, 136618. https://doi.org/10.1016/j.electacta.2020.136618
30. Wang, P., Zou, B., Ding, S., Li, L., & Huang, C. (2020). Effects of FDM-3D printing parameters on mechanical properties and microstructure of CF/PEEK and GF/PEEK. Chinese Journal of Aeronautics. https://doi.org/10.1016/j.cja.2020.05.040
31. Liu, L., & Ciftci, O. N. (2021). "Effects of high oil compositions and printing parameters on food paste properties and printability in a 3D printing food processing model. Journal of Food Engineering, 288, 110135. https://doi.org/10.1016/j.jfoodeng.2020.110135
32. Liu, Z., Li, M., Weng, Y., Qian, Y., Wong, Teck Neng, & Tan, M. J. (2020). "Modelling and parameter optimization for filament deformation in 3D cementitious material printing using support vector machine. Composites Part B: Engineering, 193, 108018. https://doi.org/10.1016/j.compositesb.2020.108018
33. Röttger, A., Boes, J., Theisen, W., Thiele, M., Esen, C., Edelmann, A., & Hellmann, R. (2020). Microstructure and mechanical properties of 316L austenitic stainless steel processed by different SLM devices. The International Journal of Advanced Manufacturing Technology, 108, 769–783. https://doi.org/10.1007/s00170-020-05371-1
34. Kumar, A., Kumar, P., & Liu, Y. Industry 4.0 Driven Manufacturing Technologies. Springer Nature. https://doi.org/10.1007/978-3-031-68271-1
35. MacDonald, E., & Wicker, R. (2016). Multiprocess 3D printing for increasing component functionality. Science, 353(6307), aaf2093.
36. Lee, J.-Y., Jia, A., & Kai Chua, C. (2017). Fundamentals and applications of 3D printing for novel materials. Applied Materials Today, 7, 120–133.

37. Mpofu, T. P., Mawere, C., & Mukosera, M. (2014). The impact and application of 3D printing technology.
38. Bi, G., Gasser, A., Wissenbach, K., Drenker, A., & Poprawe, R. (2006). Characterization of the process control for the direct laser metallic powder deposition. Surface and Coatings Technology, 201, 2676–2683.
39. Kobryn, P. A., Moore, E. H., & Semiatin, S. L. (2000). Effect of laser power and traverse speed on microstructure, porosity, and build height in laser-deposited Ti-6Al-4V. Scripta Materialia, 43, 299–305. https://doi.org/10.1016/S1359-6462(00)00408-5
40. Yu, J., Rombouts, M., & Maes, G. (2013). Cracking behavior and mechanical properties of austenitic stainless steel parts produced by laser metal deposition. Materials & Design, 45, 228–235.
41. Zheng, B., Zhou, Y., Smugeresky, J. E., Schoenung, J. M., & Lavernia, E. J. (2008). Thermal behavior and microstructure evolution during laser deposition with laser-engineered net shaping: Part II. Experimental investigation and discussion. Metallurgical and Materials Transactions A, 39, 2237–2245.
42. Zhang, K., Wang, S., Liu, W., & Shang, X. (2014). Characterization of stainless steel parts by laser metal deposition shaping. Materials & Design, 55, 104–119.
43. Ravi, G. A., Hao, X., Wain, N., Wu, X., & Attallah, M. M. (2013). Direct laser fabrication of three di-dimensional components using SC420 stainless steel. Materials & Design, 47, 731–736.
44. Smugeresky, J., Keicher, D., Romero, J., Griffith, M., & Harwell, L., Laser engineered net shaping (LENSTM) process: optimization of surface finish and microstructural properties, Proc. of the World Congress on Powder Metallurgy and Particulate Materials, Princeton, NJ, 1997.
45. El Cheikh, H., Courant, B., Branch, S., Huang, X., Hascoët, J.-Y., & Guillén, R. (2012). Direct laser fabrication process with coaxial powder projection of 316L steel. Geometrical characteristics and microstructure characterization of wall structures. Optics and Lasers in Engineering, 50, 1779–1784.
46. Milewski, J. O., Lewis, G. K., Thoma, D. J., Keel, G. I., Nemec, R. B., & Reinert, R. A. (1998). Directed light fabrication of a solid metal hemisphere using 5-axis powder deposition. Journal of Materials Processing Technology, 75, 165–172.
47. Ma, M., Wang, Z., Wang, D., & Zeng, X. (2013). Control of shape and performance for direct laser fabrication of precision large-scale metal parts with 316 l stainless steel. Optics & Laser Technology, 45, 209–216.
48. Mahatme, C., Giri, J., Mohammad, F., Ali, M. S., Sathish, T., Sunheriya, N., & Chadge, R. (2024). Experimental and numerical investigation of PLA based different lattice topologies and unit cell configurations for additive manufacturing. The International Journal, Advanced Manufacturing Technology. https://doi.org/10.1007/s00170-024-13882-4
49. Giri, J., Sathish, T., Sheikh, T., Sunehriya, N., Giri, P., Chadge, R., Mahatme, C., & Parthiban, A. (2024). Automatic liver segmentation using u-net deep learning architecture for additive manufacturing. Interactions, 245(1). https://doi.org/10.1007/s10751-024-01927-9
50. Giri, J., Sunheriya, N., Sathish, T., Kadu, Y., Chadge, R., Giri, P., Parthiban, A., & Mahatme, C. (2024). Optimization of process parameters to improve mechanical properties of fused deposition method using Taguchi method. Interactions, 245(1). https://doi.org/10.1007/s10751-024-01925-x
51. Tufail, M. S., Giri, J., Makki, E., Sathish, T., Chadge, R., & Sunheriya, N. (2024). Machinability of different cutting tool materials for electric discharge machining: A review and future prospects. AIP Advances, 14(4). https://doi.org/10.1063/5.0201614

52. Natarajan, M., Pasupuleti, T., Giri, J., Al-Lohedan, H. A., Katta, L. N., Mohammad, F., Sunheriya, N., Chadge, R., Mahatme, C., Giri, P., Mallik, S., & Sathish, T. (2024). Optimization of wire spark erosion machining of grade 9 titanium alloy (grade 9) using a hybrid learning algorithm. AIP Advances, 14(1). https://doi.org/10.1063/5.0177658
53. Srivastava, A.K., Kumar, A., Kumar, P. et al. (2023). Research progress in metal additive manufacturing: Challenges and opportunities. International Journal on Interactive Design and Manufacturing. https://doi.org/10.1007/s12008-023-01661-6

14 Overview of Burnishing Process as a Surface Improvement Technique

Nitin Jalindar Varpe and Ravindra Tajane

14.1 INTRODUCTION

Modern machinery's components must be built not only with great geometrical and dimensional correctness but also with a good surface polish, to confirm dependable performance and a long service life. Surface finish has a vital impact on functional properties like fatigue strength, wear and corrosion resistance and power loss owing to friction. Unlikely, traditional machining procedures such as milling, turning and even traditional grinding cannot achieve this demanding necessity. As a result, burnishing is used to obtain and enhance the functional qualities of a component or equipment.

Burnishing is a surface modification procedure that improves both the surface qualities and surface finish of the component. This effective method is frequently applied in the biomedical, automotive and aerospace industries to enhance the performance and dependability of the element. Burnishing is simple yet effective technique of improving surface qualities that may be performed on current equipment. Because of its great efficiency in comparison to conventional techniques like super finishing, honing and grinding, it also reduces production costs. By contrast, a burnished surface has an extended fatigue life and is incredibly wear resistant.

The burnishing technique involves rubbing hardened steel rollers or balls on workpiece's surface and providing a feed motion to it. While burnishing, significant compressive residual stress is created in the workpiece's surface, increasing the wear resistance and fatigue strength of surface layer. This technique smoothens and hardens the surface, resulting in a finish that will remain longer than one which has not been burnished.

14.2 CLASSIFICATION OF BURNISHING PROCESSES

14.2.1 Classification Based on Deformation Element

In practice, the burnishing tools have been designed to have deformation element in the form of roller or ball. Thus it's been termed roller or ball burnishing.

14.2.1.1 Ball Burnishing

The deformation agent in this method is a hard ball. Balls are made of tungsten carbide, cemented carbide, bearing steel etc. When a ball is used to distort the upper

DOI: 10.1201/9781032725086-16

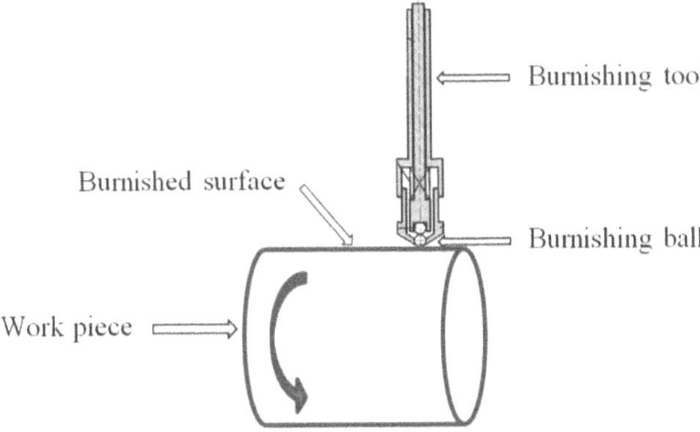

FIGURE 14.1 Scheme of ball burnishing.

layers, it produces a higher specific pressure, greater fatigue strength; hardness and deep work hardening layer because of point and rolling friction in ball and workpiece. Figure 14.1 shows a ball burnishing process.

14.2.1.2 Roller Burnishing

The deformation agent in this method is a hard roller. Roller burnishing enables individuals to avoid secondary processes, save time and cost remarkably while also enhancing product quality. Roller burnishing is a technique for creating a wear-resistant surface that is precisely formed and highly polished. Steel rollers that have been hardened and polished are pressed on a soft work item (Figure 14.2).

14.2.2 Classification Based on Motion of the Tool

Burnishing process is classified in four groups based on motion of burnishing tool and oscillation frequency; ordinary burnishing, impact burnishing, vibrating burnishing and ultrasonic burnishing.

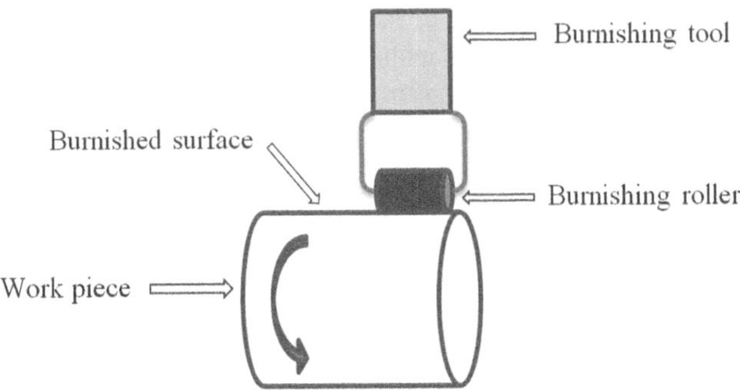

FIGURE 14.2 Schematic illustration of roller burnishing mechanism.

14.2.2.1 Normal or Ordinary Burnishing

In this, a hard roller or ball is engaged against workpiece which rotate between the two center of the lathe and feed is given parallel to the workpiece axis. The configuration of process is similar to the single point cutting process. Normal burnishing can also be used for flat surface.

14.2.2.2 Impact Burnishing

It is also known as bearingizing. This process combines rolling and penning action. In this process, harden rollers are rotating around the periphery of the cams. As these cams rotate, a rise and fall of the cam deliver 20,000 blows/min [1]. The action produces smooth surface finish improves roundness and straightness and increase surface hardness due to cold working.

14.2.2.3 Vibrating Burnishing

In the case of vibrating burnishing process in addition to the movement in direction of feed, burnishing tool is vibrated in a direction parallel to workpiece axis. The configuration is similar to the detention capacity of the surface is improved by vibrating burnishing.

14.2.2.4 Ultrasonic Burnishing

It is similar to the vibrating burnishing, in which surface is burnished with deformation tool having additional oscillatory motion along with feed motion in the feed direction. In this process a tool is oscillated with ultrasonic frequency.

14.3 SUMMARY OF PROCESS VARIABLES AND MEASURED RESPONSES

The purpose of the review is to give some insight of the subject and the theoretical framework. The literature used in the review consists of scientific journals.

14.3.1 Workpiece Materials, Their Properties and Application

Burnishing is a finishing process applied to different types of metal and their alloy for industrial application. Researcher has conducted experiment on various material used to form different components in industrial application. Table 14.1 and Figure 14.3 give detail information of material tested by researcher and their application in industries. Bar chart gives at the glance the picture of material tested.

14.3.2 Analysis of Process Parameters

Burnishing process has various process parameters for which the literature review was taken which are tabulated in Table 14.2 and summarized in Figure 14.4.

Burnishing force, speed, feed and no. of passes are commonly used parameter for study of its effect on burnishing process. Burnishing force (74%), speed (68%), feed (56%) and no. of passes (35%) are the most important parameter under review.

TABLE 14.1
Materials, Their Different Alloys and Studies Appeared in Literature

Work Material	Specification	Ref. No.	Material Application
Al, Cu, Brass	Aluminum and brass	[2]	Aluminum is used in a variety of industries, including disc brakes, truck wheels, aviation fittings, hinge pins, connectors, brake and hydraulic pistons [6, 10]. Aluminum alloys are extremely suitable for parts and systems that require a high strength-to-weight ratio and are undoubtedly the most well-known materials utilized widely in aviation, space exploration and other applications. Although pure aluminum cannot be successfully heat treated, some aluminum alloys may be heat treated to enhance its mechanical qualities up to a specific amount.
	Al 7075 T6	[5]	
	Aluminum alloy 2014	[6]	
	7178 aluminum alloy	[10]	
	AA 7075 and AISI 5140	[11]	
	Aluminum alloy AA 7075	[15]	
	6061-T6 aluminum alloy	[26]	
	2024-T3 aluminum alloy	[17]	
	Aluminum 6061	[30]	
	Aluminum 6063	[74]	
	Aluminum	[35]	
	AA2024	[79]	
	Brass; Al+Cu	[3]	
	Al, Brass and Copper	[31]	
	Al5083-AZ31B	[77]	
	Aluminum-based alloy	[37]	
Plastic	Polyoxymethylene (POM thermoplastic) and polyurethane (PUR thermosets)	[12]	Polymers can be treated for surface improvement by using ball burnishing process and produces good surface quality.
Steel	AISI 1045	[1]	Machine tool spindles, gears, bolts and nuts, and shafts in naval and automobile manufacturing are examples of AISI 1045 steels popular uses [1]. Molds of plastic injection products are manufactured by PDS5 tool steel. Key advantage of this steel is, its mold can be used without heat treatment [4]. Pre-hardened steel (AISI P20) is used for the manufacturing of tools (molds and dies) because of its toughness and high wear resistance due to 5% Cr [7]. This steel is selected because it's most commonly used in industry and produces very good surface after burnishing [9]. Used in manufacturing of molds and dies where wear resistance is desirable [16]. Austenitic X6CrNiMoTi17-12-2 alloy is the suitable material for application in corrosive media. It finds many applications in automotive industry and control valves used to control fluids, gases etc. Steel requires heat treatment before its use [20]. An aerospace material [24]. This steel is carefully chosen by considering its industrial importance and it
	PDS5 tool steel (AISI P20)	[4]	
	Pre-hardened steel (AISI P20); H13 tempered tool material	[7]	
	Hardened steel (64HRC)	[8]	
	mild steel	[9]	
	41Cr4 (AISI 5140) steel	[13]	
	PDS5 plastic injection mold steel	[14]	
	AISI 304	[16]	
	41Cr4 (AISI 5140) steel	[18,19]	
	Austenitic X6CrNiMoTi17-12-2 alloy steel	[20]	
	17-4 PH stainless steel	[24]	
	Steel-37 of hardness 220 Hv	[27]	
	En-8	[28]	
	MP35N, annealed and hardened AISI 4140	[29]	
	AISI 4340 (EN24) steel	[32]	
	Stainless steel X5CrNiMo17-12-2 for centrifugal pumps.	[34]	
	O_1 steel	[36]	
	AISI 52100	[76]	

(Continued)

TABLE 14.1 *(Continued)*
Materials, Their Different Alloys and Studies Appeared in Literature

Work Material	Specification	Ref. No.	Material Application
	ASTM 2017; (carbon steel, ASTM 1055 (1055)	[37]	degrades after burnishing through surface destruction [27]. Hardening of MP35N material is occurs due to phase transformation to hexagonal-close packed structure from face-centered cubic by plastic deformation. AISI 4140 is selected as testing material because of its wide applications [29]. Stainless steel X5CrNiMo17-12-2 for centrifugal pumps. Shafts of pumps used for marine power plants under sea [34].
Titanium	Titanium alloy (Ti-6Al-4V) alloy of aerospace Grade 5	[21]	Titanium alloy finds wide applications in aerospace engineering because of its high strength, low density, resistance to corrosion and fracture. It is commonly used for commercial and industrial applications like oil refinery, propeller blades, biomedical implants, nuclear reactors, missile cases and under water applications. Due to bio compatibility of titanium and its alloys, they are preferred for biomedical implants [21]. As nitinol's mechanical properties are similar to biological tissues, it finds wide applications in biomedical field [23]. Ti-6Al-4V is favorite material for airframe and components of engine [25].
	Nitinol is a nickel–titanium alloy-super elastic Ni50.8Ti49.2 (50.8 at.% Ni–49.2 at.% Ti)	[23]	
	Ti-6Al-4V	[25]	
	Ti-6Al-4V	[33]	

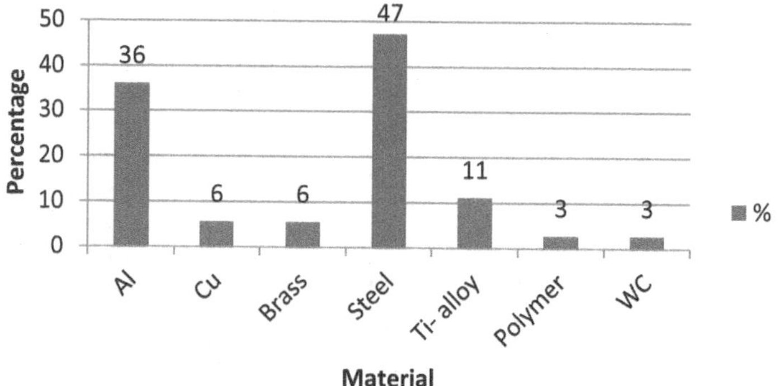

FIGURE 14.3 Literature review on work material.

TABLE 14.2
Analysis of Process Parameters of Burnishing Process

Sr. No.	Process Parameter	Reference Paper No.
1.	Burnishing Force	[1, 2, 4, 5, 7–12, 14–16, 21, 22–25, 27, 30, 32, 35–37]
2.	Burnishing Speed	[1, 4–10, 12, 15, 16, 18, 21, 24, 26–28, 30–34, 37]
3.	Feed Rate	[4–6, 8–10, 12, 13, 15, 16, 18, 21, 24, 27, 28, 31–33, 37]
4.	No. of Tool Passes	[1, 5–7, 9, 10, 15, 21, 26–28, 32]
5.	Ball Diameter	[1, 3, 12, 13, 30, 34]
6.	Depth of Burnishing	[6, 7, 26, 31, 33]
7.	Lubricant Used	[3, 13, 28, 35]
8.	Ball Material	[4]
9.	Initial S/f Hardness	[3]
10.	Initial S/f Roughness	[3]
11.	Direction of Burnishing	[7, 37]

14.3.3 Lubricants and Their Performance

Various lubricants that have been mentioned in the literature, as well as the effects they have on surface quality, are mentioned in this section. At distinct levels of process variables, kerosene produces good surface quality in comparison with SAE-30 oil. SAE-30 oil having graphite 5%–10% by weight [1].

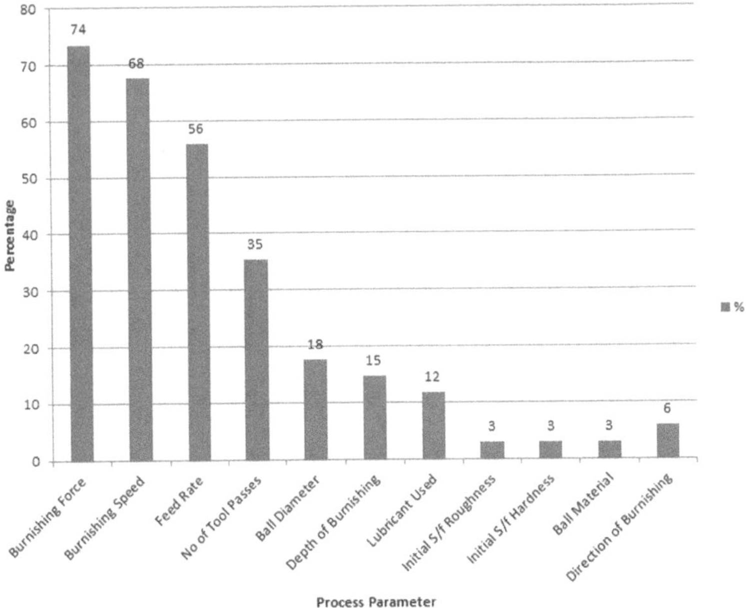

FIGURE 14.4 Literature review on process parameter.

TABLE 14.3
Commercial Lubricants

Sr. No.	Types of Lubrication	Viscosity Gread	Viscosity (mm^2/s at 40°C)
1	Gear oil	85w/140 (SAE)	413
2	Industrial gear oil	320 (ISO)	320
3	Diesel engine oil	S3 50 (SAE)	210
4	Gasolin engine oil	20w/50 (SAE)	157
5	Mixed fleet engine oil	SX 30 (SAE)	96
6	Hydraulic oil	37 (ISO)	37
7	Braek oil	A5 (SAE)	8

To explore lubricants effect on ball burnishing process, commercial lubricants of various types were utilized by A.M. Hassan and A.M. Maqableh (2000) [3], these lubricants being shown in following Table 14.3.

In burnishing process, the machining coolant is used as lubricant, which avoids adhesion in between workpiece and burnishing ball [7]. At the time of burnishing operation, external reservoir is used to supply constant lubrication [12]. Burnishing was performed under cooling medium which having combination of machine oil and kerosene as 15% and 85%, respectively [13] (Figure 14.5).

Lubricant shows significant effect during roller burnishing process. Surface finish increases considerably as burnishing force increases, if lubricant is used during burnishing as compared to dry condition. Aluminum material achieves good surface finish in presence of kerosene as compared to SAE 40 oil [35].

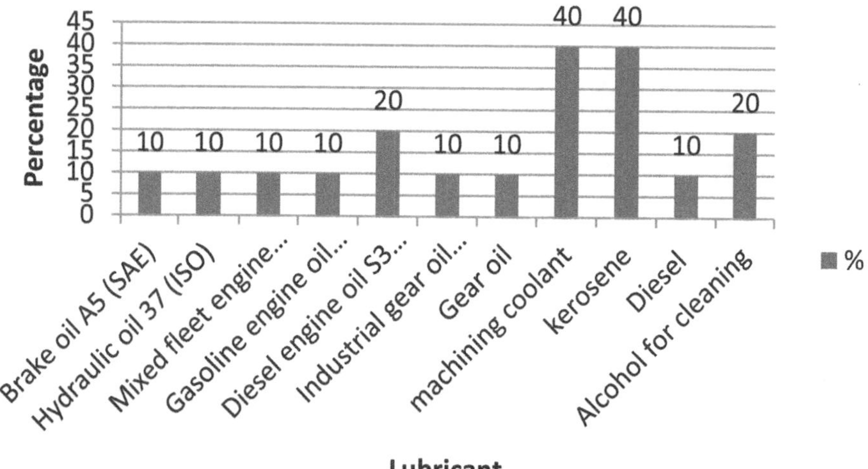

FIGURE 14.5 Literature review on lubricant.

TABLE 14.4
Ball Burnishing Tools

Ref. No.	Tool Used	Tool Material	Machined Surface
[1]	Simple and rigid ball burnishing tool	Steel balls	Aluminum specimen with 25 mm diameter
[4]	Spring loaded ball burnishing tool	Titanium nitride coated ball (TiN), Chromium coated steel ball (CrC), Tungsten carbide (WC)	Plastic injection molding steel PDS5
[5]	Advanced helical spring incorporated ball burnishing tool	13 mm diameter ball (WC)	Aluminum Alloy (Al 7075 T6) 80 x120 mm
[7]	Hydrostatic ball burnishing tool	Ceramic ball	Heat treated and Tempered steel
[10]	Screw jack type ball burnishing tool	Tungsten carbide (WC)	30 mm Al alloy 7178 bars
[11]	Ball burnishing tool with backup balls	Si3N4	AISI 5140-solid bars of Ø35 mm AA 7075 solid bars of Ø30 mm
[15]	Screw jack type ball burnishing tool with ball retaining cap	18 mm diameter ball	AA7075 with Ø30 bar

14.3.4 Burnishing Tools

In burnishing process, a hard roller or ball is engaged against workpiece. It's cold working process where plastic deformation occurs on workpiece.

The present literature review highlights experiments done with various burnishing tool. A study of various burnishing tools, with respect to its material, design and the surface for which it was utilized by different researchers, is presented here. Table 14.4 summarizes ball burnishing tools and Table 14.5 summarizes roller burnishing tools.

TABLE 14.5
Roller Burnishing Tools

Ref. No.	Tool Used	Tool Material	Machined Surface
[17]	Spring loaded roller burnishing tool	Steel roller	Steel-37 with hardness 220 *Hv*
[20]	Spring loaded roller burnishing tool with roller retaining cap	Carbon chromium rollers	Aluminum 6061 cylindrical rod
[21]	Multi roller burnishing tool for vertical machining	Multi roller burnishing tool	100 mm long and 45 mm square Al, Brass, Copper bars
[22]	Adjustable type roller burnishing tool	Tin-coated EN31 rollers	Steel En24
[27]	Bearing-type roller burnishing tool	Hardened alloy steel	Ø27.8 mm Mild Steel

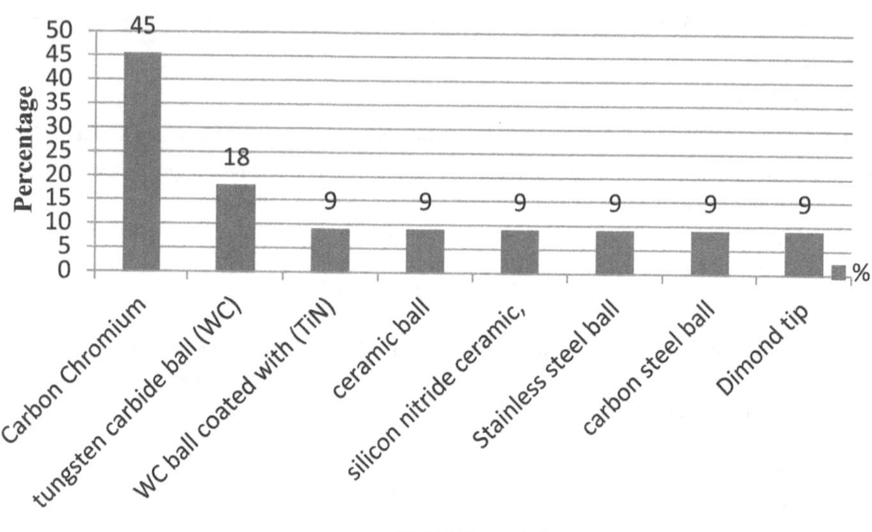

FIGURE 14.6 Literature review on ball material.

14.3.5 Analysis of Ball Material and Ball Diameter

Burnishing operation is performed by hard ball or roller. Ball or roller diameter and material shows significant impact on burnished components surface roughness and hardness. Ball material and ball diameter selected by scientist for burnishing operation is summarized in Figures 14.6 and 14.7, respectively. Diameter of 4 mm–18 mm [10, 15] for round diamond sliding burnishing tool [20] are used in

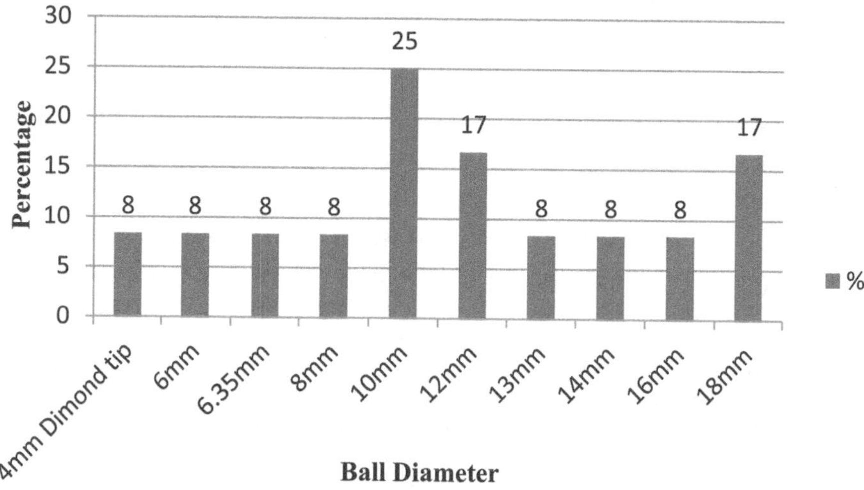

FIGURE 14.7 Literature review on ball diameter.

burnishing process under review. Commonly used ball material is carbon chromium steel, as it is freely available from ball bearing. Although coting of TiN can be given on tungsten carbide [75] or carbon chromium steel ball for improvement in the performance of burnishing process.

14.3.6 Literature Review of Machine Tool Used

Burnishing operation is performed on lathe, milling machine three axes and four axes machining center. Figure 14.8 shows the summary of machine used in ball and roller burnishing. One can conclude that ball burnishing process mostly performed on cylindrical component CNC lathe though little work has been conducted on plane surface burnishing with milling machine and CNC three axes and four axes machining center.

14.3.7 Methodology Used

Surface response method, factorial design, design of experiment, artificial neural network (ANN) [71], Taguchi [72] etc. are the method used by researcher to analyze the impact of process parameter on responses. Only one researcher has attempted the multi-objective optimization using gray relation analysis [15]. Figure 14.9 gives overall idea of methodology used in analyzing the burnishing process.

14.3.8 Measured Variable (Response) and Results Obtained

Burnishing process is affected by various parameters/variables. These variables have its influence on the surface roughness, microhardness and other responses. Table 14.6 and Figure 14.10 investigate different measured responses. Table 14.7 shows the summary of literature reviewed.

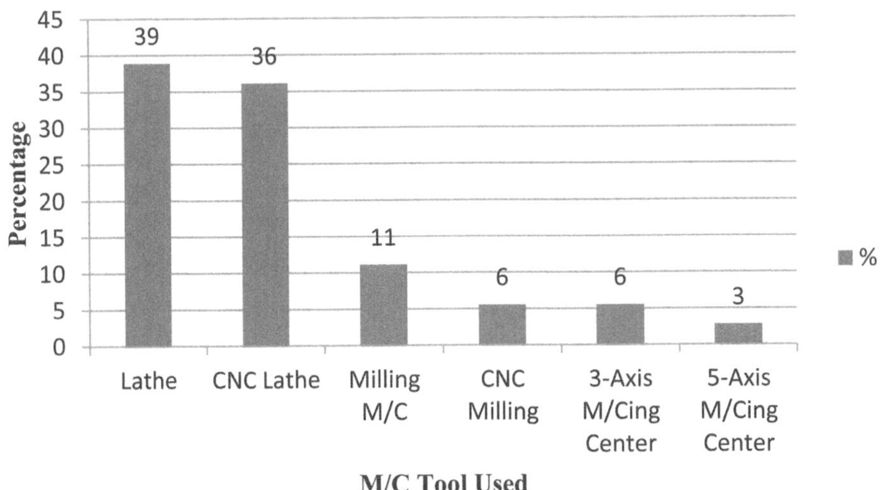

FIGURE 14.8 Literature review on M/C used.

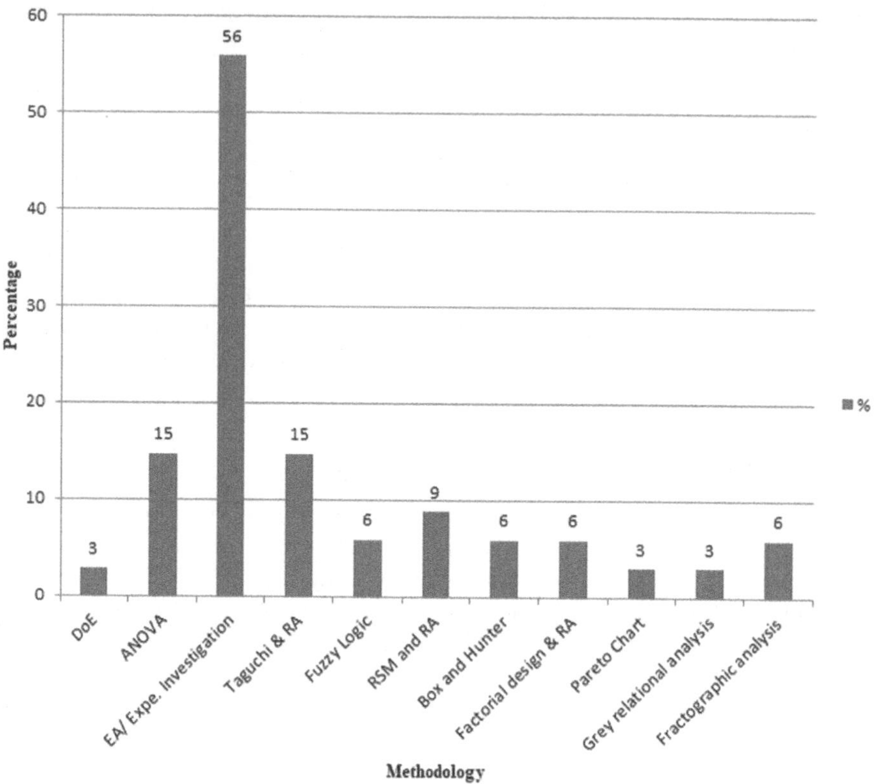

FIGURE 14.9 Literature review on methodology.

TABLE 14.6
Measured Responses

Sr. No.	Process Parameter	Reference Paper No.
1	Surface Roughness	[1, 2, 3, 4, 5, 7, 8, 9, 10, 11, 12, 13, 14, 15, 16, 18, 21, 22, 24, 25, 25, 27, 28, 29, 30, 31, 32, 33, 34, 35, 36, 37, 73, 78, 79, 80, 81, 82, 83, 84]
2	Microhardness	[1, 2, 3, 5, 6, 7, 15, 16, 18, 20, 21, 25, 26, 27, 28, 29, 31, 33, 34]
3	Fatigue Life	[1, 2, 16]
4	Out-of-Roundness	[6, 9]
5	Micro Structure	[7, 18, 21]
6	Residual Stresses	[16, 24, 25, 26]
7	Tension Test	[16, 36]
8	Corrosion Resistance	[2, 17, 20]
9	Austenite-To-Martensitic Phase Transformation	[23]
10	Refined Grain Structure	[25]
11	Tribological Properties	[29]

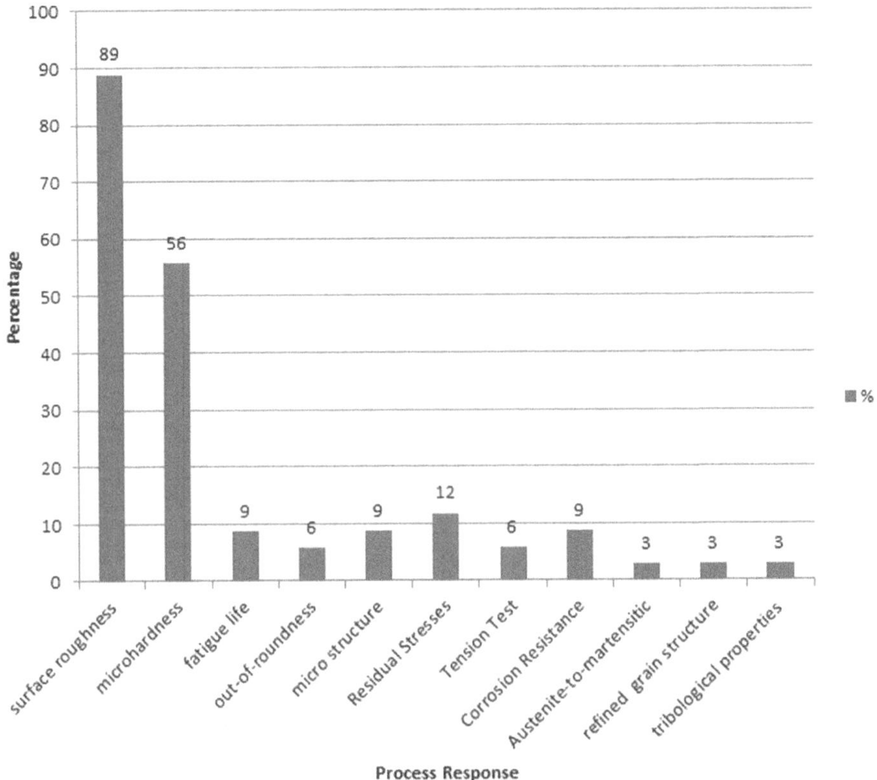

FIGURE 14.10 Literature review on measured responses.

14.4 SUMMARY OF LITERATURE REVIEW

TABLE 14.7
Literature Review Summary

Ref. No.	Work Material	Machine Tool	Process Parameter	Response	Methodology	Conclusion
[1]	AISI 1045	General-purpose lathe machine	Pressure, speed, diameter of ball, no. of passes	Microhardness, fatigue life, surface roughness, surface integrity	DOE, ANOVA	Optimum conditions are predicted for surface finish, hardness and process economy. The database created by the EA might be very helpful in determining the most suitable process variables and factors to finish different machine parts at a reasonable cost.
[2]	Al (99.85% Al) and brass	Lathe machine	Force; constant Feed: 0.1 Speed: 25.8 Number of passes: 1 Balls dia.:10	Roughness; subsurface hardness; fatigue strength; corrosion Resistance.	EA experimental analysis;	On the metallic surface of elements, shot peening is utilized to enhance specific characteristics. These qualities will further enhance when burnishing techniques are used on shot-peened elements.
[3]	Brass; Al and copper were used to obtain Al+Cu	Colchester master 2500 lathe	Lubricants, surface hardness, roughness, ball dia.	Surface roughness; surface hardness	EA experimental analysis;	The findings indicate that the majority of the initial burnishing factors having significant effects on the burnishing procedure. Both, the overall amount of hardness and ultimate surface finish will increase as the ball diameter increases up to certain size.
[4]	PDS5 tool steel (equivalent to AISI P20)	Three-axis machining center	Ball material; speed; force; feed.	Surface finish	Taguchi's orthogonal array L_{18}; ANOVA	By using ideal burnishing variables on plastic injection mold's freeform surface, surface roughness Ra possibly improved from 0.842 to 0.187μm.
[5]	Al 7075 T6	Taksan 40T1500 CNC milling machine	Force (N); feed; speed; no. of passes.	Surface roughness and hardness	Fuzzy logic	This investigation demonstrates that increasing the applied force has an adverse effect on surface finish. With 200 N, the best surface finish was achieved.
[6]	Al alloy 2014	CNC lathe machine	Speed, feed, depth of penetration and no. of passes.	Out-of-roundness and surface microhardness	RSM, with Box and Hunter method	It is possible to effectively finish non-ferrous metals interior surface that are challenging to grind using traditional grinding techniques by using the recommended internal ball burnishing tool.

[7]	Pre-hardened steel AISI P20	Five axis machining center	Force; velocity; radial depth; no. of passes; Direction;	Roughness; microhardness; microstructure photographs	EA experimental analysis;	A recently developed improved surface treatment method for free-form objects is ball burnishing; it replaced high-speed ball-end milling.
[8]	Hardened steel (64HRC)	CNC lathe	Force, feed, speed.	Surface roughness	2^5 screening factorial experiment; Pareto chart;	The hydrostatic principle was applied by the tool. The burnishing speed and feed, hydraulic pressure, nose radius and previous hard turning feed were selected process variables. The results indicated that pressure had a significant impact on the process and that the initial roughness before forceful turning had a significant impact as well.
[9]	Mild steel	Regular lathe	Speed, force, feed and no. of balls used in single pass	Surface roughness, out-of-roundness, change in dia. of workpiece	Fuzzy logic model	A fuzzy model shown that center rest ball burnishing equipment can provide acceptable surface qualities. To regulate the levels of all surface attributes, the three most crucial factors are burnishing feed, force and no. of ball passes.
[10]	7178 Al alloy	CNC Lathe (Industrial type)	Force, no. of passes, feed and speed	Surface roughness	Desirability function approach (DFA) together with (RSM) with CCD; ANOVA	The findings showed that no. of passes and burnishing force were important variables influencing the surface finish.
[11]	AA 7075 and AISI 5140	C6132A machine tool;	Burnishing force	Surface roughness	Analytical model (AM)	Researcher derives and validates relationship between burnishing force and surface roughness.
[12]	Thermoplastic and thermosets	Lathe machine	Force, speed, feed rate, ball diameter.	Surface roughness;	EA experimental analysis;	The ball burnishing procedure has a strong chance of being adopted as a novel surface treatment technique for polymers, according to the findings of the surface quality and tribology tests.
[13]	41Cr4 (AISI 5140) steel	Lathe machine	CONSTAN Ball dia, force, feed, cooling medium	Surface roughness	EA experimental analysis;	Cutting-burnishing and cutting-abrasive are hybrid techniques as a substitute to conventional turning and grinding.

(Continued)

TABLE 14.7 *(Continued)*
Literature Review Summary

Ref. No.	Work Material	Machine Tool	Process Parameter	Response	Methodology	Conclusion
[14]	PDS5 plastic injection mold steel	Three-axis machining center	Force (N) and constant Feed (mm/min) 800;	Surface roughness	EA experimental analysis;	This study proposes an invention of a novel ball burnishing integrated with a load cell combined with a machining center to enhance the PDS5 plastic injection mold steel's surface finish.
[15]	7XXX series wrought aluminum alloy AA 7075	FANUC GT- 250B CNC lathe.	Force(F), feed rate (f), no. of passes (N) and speed (V)	Surface roughness and microhardness	Grey relational analysis and Taguchi method $L_{16}(4^4)$ orthogonal array; ANOVA	The Taguchi is a highly useful technique for process optimization when there aren't as many trial runs. Smoother surfaces with better surface hardness are outputs of all sorts of burnishing methods.
[16]	AISI 304	CMZ turning center (model TC/25 BTY)CN C	Constant-fluid burnishing pressure, speed, and feed	Roughness; hardness; grain size; residual stresses; tensile tests; fatigue tests	EA experimental analysis; fractographic analysis;	Surface roughness, fractograph, cyclic relaxation effects and deep residual stresses are analyzed and experimental results are presented in this chapter.
[17]	2024-T3 Aluminum alloy	Lathe	Speed; feed	Corrosion resistance	EA experimental analysis	By burnishing the surface, the corrosion performance measured by electrochemical impedance spectroscopy was enhanced. Furthermore, impact of burnishing conditions on donor concentration of surface passive film and its correlation with corrosion resistance were examined. SEM data following corrosion immersion provided more evidence for these findings.
[18]	41Cr4 (AISI 5140) steel	CNC turning center,	Burnishing speed burnishing feed	Surface roughness; microhardness-microstructure	EA experimental analysis; fractographic analysis	The workpiece can be cryogenically pre-cooled to further regulate the burnishing effects.
[20]	Austenitic X6CrNiMoTi17-12-2 alloy steel	Engine lathe;	4 mm round tip tool, burnishing force 150 N, feed 0.11 mm/rev.	Microhardness; stress; Corrosion resistance measure using the electrochemical method	Electrochemical investigation, corrosion resistance were determined	Enhancing corrosion resistance by slide burnishing is possible as long as the machined surface has the proper geometry. Surface layer condition parameters and other elements of the surface GEOMETRICAL structure are important considerations in addition to minor surface roughness.

[21]	Titanium alloy (Ti–6Al–4V) Grade 5	CNC Ace turn mill	Speed; feed; force; no. of passes.	Surface roughness, hardness; microstructure	Taguchi's L^{25} orthogonal array; analysis of means (ANOM); (ANOVA); Microstructure and SEM analysis	For good surface finish, a burnishing speed in middle range, less feed and small force with three passes work well together. However, surface hardness might be increased by burnishing at a moderate speed and feed, using greater force and making more tool passes.
[22]	Thermally coated surfaces WC12Co coating	Milling machine	Burnishing force 150 N, feed 0.11 mm/rev.	Surface finish	Analytical model-verified on the basis of experimental tests	A rolling process may be created without the need for substantial preparatory testing since the roughness can be estimated using the existing model.
[23]	Nitinol is a nickel–titanium alloy	CNC milling machine;	Force (load)	Austenite-to-martensitic phase transformation	Hertz theory of elastic contact; finite element simulation	To examine austenite to martensitic phase transition both experimentally and theoretically, nitinol (Ni50.8Ti49.2) was burnished under various loads.
[24]	17-4 PH stainless steel	CNC lathe	Burnishing pressure, speed and feed	Surface roughness; residual stresses.	3^3 factorial design by using Minitab software and regression analysis	Rougher surfaces result from increased pressure. Although feed reduces surface roughness but, more feed will result in increased surface roughness.
[25]	Ti-6Al-4V	HAAS TL-2 CNC lathe	Force and. constant: feed rate = 0.05 mm/rev, and speed = 150 m/min	Roughness; Microhardness; refined grain structure, residual stresses	Experimental analysis	To enhance Ti-6Al-4V alloy properties, cryogenic burnishing was used to create SPD (severe plastic deformation) layer. Ti-6Al-4V burnishing is governed by two main processes: thermal softening and strain hardening.
[26]	6061-T6 aluminum alloy	Vertical milling machine	Speed, depth of penetration and no. of passes	Roughness, microhardness and residual stress	RSM with Group Method of Data Handling Technique	Surfaces with low burnishing rates and high penetration depths are significantly smoother than those with high burnishing speeds and deep penetration due to chatter.
[27]	Steel-37 of hardness 220 Hv	Center lathe model DIZ	Burnishing speed, force, feed and number of passes	Surface microhardness and roughness	Experiments design proposed by Box and Hunter	Residual stress distribution in the orthogonally burnished surface region is found by deflection etching method. Surface roughness and microhardness of St-37 resulting from roller burnishing under lubricated circumstances are predicted using mathematical models.

(Continued)

TABLE 14.7 ***(Continued)***
Literature Review Summary

Ref. No.	Work Material	Machine Tool	Process Parameter	Response	Methodology	Conclusion
[28]	En-8	HMT NH-26 lathe	Speed; feed; no. of passes; lubricant	Surface finish and surface microhardness	Taguchi L9 orthogonal array	One of the crucial procedures for precision component machining is the roller burnishing process, results of which are presented.
[29]	MP35N, annealed and hardened AISI 4140	CNC lathe	Speed; feed; no. of passes;	Surface finish and surface microhardness	Experimentally investigated	A novel hybrid method for polishing metals has been offered: laser-assisted burnishing (LAB). To assess the suggested method, traditional burnishing and laboratory experiments were conducted on annealed and hardened AISI 4140 and MP35N.
[30]	Aluminum 6061	LA430 lathe	Speed, force and burnishing tool dimension	Surface qualities and tribological properties	Experimentally investigated	Investigated how surface characteristics and tribological properties were affected by force, speed and tool dimension. It was discovered that there is a 40% improvement in surface finish. On the other hand, the surface finish improves less as roller contact width increases.
[31]	Aluminum, brass and copper	PBM – VS 300 milling machine;	Spindle rotations (speed)rpm, feed rate and depth of penetration.	Surface roughness and surface hardness	Experimentally investigated	By choosing the right input parameters, the trials help to improve the burnished surface's quality.
[32]	AISI 4340 (EN24) steel	Lathe machine	Speed, feed, burnishing force, no. of passes	Surface roughness	Taguchi's OA; ANOVA	This effort attempts to coat rollers with TiN using a DC magnetron sputtering machine, which is subsequently utilized for burnishing. There is a performance comparison between TiN-coated and untreated rollers.

[33]	Ti-6Al-4V	Pinnacle vertical milling machine	Spindle speed; feed rate; depth of penetration	Surface roughness; microhardness	Experimentally investigated	The hardness and roughness of the surface are enhanced by roller burnishing. Compressive stress will be imparted and fatigue life will be enhanced. The burnishing procedure is challenging for this grade of material because titanium alloy is a challenging material to machine. For same spindle rotation, feed rate, and penetration depth, a low surface roughness and high hardness were achieved.
[34]	Stainless steel X5CrNiMo17-12-2 for centrifugal pumps.	Universal lathe	Burnishing speed (vn), Radius of a roller's ; and feed (fn) is constant	Strengthening of the surface layer shafts neck i.e. microhardness and surface roughness.	Experimentally investigated	Many complexly shaped parts used in machinery such as camshafts, drive shafts and crankshafts are subject to varying stresses. Burnishing can increase the fatigue resistance of such materials used in shipbuilding. Analyze the impact of the curve burnishing radius on the strain hardening of centrifugal pumps shaft neck's surface layer made of stainless steel X5CrNiMo17-12-2.
[35]	Al	Lathe	Burnishing force; lubricant	Surface roughness	Experimentally investigated	In an experiment using roller burnishing on an aluminum workpiece under various circumstances, a new design of dynamometer is employed. For various lubricant applications, ideal burnishing force values and surface roughness value (Ra) are found.
[36]	O_1 steel	Lathe	Force constant feed rate 0.02 m/s and speed 100 rpm.	True stresses; U.T.S; maximum % elongation and surface roughness	Experimentally investigated	This research examines the impact of varying forces (105, 140, 175, 210)N and presents the findings of the trials. The study's main conclusions are that the material's real stress has increased by around 150 MPa, the U.T.S. has grown by 166 MPa and the surface quality has improved by 12.5%.
[37]	ASTM Al 2017, ASTM 1055	(CNC) lathe	Speed, thrust force, feed and angle of inclination	Surface roughness	Experimentally investigated	This research proposes a novel roller burnishing technique that combines a sliding and rolling action.

(*Continued*)

TABLE 14.7 ***(Continued)***
Literature Review Summary

Ref. No.	Work Material	Machine Tool	Process Parameter	Response	Methodology	Conclusion
[38]	AISI 1045	Lathe	Speed; feed; no. of passes	Fatigue strength, surface roughness	Experimentally investigated	This work presents analysis and experimental data related to fractography, cyclic relaxation effects, in-depth residual stresses and surface roughness.
[21]	Ti-6Al-4V	Lathe	burnishing speed, feed, burnishing force, no. of passes	Surface hardness and roughness	Taguchi L25 orthogonal array	The optimization findings in this research showed that while no. of passes and burnishing force are essential factors in maximizing hardness, burnishing speed and feed are critical factors for improving surface finish.
[39]	Turbine compressor and pump parts	Lathe	Feed rate	Surface residual stress	2D finite element model	This study examined the interference effects caused by adjacent burnishing points in relation to an elastic burnishing tool.
[40]	TA2	Lathe	Burnishing speed, feed, burnishing force, no. of passes	Microhardness, roughness, residual stresses and topography	Single factor experiments	Surface topography of burnished part is mostly influenced by the burnishing feed and force, as indicated by the power spectral density profile analysis results presented in this research. Following the burnishing process, the surface residual stress increases from −67.7 MPa to −400.5 MPa. At a feed rate of 0.05 mm, surface finish achieves its minimal value of 0.057 μm.
[41]	AISI 316Ti	Lathe	Burnishing speed, feed, burnishing force, no. of passes	Microhardness, roughness, wear resistance, residual stress, fatigue strength (life)	Experimental and FEM analysis	This research examines impact of burnishing variables on produced fatigue strength, residual stress, wears resistance, microhardness and achieved roughness.
[42]	AISI 1040	Lathe	Speed, feed rate, penetration depth, no. of passes	Surface roughness and hardness	Grey relational analysis (GRA)	The process significantly improved experimental results, which were optimized using multi-parametric optimization. This would help the steel industry increase productivity while maintaining the quality of high-carbon steel.

[43]	FSW components	Lathe	Speed, feed rate, penetration depth, no. of passes	Microhardness, roughness, residual stresses and topography	Experimentally investigated	An overview of effects of burnishing variables on resultant wear resistance, fatigue strength (life), microhardness and roughness is provided in this study.
[44]	Aluminum alloy	Lathe	Burnishing speed, force, feed rate, no. of passes	Surface hardness	Taguchi, fuzzy logic and regression models	This work optimized the surface hardness by using Taguchi's S/N ratio. Through ANOVA, influence of burnishing variables on hardness was determined.
[45]	Ti-6Al-4V	Lathe	Burnishing speed, feed, force and no. of pass	Wear resistance	Taguchi L16 OA	The findings of this chapter's Taguchi optimization showed that, while burnishing speed and feed are crucial for reducing coefficient of friction; burnishing force and no. of passes are most essential factors to minimize specific wear rate.
[46]	AISI 4140	Lathe	Burnishing force, ball burnishing diameter and depth of cut	surface roughness (Ra), compressive residual stress and microhardness (μH)	Taguchi optimization	This chapter addresses Combined Turning-Burnishing (CoTuB), an integrated manufacturing method that uses a single machine tool to concurrently execute turning and ball burnishing. This procedure was designed to take advantage to enhance surface integrity and quality.
[47]	EN31 Steel	Lathe	Burnishing speed, feed, force and no. of pass	Surface roughness, hardness, corrosion, wear resistance and residual stress	Principal Component Analysis (PCA) integrated Taguchi	The results of the optimization showed a 52% increase in wear resistance over the turned surface. Hardness improved from 178.5 Hv to 265 Hv and surface roughness decreased from 0.439 μm to 0.091 μm after burnishing. Compressive residual stress was created just beneath the burnished surface, as demonstrated by X-ray diffraction. Surface micrographs and recorded polarization curves show better and stable corrosion resistance.

(Continued)

TABLE 14.7 *(Continued)*
Literature Review Summary

Ref. No.	Work Material	Machine Tool	Process Parameter	Response	Methodology	Conclusion
[48]	AA6061-T6 alloy	Lathe	Burnishing speed, feed, force	Surface finish	Taguchi, ANOVA	The findings demonstrate that an optimal burnishing setting greatly improves the specimens' surface quality. The examination of the data reveals that, at 54.51%, the burnishing feed rate has the largest effect on surface roughness, followed by load and burnishing speed.
[49]	Ti-6Al-4V alloy	Lathe	Burnishing speed, feed, force, number of pass, lubricant	Surface integrity	Experimental analysis	All of the investigated burnishing parameters except speed were found to be crucial in regulating the treated surface integrity and, therefore, the functional performance of Ti-6Al-4V aerospace components.
[20]	X6CrNiMoTi1 7-12-2 stainless steel	Milling machine	Burnishing, drawing, polishing	Corrosion resistance	Experimental analysis	The article introduces diamond slide burnishing, an unconventional finishing technique. Through expedited electrochemical experiments employing the potentiodynamic approach, the impact of burnishing, polishing and drawing on corrosion resistance was investigated.
[1]	AISI 1045	Machining center	Burnishing speed	Roughness, microhardness, surface integrity, fatigue life	Mathematical expression	Surface quality features of work material were assessed in this chapter to investigate effects of parameters, determining which selected parameter is prevailing factor, ranking the factors and using the factors at different levels to improve fatigue life, surface hardness and finish.

[50]	7178 aluminum alloy	Lathe	Burnishing speed, feed, force, no. of pass	Surface roughness	Desirability function approach (DFA) together with RSM	The findings of this study showed that no. of passes and burnishing force had a substantial effect on surface roughness. The correctness of the anticipated model was validated by the expected surface roughness values followed by validation tests conducted at ideal conditions.
[51]	Stainless steel	Lathe	Burnishing speed, feed, force, no. of pass	Surface roughness, microhardness	Taguchi's orthogonal array; ANOVA	According to the results analysis, burnishing force had greatest impact on both microhardness and surface roughness, accounting for 39.87% of the surface roughness and 42.85% of the microhardness.
[52]	AISI 1045 steel	Lathe	Burnishing speed, feed and force	Surface roughness, hardness and corrosion resistance	3^3 factorial design and the response surface methodology	The results indicate that ball burnishing process may enhance the corrosion resistance of AISI-1045 steel by increasing hardness from 202 HB to 236 HB and reducing surface roughness from 3.51 μm to 0.61 μm.
[53]	Mg–Zn–Ca alloy	Lathe	Burnishing feed, depth of press, force, number of pass	Surface roughness, surface microhardness and corrosion resistance	Taguchi's orthogonal array	Based on ball burnishing trials, a significant increase in surface finish of 129 nm and microhardness of 107 Hv was attained with a press depth of 0.45 mm, feed rate of 450 mm min–1, burnishing force of 250 N and no. of passes: 2.
[8]	Hardened steel	Lathe	Burnishing feed and speed	Surface roughness, fatigue strength, corrosion resistance and bearing ratio	3^3 factorial design	Heat-treated steel parts may now be burnished to 65 HRC thanks to recent advancements. Because compressive stresses are incorporated into the surface layer, burnishing results in a good surface finish that is equivalent to grinding as well as an increase in the mechanical properties of the surface.
[54]	AISI P20	Five axis milling machine	Burnishing feed	Surface roughness	Experimental analysis	The burnishing method is especially appealing to obtain the necessary quality of dies and molds because of the good final surface quality. Using the maximum linear feed, this method may be easily implemented on the same high-speed equipment used for sculptured milling.

(Continued)

TABLE 14.7 ***(Continued)***
Literature Review Summary

Ref. No.	Work Material	Machine Tool	Process Parameter	Response	Methodology	Conclusion
[55]	AISI 4140	laser-assisted burnishing (LAB) setup	Burnishing feed and speed	Surface roughness, hardness and compressive residual stress	Experimental analysis	Experiments demonstrate that laser assisted burnishing (LAB) can generate considerably better surface finish, greater hardness, and deep compressive residual stress because it causes more plastic deformation than conventional burnishing since it temporarily softens the workpiece material before burnishing.
[56]	A92017	Lathe	Burnishing feed and speed	Surface roughness	Experimental analysis	Conclusions on the enhancement of surface finish achieved on workpieces through the ball burnishing method are made in this research. The principal contribution of this study is the handling of convex and concave shapes.
[57]	AISI420	Industrial type of CNC lathe	Burnishing speed, feed, force, no. of pass, lubricant	Surface roughness, hardness	Taguchi's L18 OA, ANOVA	With the right treatment settings, the test specimen's fine turned surface finish could be enhanced from Ra 1.1 μm to Ra 0.025 μm, and its surface hardness could be raised from HRc 51 to HRc 52.5.
[32]	EN24 steel	Machining center	Burnishing speed, feed, force, number of pass	Surface roughness	Mathematical expression	In burnishing operations, it was found that the TiN-coated roller performs better than the uncoated rollers. The effectiveness of the roller during burnishing is largely dependent on burnishing speed, force and no. of passes especially when considering the components surface finish that are generated.
[58]	AISI 1010 Steel	Vertical machining center	Burnishing speed, feed, force	Surface roughness, surface hardness and microstructure	Taguchi and response surface methodology (RSM)	It was discovered that burnishing force, speed and feed rate had the greatest and least respective impact on surface hardness and roughness. Furthermore, burnished surface microstructural analyses show that burnishing forces greater than 400N result in flaking of the burnished surfaces.

[59]	17-4 PH stainless steel	Industrial type of lathe	Burnishing speed, feed	Surface roughness, hardness and compressive residual stress	Second-order empirical model	This study demonstrated how to modify turned surfaces of an aerospace material using ball burnishing technique. Research is done to investigate impact of burnishing parameters on responses.
[60]	6061-T6 aluminum alloy	Vertical machining machine	Burnishing speed, depth of penetration and no. of passes	Surface roughness, hardness and compressive residual stress	Group Method of Data Handling Technique (GMDH)	Research indicates that surfaces with minimum burnishing speeds and maximum penetration depths are significantly smoother than those with high speeds and deep penetrations, which result in rougher surfaces due to chatter. As burnishing speed increases, compressive residual stress value drops. As the no. of passes or burnishing depth of penetration increases, the maximum compressive residual stress also rises.
[61]	EN 31 steel	Industrial type of lathe	Burnishing speed, feed, force, no. of pass	Surface roughness, hardness	Taguchi's orthogonal array L_{25}; ANOVA	The Taguchi technique is used to determine which combination of process variables will result in lesser surface roughness and high hardness. Following burnishing, hardness rises from 179.5 Hv to 266.5 Hv and surface roughness falls from 0.446 µm to 0.089 µm.
[62]	Mg Ze41A alloy	Lathe	Burnishing speed, feed, force, no. of pass	Surface roughness, microhardness	Experimental analysis	Aim of current study is to use ball burnishing technique to increase Mg Ze41A alloy's surface integrity. With optimal parameters setting surface roughness decreased by 94.90% from 0.941 µm to 0.048 µm. Similarly, hardness improved from 75.2 Hv to 113.27 Hv resulted in a 50.62% improvement in microhardness.
[63]	Mg Ze41A alloy	Lathe	Burnishing speed, feed, force, number of pass	Surface integrity	FIS, ANN and ANFIS	This work uses soft computing methods to estimate surface quality of ball-burnished magnesium Ze41A alloy, including ANFIS, ANN and FIS.

(Continued)

TABLE 14.7 ***(Continued)***
Literature Review Summary

Ref. No.	Work Material	Machine Tool	Process Parameter	Response	Methodology	Conclusion
[64]	Magnesium Ze41A alloy	Milling machine	Burnishing speed, feed, force, no. of pass	Wear rate	Fuzzy interface system	Enhancement in microhardness and surface finish during the ball burnishing process appears to be the cause of the notable improvement in tribological performance of the magnesium Ze41A alloy. There has been a 51.22% reduction in the wear rate and a 72.58% CoF.
[65]	Ti60 alloy	Lathe	Burnishing speed, feed, force, no. of pass	Surface hardness, roughness, residual stress, microstructure and microhardness	Experimental analysis	The combined turning and ball burnishing test was conducted in order to study the surface characteristics of the Ti60 alloy. Surface roughness, morphology, microhardness, residual stress and microstructure were among the surface characteristics that were examined.
[66]	β-Ti alloy	Lathe	Burnishing speed, feed, force, no. of pass	Surface hardness, roughness	Experimental analysis	In order to enhance the surface finish and hardness for possible applications in biomedical implants, a comparative study on the surface modification of β-Phase titanium alloy using electric discharge machining (EDM) and ball burnishing assisted electrical discharge cladding (BB-EDC) process has been conducted in this work.
[67]	Stavax	SIEMENS CNC	Burnishing force, depth of cut, number of pass	Surface roughness	Mathematical modeling	This work contributes to a deeper understanding of the surface creation mechanism during the burnishing process. Furthermore, the obtained findings demonstrate a notable 95% surface finish increase, demonstrating the potential of the burnishing process as well as its speed and economy.

[68]	EN 31 Steel	Industrial lathe	Burnishing speed, feed, force, no. of pass	Surface roughness, surface hardness	Principal Component Analysis (PCA) coupled with Taguchi	Principal component analysis (PCA) integrated Taguchi optimization approach are used in this study to solve a multi-response challenge and find the ideal burnishing process variable that will provide the necessary surface hardness and finish to meet all of the shank's functional criteria.
[69]	CuAl8Fe3 bronze	Vertical machine center Haas MiniMill	Burnishing speed, feed	Surface roughness, microhardness	Non-dominated sorting genetic algorithm (NSGA-II)	This work describes a new two-transition, single-operation method for machining holes in CuAl8Fe3 bronze sliding bearing bushings. Regression analysis and experiments were used to examine the 2D and 3D surface texture characteristics of the processed holes that were created following the first and second transitions.
[70]	Al7075 alloy	A 2D ultrasonic-assisted surface burnishing system	Burnishing force	Surface roughness, surface hardness	Mathematical modeling	The stability and sophistication of the burnishing mechanism are noteworthy. The Al7075 alloy specimen that has been burnished has a surface that is very smooth and hard.

14.5 FUTURE SCOPE

- The hardness of rotary friction welded parts is dropped at weld line and in heat affected zone (HAZ) which can be improved by burnishing. However, improved hardness in this zone is still below the hardness of burnished base metal. Therefore the burnishing process can be investigated to adjust and optimize the burnishing parameters online so as to match hardness of HAZ with that of base metal.
- Burnishing process can be investigated and optimized for wear resistance parts of die and mold such as guide post, guide bush, ejector pin etc. to enhance their surface finish and surface hardness and consequently increase the wear, friction and corrosion resistance. Thus proposing new method of processing these parts that incorporate the burnishing process to replace heat treatment and abrasive finishing process of existing method. The proposed method can be investigated on the performance and saving in processing time, cost and health and risk associated with existing hazardous processes.
- An attempt can be made to investigate the burnishing process to eliminate any of the hazardous process like buffing, coating, plating etc.

14.6 CONCLUSIONS

Burnishing process review gives brief idea about materials, lubricant comparison along with process parameter, measured responses and results obtained.

- As observed, authors prefer aluminum and steel materials, but there is a lot of room to work on polymers.
- Burnishing speed, feed and force were most employed process variables as compared to ball material, no. of rotations and burnishing direction.
- SAE-oils are commonly utilized lubricants but use of kerosene gives increased hardness after soluble oil.
- Ball burnishing tool gives rolling friction due to point contact and roller burnishing tool gives sliding friction due to line contact.
- Surface roughness and microhardness are favorite measured responses.
- Mathematical modeling and DOE are mostly preferred methodologies. Now a day, investigators are favoring Taguchi, RSM and fuzzy logic to optimize burnishing process variables.
- Some of the applications where the potentials of burnishing process can be well explored are-
  - i. The friction stir welding (FSW) Al and Al alloy component
  - ii. Aluminum matrix composite (AMCs).

REFERENCES

1. C. Y. Seemikeri, P. K. Brahmankar and S. B. Mahagaonkar, "Investigations on surface integrity of AISI 1045 using LPB tool", *Tribology International* 41, 8 (2008): 724–734.

2. A. M. Hassan and M. S. Momani, "Further improvements in some properties of shot peened components using the burnishing process", *International Journal of Machine Tools & Manufacture* 40 (2000): 1775–1786.
3. A. M. Hassan and A. M. Maqableh, "The effects of initial burnishing parameters on non-ferrous components", *Journal of Materials Processing Technology* 102 (2000): 115–121.
4. F. J. Shiou and C. H. Chen, "Freeform surface finish of plastic injection mold by using ball-burnishing process", *Journal of Materials Processing Technology* 140 (2003): 248–254.
5. H. Basak and H. H. Goktas, "Burnishing process on al-alloy and optimization of surface roughness and surface hardness by fuzzy logic", *Materials and Design* 30 (2009): 1275–1281.
6. S. P. Dwivedi, N. K. Maurya and M. Maurya, "Assessment of hardness on AA2014/eggshell composite produced via electromagnetic stir casting method", *Evergreen Joint Journal of Novel Carbon Resource Sciences & Green Asia Strategy* 06, 04 (2019): 285–294.
7. L. N. López de Lacalle, A. Lamikiz, J. Munoa and J. A. Sánchez, "Quality improvement of ball-end milled sculptured surfaces by ball burnishing", *International Journal of Machine Tools & Manufacture* 45, (2005): 1659–1668.
8. L. Luca, S. Neagu-Ventzel and I. Marinescu, "Effects of working parameters on surface finish in ball-burnishing of hardened steels", *Precision Engineering* 29, 2 (2005): 253–256.
9. A. A. Ibrahim, S. M. Abd Rabbo, M. H. El-Axir and A. A. Ebied, "Center rest balls burnishing parameters adaptation of steel components using fuzzy logic", *Journal of Materials Processing Technology* 209 (2009): 2428–2435.
10. A. Sagbas, "Analysis and optimization of surface roughness in the ball burnishing process using response surface methodology and desirability function", *Advances in Engineering Software* 42 (2011): 992–998.
11. F. L. Lin, W. Xia, Z. Y. Zhou, J. Zhao and Z. Q. Tang, "Analytical prediction and experimental verification of surface roughness during the burnishing process", *International Journal of Machine Tools & Manufacture* 62 (2012): 67–75.
12. K. O. Lowa and K. J. Wong, "Influence of ball burnishing on surface quality and tribological characteristics of polymers under dry sliding conditions", *Tribology International* 44 (2011): 144–153.
13. W. Grzesik and K. Zak, "Modification of surface finish produced by hard turning using superfinishing and burnishing operation", *Journal of Materials Processing Technology* 212 (2012): 315–322.
14. F. J. Shiou and C. H. Chuang, "Precision surface finish of the mold steel PDS5 using an innovative ball burnishing tool embedded with a load cell", *Precision Engineering* 34 (2010): 76–84.
15. Procesu, Uporaba Grey-TaguchijeveMetodePri. "Use of grey based Taguchi method in ball burnishing process for the optimization of surface roughness and microhardness of AA 7075 aluminum alloy." Materiali in tehnologije 44, 3 (2010): 129–135.
16. J. Triyono, R. Rahajeng and E. Surojo, "Surface modification and hardness behavior of AISI 304 as an artificial hip joint using ammonia and scallop shell powder as a nitriding agent", *Evergreen Joint Journal of Novel Carbon Resource Sciences & Green Asia Strategy* 8, 2 (2021): 335–343.
17. L. Jinlong and L. Hongyun, "Effect of surface burnishing on texture and corrosion behavior of 2024 aluminum alloy", *Surface & Coatings Technology* 235 (2013): 513–520.
18. W. Grzesik and K. Zak, "Producing high quality hardened parts using sequential hard turning and ball burnishing operation", *Precision Engineering* 37 (2013): 849–855.

19. P. Ballanda, L. Tabourota, F. Degrea and V. Moreaub, "Mechanics of the burnishing process", *Precision Engineering* 37 (2013): 129–134.
20. K. Konefal, M. Korzynski, Z. Byczkowskab and K. Korzynskac, "Improved corrosion resistance of stainless steel X6CrNiMoTi17-12-2 by slide diamond burnishing", *Journal of Materials Processing Technology* 213, 11 (2013): 1997–2004.
21. G. D. Revankar, R. Shetty, S. S. Rao and V. N. Gaitonde, "Analysis of surface roughness and hardness in ball burnishing of titanium alloy", *Measurement* 58 (2014): 256–268. DOI: 10.1016/j.measurement.2014.08.043.
22. L. Hiegemann, C. Weddeling, N. B. Khalifa and A. E. Tekkaya, "Analytical prediction of roughness after ball burnishing of thermally coated surfaces", *Procedia Engineering* 81 (2014): 1921–1926.
23. C. H. Fua, M. P. Sealya, Y. B. Guoa and X. T. Wei, "Austenite–martensitic phase transformation of biomedical Nitinol by ball burnishing", *Journal of Materials Processing Technology* 214 (2014): 3122–3130.
24. T. Zhang, N. Bugtai and I. D. Marinescu, "Burnishing of aerospace alloy: A theoretical–experimental approach", *Journal of Manufacturing Systems* 37, 2 (2015): 472–478.
25. J. Caudill, B. Huanga, C. Arvina, J. Schoopa, K. Meyer and I. S. Jawahir, "Enhancing the surface integrity of Ti-6Al-4V alloy through cryogenic burnishing", *Procedia CIRP* 13 (2014): 243–248.
26. M. M. El-Khabeery and M. H. El-Axir, "Experimental techniques for studying the effects of milling roller-burnishing parameters on surface integrity", *International Journal of Machine Tools & Manufacture* 41 (2001): 1705–1719.
27. M. H. El-Axir, "An investigation into roller burnishing", *International Journal of Machine Tools & Manufacture* 40 (2000): 1603–1617.
28. C. S. Jawalkar and R. S. Walia, "Study of roller burnishing process on En-8 specimens using design of experiments", *Journal of Mechanical Engineering Research* 1, 1 (2009): 038–045.
29. Y. Tian and Y. C. Shin, "Laser-assisted burnishing of metals", *International Journal of Machine Tools & Manufacture* 47 (2007): 14–22.
30. N. S. M. El-Tayeb, K. O. Low and P. V. Brevern, "Influence of roller burnishing contact width and burnishing orientation on surface quality and tribological behavior of Aluminium 6061", *Journal of Materials Processing Technology* 186 (2007): 272–278.
31. S. Thamizhmanii, B. Saparudin and S. Hasan, "A study of multi-roller burnishing on non- ferrous metals", *Journal of Achievements in Materials and Manufacturing Engineering* 22, 2 (2007): 95–98.
32. B. C. Yeldose and B. Ramamoorthy, "An investigation into the high performance of TiN-coated rollers in burnishing process", *Journal of Materials Processing Technology* 207 (2008): 350–355.
33. S. Thamizhmnaii, B. B. Omar, S. Saparudin and S. Hasan, "Surface roughness investigation and hardness by burnishing on titanium alloy", *Journal of Achievements in Materials and Manufacturing Engineering* 28, 2 (2008): 139–142.
34. T. Dyl, "The burnishing strengthen shafts neck of centrifugal pumps", *Journal of KONES Powertrain and Transport* 18, 2 (2011): pp. 123–128.
35. J. N. Malleswara Rao, A. C. Kesava Reddy and P. V. Rama Rao, "Design and fabrication of new type of dynamometer to measure radial component of cutting force and experimental investigation of optimum burnishing force in roller burnishing process", *Indian Journal of Science and Technology* 3, 7 (2010), ISSN: 0974-6846, pp. 737–742.
36. K. S. Rababa and M. M. Al-mahasne, "Effect of roller burnishing on the mechanical behavior and surface quality of O1 alloy steel", *Research Journal of Applied Sciences, Engineering and Technology* 3, 3 (2011): 227–233.

37. M. Okada, S. Suenobu, K. Watanabe, Y. Yamashita and N. Asakawa, "Development and burnishing characteristics of roller burnishing method with rolling and sliding effects", *Mechatronics* 29 (2014): 110–118.
38. R. Aviles, J. Albizuri, A. Rodriguez and L. N. Lopez De Lacalle, "Influence of low-plasticity ball burnishing on the high-cycle fatigue strength of medium carbon AISI 1045 steel", *International Journal of Fatigue* 55 (2013): 230–244. DOI: 10.1016/j.ijfatigue.2013.06.024.
39. D. He, B. Wang, J. Zhang, S. Liao and W. J. Deng, "Investigation of interference effects on the burnishing process", *International Journal of Advanced Manufacturing Technology* 95 (2018): 1–10. DOI: 10.1007/s00170-017-0640-3.
40. X. Yuan, Y. Sun, C. Li and W. Liu, "Experimental investigation into the effect of low plasticity burnishing parameters on the surface integrity Of TA2", *International Journal of Advanced Manufacturing Technology* 88, 1–4 (2017): 1089–1099. DOI: 10.1007/s00170-016-8838-3.
41. J. T. Maximov, G. V. Duncheva, A. P. Anchev, N. Ganev, I. M. Amudjev and V. P. Dunchev, "Effect of slide burnishing method on the surface integrity of AISI 316Ti Chromium – Nickel steel", *Journal of the Brazilian Society of Mechanical Sciences and Engineering* 40, (2018): 194. DOI: 10.1007/s40430-018-1135-3.
42. S. Kumar, B. Mitra and N. Kumar, "Application of GRA method for multi-objective optimization of roller burnishing process parameters using a carbide tool on high carbon steel (AISI-1040)", *Grey Systems: Theory and Application* 9, 4 (2019): 449–463. DOI: 10.1108/GS-03-2019-0006.
43. N. J. Varpe, A. Hamilton and U. Gurnani, "Review of effect of burnishing process on surface integrity of friction stir welded components", *Design Engineering* 4 (2021):651–666.
44. P. S. Kumar, B. S. Babu and V. Sugumaran, "Comparative modeling on surface roughness for roller burnishing process, using fuzzy logic", *International Journal of Mechanical and Production Engineering Research and Development* 8, 1 (2018): 43–64.
45. G. D. Revankara, R. Shetty, S. S. Rao and V. N. Gaitonded, "Wear resistance enhancement of titanium alloy (Ti–6Al–4V) by Ball burnishing process", *Journal of Materials Research and Technology* 6, 1 (2017): 13–32. DOI: 10.1016/j.jmrt.2016.03.007.
46. A. Rami, F. Gharbi, S. Sghaier and H. Hamdi, "Some insights on combined turning-burnishing (Cotub) process on workpiece surface integrity", *International Journal of Precision Engineering and Manufacturing* 19, 1 (2018): 67–78. DOI: 10.1007/s12541-018-0008-0.
47. N. J. Varpe and A. Hamilton, "Investigation into burnishing process to examine effect on surface integrity, wear and corrosion resistance of carbon alloy (EN31) steel", *Journal of Materials Engineering and Performance* (2023). DOI: 10.1007/s11665-023-08524-x.
48. V. Barahate, A. R. Govande, G. Tiwari, B. R. Sunil and R. Dumpala, "Parameter optimization during single roller burnishing of AA6061-T6 alloy by design of experiments", *Materialstoday Proceedings* 50, 5 (2022): 1967–1970.
49. J. Caudill, J. Schoop and I. S. Jawahir, "Correlation of surface integrity with processing parameters and advanced interface cooling/lubrication in burnishing of Ti-6Al-4V alloy", *Advances in Materials and Processing Technologies* 5, 1 (2018): 53–66. DOI: 10.1080/2374068X.2018.1511215.
50. S. Aysun, "Analysis and optimization of surface roughness in the ball burnishing process using response surface methodology and desirability function", *Advances in Engineering Software* 42, 11 (2017): 992–998.
51. T. A. El-Taweel and M. H. El-Axir, "Analysis and optimization of the ball burnishing process through the Taguchi technique", *International Journal of Advanced Manufacturing Technology* 41 (2009): 301–310.

52. A. Saldaña-Robles, H. Plascencia-Mora, A. Saldaña-Robles, A. Marquez-Herrera and J. A. Diosdado-De la Peña, "Influence of ball-burnishing on roughness, hardness and corrosion resistance of AISI 1045 steel", *Surface and Coating Technology* 339 (2018): 191–198.
53. B. B. Buldum and S. C. Cagan, "Study of ball burnishing process on the surface roughness and microhardness of AZ91D Alloy", *Experimental Techniques* 42, 2 (2018): 233–241.
54. L. N. López De Lacalle, A. Lamikiz, J. Muñoa and J. A. Sánchez, "Quality improvement of ball-end milled sculptured surfaces by ball burnishing", *International Journal of Machine Tools and Manufacture* 45, 15 (2005): 1659–1668.
55. Y. Tian and Y. C. Shin, "Laser-assisted burnishing of metals". *International Journal of Machine Tools and Manufacture* 47, 1 (2007): 14–22.
56. J. A. Travieso-Rodríguez, G. Dessein and H. V. Gonzalez-Rojas, "Improving the surface finish of concave and convex surfaces using a ball burnishing process", *Materials and Manufacturing Processes* 26 (2011): 37–41. DOI: 10.1080/10426914.2010.544819.
57. F. J. Shiou, S. J. Huang, A. J. Shih, J. Zhu and M. Yoshino, "Fine surface finish of a hardened stainless steel using a new burnishing tool", *Procedia Manufacturing* 10 (2017): 208–217.
58. F. Gharbi, S. Sghaier, K. J. Al-Fadhalah and T. Benameur, "Effect of ball burnishing process on the surface quality and microstructure properties of AISI 1010 steel plates", *Journal of Materials Engineering and Performance* 20, 6 (2011): 903–910.
59. T. Zhang, N. Bugtai and I. D. Marinescu, "Burnishing of aerospace alloy: A theoretical-experimental approach", *Journal of Manufacturing Systems* 42 (2014): 2–8.
60. M. M. El-Khabeery and M. H. El-Axir, "Experimental techniques for studying the effects of milling roller-burnishing parameters on surface integrity", *International Journal of Machine Tools and Manufacture* 41, 12 (2001): 1705–1719.
61. N. J. Varpe, A. Hamilton and U. Gurnani, "Optimization of Burnishing process by Taguchi method for surface enhancement of EN31 steel", *Surface Topography: Metrology and Properties* 10 (2022): 015017. DOI:10.1088/2051-672X/ac4f37.
62. G. V. Jagadeesh and S. G. Setti, "Surface integrity of ball burnished bioresorbable magnesium alloy", *Advances in Manufacturing* 11, 5 (2022): 342–362. DOI:10.1007/s40436-021-00387-6.
63. G. V. Jagadeesh and S. G. Setti, "Modeling the surface integrity of ball burnished biocompatible magnesium alloy by soft computing techniques", *Transactions of the Indian Institute of Metals* 75, 6 (2022):1603–1618.
64. G. V. Jagadeesh and S. G. Setti, "Tribological performance evaluation of ball burnished magnesium alloy for bioresorbable implant applications", *Journal of Materials Engineering and Performance* 31, 2 (2022): 1170–1186.
65. L. K. Han, Y. Changfeng and Z. Dinghua, "Studies on the surface characteristics of Ti60 alloy induced by turning combined with ball burnishing", *Journal of Manufacturing Processes* 76 (2022): 349–364.
66. R. Wandra, C. Prakash and S. Singh, "Investigation on surface roughness and hardness of β-Ti alloy by ball burnishing assisted electrical discharge cladding for bio-medical applications", *Materialstoday* 50, 5 (2022): 848–854.
67. R. Vaishya, V. Sharma and A. Gupta, "Mathematical modeling and experimental validation of surface roughness in ball burnishing process", *Coatings* 12, 10 (2022): 1506. DOI: 10.3390/coatings12101506.
68. N. J. Varpe, R. Tajane, U. Gurnani and A. Hamilton, "Multi objective optimization of burnishing to eliminate heat treatment in reamer shank manufacturing by using Taguchi coupled principal component analysis (PCA)", *Advances in Materials and Processing Technologies* 9, 4 (2022): 1–17. DOI: 10.1080/2374068X.2022.2118931.

69. G. V. Duncheva, J. T. Maximov, A. P. Anchev, V. P. Dunchev and Y. B. Argirov, "Multi-objective optimization of the internal diamond burnishing process", *Materials and Manufacturing Processes* 37, 1 (2021): 428–436.
70. C. Ding, Z. Zhou, Z. Piao and P. Mao, "Influence of the ultrasonic vibration on system dynamic responses in the multi-ball surface burnishing process", *Journal of Manufacturing science and Engineering* 144, 5 (2022): 051002. DOI: 10.1115/1.4052392.
71. R. Kumar, B. Kumar Singh, A. Kumar and A. Kumar, "Integrating selective flocculation techniques for enhanced efficiency in manufacturing processes: A novel approach through artificial neural network modeling", *Journal of Alloys and Metallurgical Systems* (2024): 100088. DOI: 10.1016/j.jalmes.2024.100088.
72. R. Ranjan, S. Kumar Jha and A. Kumar, "Enhancing hardness in SMAW welded joints with low-frequency vibration assistance: An experimental and analytical study", *Journal of Adhesion Science and Technology* (2024). DOI: 10.1080/01694243.2024.2371227.
73. S. Kumar Murmu, S. Chattopadhayaya, R. Cep and A. Kumar, "Exploring tribological properties in the design and manufacturing of metal matrix composites: An investigation into the AL6061-SiC-fly ASH alloy fabricated via stir casting process", *Frontiers in Materials* 11 (2024). DOI: 10.3389/fmats.2024.1415907.
74. I. Chowdhury, K. Sengupta, P. Chandra Chandra and A. Kumar, "Progressive effect of dual-hybridization in friction stir welding by ultrasonic vibration and resistive heating for joining dissimilar material Al6063 aluminium alloy and C26000 copper alloy", *Journal of Adhesion Science and Technology* (2024). DOI: 10.1080/01694243.2024.2359842.
75. S. Gangwar, S. Chandra Mondal and A. Kumar, "Performance analysis and optimization of machining parameters using coated tungsten carbide cutting tool developed by novel S3P coating method", *International Journal on Interactive Design and Manufacturing* (2024). DOI: 10.1007/s12008-024-01852-9.
76. G. S. Ghule, S. Sanap, S. Chinchanikar, R. Cep and A. Kumar, "Investigation of conventional and ultrasonic vibration-assisted turning of hardened steel using a coated carbide tool", *Frontiers in Mechanical Engineering* 10 (2024). DOI: 10.3389/fmech.2024.1391315.
77. A. Kumar Srivastava, S. Tiwari, P. Pachauri, N. Gupta, B. Sunil and A. Kumar, "Bonding strength and microstructural features of Al5083-AZ31B alloys laminated sheet through friction stir additive manufacturing", *Journal of Adhesion Science and Technology* 38, 4 (2024): 583–596. DOI: 10.1080/01694243.2023.2240637.
78. R. Kumar, V. Kumar and A. Kumar, "A review on effect of computer-aided machining parameters in incremental sheet forming", *Handbook of Flexible and Smart Sheet Forming Techniques: Industry 4.0 Approaches* (2023). DOI: 10.1002/9781119986454.ch3.
79. L. Rathee, R. Rathee and A. Kumar, "Experimental study of impact parameters on the deforming loads in incremental forming", *Modeling, Characterization, and Processing of Smart Materials* (2023): 140–150. DOI: 10.4018/978-1-6684-9224-6.ch006.
80. T. Sathish, R. Saravanan, S. J. Arunachalam, A. Parthiban and J. Giri, "Comparing Kevlar fiber influence on PP/sisal/SiO2 nanoparticle fillers/Kevlar hybrid nanocomposites with plain nanocomposite for enhanced mechanical properties", *Interactions* 245, 1 (2024). DOI: 10.1007/s10751-024-01935-9.
81. P. A. Thakare, N. Kumar, V. B. Ugale, J. Giri, N. Sunheriya and H. A. Al-Lohedan, "Effect of impact and flexural loading on hybrid composite made of Kevlar and natural fibers", *AIP Advances* 14, (4) (2024). DOI: 10.1063/5.0195907.
82. M. Natarajan, T. Pasupuleti, J. Giri, H. A. Al-Lohedan, L. N. Katta, F. Mohammad, N. Sunheriya, R. Chadge, C. Mahatme, P. Giri, S. Mallik and T. Sathish, "Optimization of wire spark erosion machining of Grade 9 titanium alloy (Grade 9) using a hybrid learning algorithm", *AIP Advances* 14, (1) (2024). DOI: 10.1063/5.0177658.
83. K. L. Narasimhamu, M. Natarajan, P. Thejasree, E. Makki, J. Giri, N. Sunheriya, R. Chadge, C. Mahatme, P. Giri and T. Sathish, "Development of hybrid optimization

model using grey-ANFIS-Jaya algorithm for CNC drilling of aluminium alloy", *Journal of Engineering* 2024 (2024): 1–12. DOI: 10.1155/2024/1476770.

84. N. Praveen, S. K. NG, C. D. Prasad, J. Giri, I. Albaijan, U. S. Mallik and T. Sathish, "Effect of pulse time (Ton), pause time (Toff), peak current (Ip) on MRR and surface roughness of Cu–Al–Mn ternary shape memory alloy using wire EDM", *Journal of Materials Research and Technology/Journal of Materials Research and Technology* (2024). DOI: 10.1016/j.jmrt.2024.03.122.

15 Laser Powder Bed Fusion Printing

Material and Process Parameter

Ashish Kumar, Ashish Kumar Srivastava, Parveen Kumar, Ajay Kumar, and Vikas Choudhary

15.1 INTRODUCTION

Additive manufacturing (AM) also referred to as 3D printing has completely revolutionized the manufacturing process from beginning to last functional parts production. Additionally, it is promoting the engineering development of the coming generation. Because AM makes it possible to manufacture extremely complex object, thin-walled parts at a cost-effective price, it has had a vast impact on a wide range of industries [1]. AM is divided into seven major types [2, 3], as shown in Figure 15.1. They are fused filament fabrication, vat photopolymerization, powder bed fusion (PBF), material jetting, binder jetting, direct light fabrication, laminated object manufacturing [4].

15.1.1 PBF Overview

PBF technology widely acceptable because of its versatility through the utilization of a wide range of materials such as metals ceramics and polymer, thereby its potential in various industries. This comprehensive analysis is centered on the PBF approach, which is intricately linked to the production of precise components. Within the realm of PBF, a heat source is harnessed to meld powdered materials, enabling the creation of complex metallic parts with a high level of accuracy in terms of dimensions. PBF methods sintering heat sintering, selective laser sintering, direct metal laser sintering, multi jet fusion, electron beam melting, selective laser melting, as illustrated in Figure 15.2. These processes play an important role in advancing the capabilities of PBF technology and expanding its applications across diverse sectors [1]. The powder particles are spherical in shape, and the particle sizes used for selective laser melting (SLM), selective laser sintering (SLS), and electron beam melting (EBM) are, respectively, 15–40 μm, 20–80 μm, and 40–100 μm [5]. Table 15.1 presents characteristics of different types of PBF processes.

DOI: 10.1201/9781032725086-17

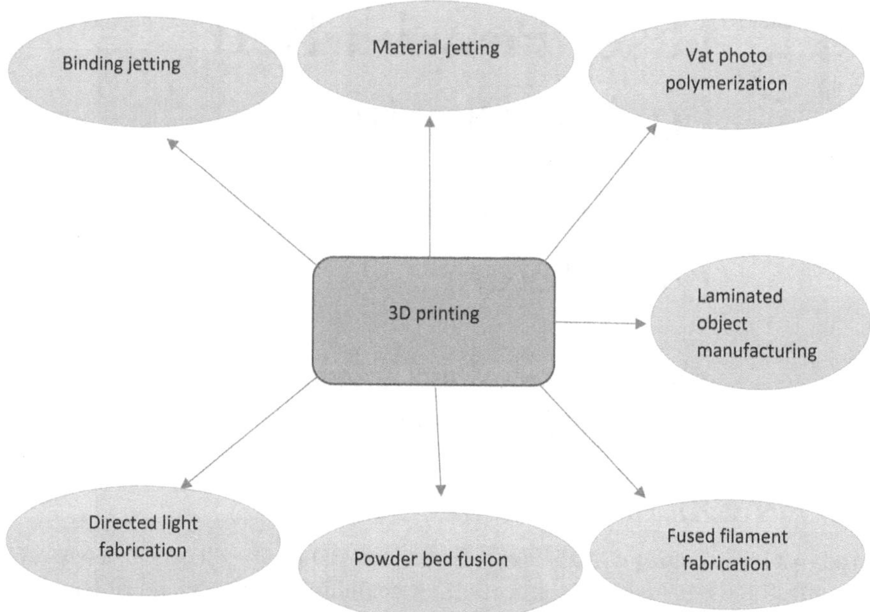

FIGURE 15.1 Types of additive-manufacturing technologies.

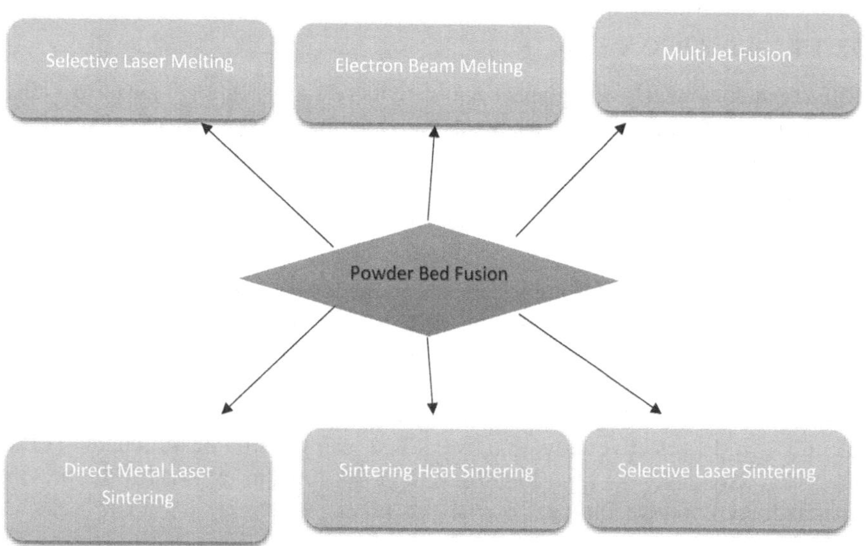

FIGURE 15.2 Different PBF technologies.

TABLE 15.1
Different Type of Powder Bed Fusion

Type of PBF	Energy Source	Specifications	History	Materials Used
Selective laser sintering (SLS) [6–8]	CO_2 laser, fiber laser	Layer thickness: 0.05–0.15 mm. Build volume: up to 700 × 380 × 580 mm³. Typical accuracy: ± 0.2 mm.	Invented in the mid-1980s by Dr. Carl Deckard and Dr. Joe Beaman at the University of Texas at Austin	Polyamide (Nylon), polystyrene, thermoplastic elastomers, composites (e.g., carbon fiber-filled)
Direct metal laser sintering (DMLS) [9]	Fiber laser	Layer thickness: 0.02–0.05 mm. Build volume: up to 500 × 280 × 325 mm. Typical accuracy: ± 0.1 mm.	Developed in the 1990	Stainless steel, aluminum, titanium, nickel alloys, cobalt-chrome
Selective laser melting (SLM) [10–12]	Fiber laser	Layer thickness: 0.02–0.05 mm. Build volume: up to 800 × 400 × 500 mm. Typical accuracy: ± 0.1 mm.	Developed in Germany in the mid-1990s by Fraunhofer Institute ILT	Aluminum, titanium, steel, nickel alloys, cobalt-chrome
Electron beam melting (EBM) [13–15]	Electron beam	Layer thickness: 0.05–0.2 mm. Build volume: up to 350 × 350 × 380 mm. Typical accuracy: ± 0.2 mm.	Developed in the late 1990s by Arcam AB (now part of GE Additive)	Titanium, cobalt-chrome, inconel
Multi jet fusion (MJF) [16]	Infrared light, fusing agents	Layer thickness: 0.08–0.12 mm. Build volume: up to 380 × 284 × 380 mm. Typical accuracy: ± 0.2 mm.	Introduced by HP Inc. in 2016	Polyamide (nylon), thermoplastic polyurethane (TPU), polypropylene
Selective heat sintering (SHS) [6]	Infrared heater or thermal print head	Layer thickness ranges from 0.1 to 0.2 mm. Build volume around 200 × 200 × 150 mm. Dimensional accuracy is typically within ±0.2 mm.	Developed in the 1990s by the company DTM Corporation	Nylon, polystyrene, and other polymers

15.1.2 LPBF: A Well-Known AM Method

The steps involved in creating a component using Laser Powder Bed Fusion (LPBF) are as follows: (i) First of all by using CAD software we design and create the 3D part, followed by cutting the model into the necessary number of layers

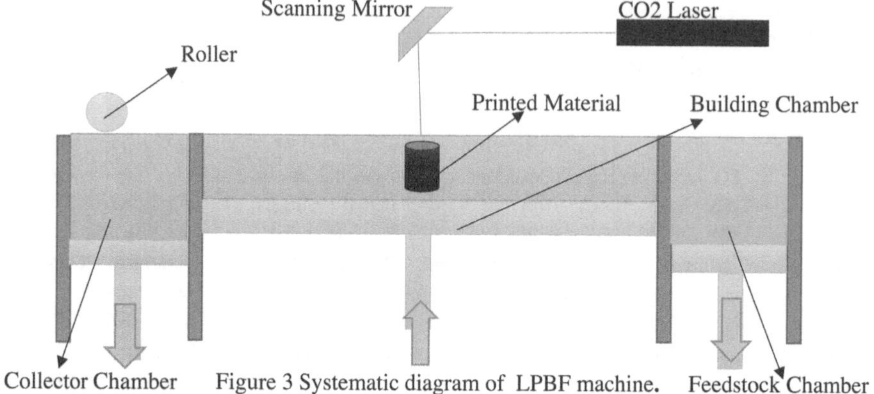

FIGURE 15.3 Systematic diagram of LPBF machine.

with a suitable thickness. (ii) A foundation is mounted on the build platform for the construction process. (iii) The first layer is applied to the build platform in accordance with the suitable layer thickness. (iv) The build chamber is placed into an inert environment, mostly composed of nitrogen and argon, to reduce the effect of oxidation. (v) The laser from laser source start to fall on powder thus powder start melting and bond formation among powder particles. (vi) The construction platform is lowered, and the final two processes of spreading the powder bed and repeatedly scanning it are repeated until the completed component is created. Figure 15.3 lists the parts of an LPBF machine [17]. Figure 15.4 represents heat transfer methods.

- **Two or three chambers are typically found inside a PBF machine**
  - Feedstock chambers (powder supply)
  - Building chambers (Where the part is built)
  - Collective chambers (Excess powder deposit)
- The feedstock chamber pushes the powder placed on a platform upward to allow the roller to redistribute the supply evenly across the building platform.

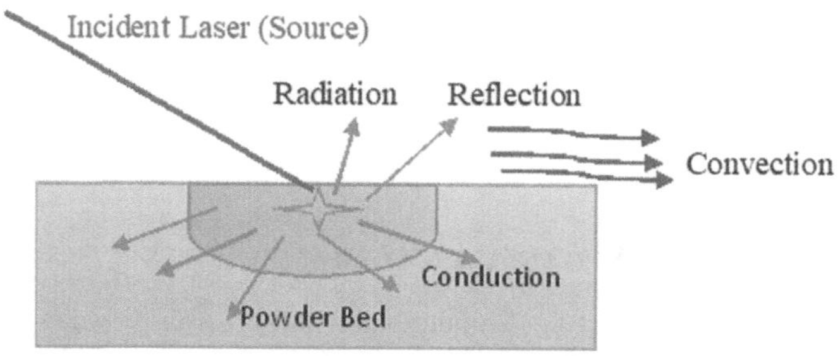

FIGURE 15.4 Shows heat transfer methods.

- The building chamber comprises of inert gas, often argon to avoid oxidization of the metal as it melts.
- The collection chamber collects the excess powder spread by the roller to ensures an even spread.

When incident laser fall on build platform then heat transfer take place in following steps:

- Radiation emission in surrounding
- Reflection or Scattering in surrounding
- Convection heat flow
- Conduction in powder bed
- Laser Absorption
- Vaporization

LPBF is similar as welding because in this process also welding bead is made because as laser beam fall on powder particles, powder particles completely melts and thus lead in formation of welding bead. It is therefore a kind of deposition method of welding [18]. In an LPBF process, a number of physical behaviors take place, which including heat transfer, absorption solidification, chemical reaction, phase transformation, movement of molten metals in welding bead [18, 19]. The transformation of light energy into heat energy forms the melt pool, and due to surface tension required shapes is obtained such as cylinder and sphere [20].

15.2 MATERIALS

This section highlights current initiatives and developments in the processing of functional materials for industrial product development using additive manufacturing.

15.2.1 Titanium

Titanium (Ti) alloys are widely used in the aerospace, automotive, and biomedical sectors because of their high-temperature tolerance, compatibility, and high strength-to-weight ratio. resistance.

Generally titanium is used in biomedical sectors to fulfill the essential requirements of low Young's modulus, high strength, low density, high wear resistance, high corrosion resistance, and high compatibility. Ti alloys possess excellent properties mentioned above and prove to be a good choice for the biomedical industry. Dental fields, joint replacements, implants, orthodontic components, surgical equipment, and artificial heart valves are among its biomedical uses. Because of the high strength of Ti alloy manufacturer has always presented challenges in terms of time, energy, and material requirements. Nevertheless, there is now more Ti component manufacture thanks to the development of additive manufacturing, particularly LPBF. Due to its superior mechanical properties, Tie6Ale4V has now supplanted commercially pure (CP) Ti in orthopedic prostheses [21–26].

15.2.2 Magnesium-Based Alloys

Magnesium is the sixth most abundant element in the earth crust. The basic property of material is light weight (even lighter than Ti and Al elements) high strength-to-weight ratio. Magnesium is widely used in different industry such as automobiles, clinical, and aviation. Magnesium is widely used in alloy to improve property such as machinability, castability, high thermal stability, mechanical property, and electrical conductivity. But the main drawbacks of magnesium are low elastic modulus, low corrosion resistance, poor creep resistance and low strength [27].

15.2.3 Aluminium-Based Alloys

Aluminium is widely used in world after steel and iron due to its property like high strength-to-weight ratio, low density and high strength, corrosion resistance [28–30]. We made the components of aluminium by using LPBF technique because by using this process we get dense product. But for some alloys such as Al/Fe_2O_3 powders, it was necessary to employ in situ formation of particle-reinforced Al matrix to overcome defects such as balling, doss formation and part distortion, which causes poor surface finish of the part produced [31–33].

15.2.4 Stainless Steel 316L

Stainless steel 316L is a famous grade of stainless steel known for its excellent corrosion resistance, mechanical properties, and versatility. Here are some of its key properties: Corrosion resistance: SS316L offers superior corrosion resistance, especially in environments containing chlorides, such as seawater or acidic conditions. This resistance is due to its high chromium (Cr) and molybdenum (Mo) content, which form a passive oxide layer on the surface, protecting it from corrosion. High temperature resistance: SS316L maintains its mechanical properties at elevated temperatures, making it suitable for applications involving high temperatures, such as in chemical processing, marine environments, and medical devices. Strength and ductility: SS316L exhibits good mechanical properties, including high tensile strength and excellent ductility. This combination of strength and ductility allows for the fabrication of complex components and structures without sacrificing integrity. Low carbon content: The "L" in SS316L denotes its low carbon content compared to standard SS316. This low carbon content minimizes carbide precipitation during welding, which reduces the risk of intergranular corrosion and sensitization [34]. Table 15.2 highlights the chemical properties of SS316L. Table 15.3 discusses the materials processed by LPBF and their advantages.

15.3 LPBF PROCESS PARAMETER

The setup conditions of a manufacturing process are greatly influenced by the process parameters of each given process. The ranges of process parameters are shown in coupled with notable impacts, such laser strength, spacing scan speed and particle size.

TABLE 15.2
Show Chemical Properties of SS316L

Element	% Present
Nitrogen (N)	0.10
Sulfur(S)	0.015
Phosphorous (P)	0.045
Manganese (Mn)	2.00
Silicon (Si)	1.00
Chromium (Cr)	16.50–18.50
Nickel (Ni)	10.00–13.00
Carbon (C)	0.01
Molybdenum (Mo)	2.00–2.50

15.3.1 Density

The density of material produced by LPBF is expressed by which is given below,

$$€ = P/U.h.d$$

Where laser power (P), scan speed (u), hatch spacing (h) and layer thickness (d) can be combined to form a single equation [31].

TABLE 15.3
Materials and Its Advantages

Material	Advantages
Titanium and its alloys	The component made by this have low allergic risk and it is biocompatibility. High strength-to-weight ratio and structural integrity [35]. Corrosion resistance and ensures the longevity of the implant in the body [2].
Co-Cr-based alloys	Have high wear resistance [36] Corrosion resistance and biocompatible [23]. As we know that load-bearing capacity is more so it used in knee and hip replacement [37]
Ni-Ti alloy	Unique shape memory and super elasticity. Corrosion resistance and biocompatible. It is widely used in stents, vascular implants, and devices for orthodontics [38].
Stainless steel	High resistance to corrosion and biocompatibility [39]. It is widely used in day-to-day life because of economical by value and implant requirement [40].
Magnesium and its alloys	Lightweight and density is similar to that of bone [41]. Biodegradable [42]. It is widely used in implantation scenarios [43], such as bone fixation devices.
AlSi10Mg	This material give favorable combined property of corrosion resistance, high specific strength, and low density [44].

Low energy density arises in LPBF due to high speed thus insufficient melting takes place which result in porosity and generating discontinuous cracks. But as the energy density increased better melting of powder takes place which result in disappearance of pores and a smooth surface finish is obtained. Again energy density increased further to 21J/mm^2, density is reduced this is because sufficient energy to make liquid pool, low scan speed causes spheroidization of material takes place [27].

15.3.2 Scanning

Poor surface quality, porosity, and residual strains are examples of LPBF defects that are caused by the creation of non-uniform temperature gradients. Utilizing a scanning approach, the issue with the temperature gradient during the cooling and heating phases. A number of researchers developed different scanning techniques to create completely dense components with essentially no flaws. As seen in, some of the typical and frequently recognized scanning include bidirectional, unidirectional, and chessboard strategies. The unidirectional strategy is the simplest but gives the low-density results among other methods [45] (Figure 15.5).

15.3.3 Powder Size

The hallmarks of the powder used in the LPBF process are important in establishing the characteristics of the components that are produced as well as keeping an eye on the process stability. We mainly discuss the property of powder by knowing the size, shape, composition, internal porosity, and surface morphology are typically used. Physical factors including flowability, or the degree to which particles tend to flow, and apparent density, or how well a powder may be packed, are also taken into account. One characteristic that keeps an eye on all the other process factors, such powder packing, flowability, and the characteristics of the heat impacts. They are also crucial in determining the process' other parameters, such as the layer thickness [46]. Figure 15.6 shows laser power, scan speed, and hatch spacing. Figure 15.7 highlights various scan strategies.

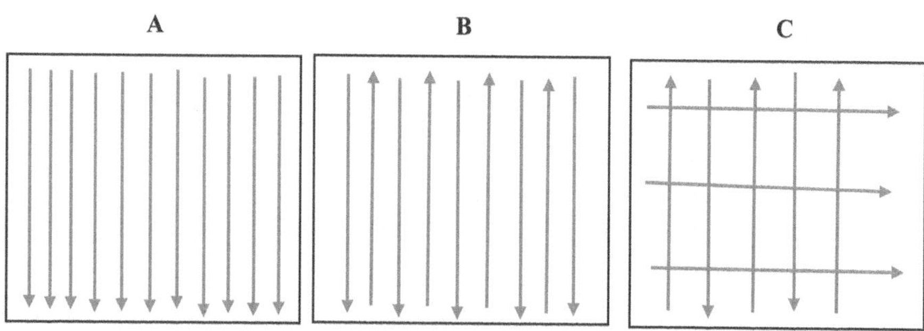

FIGURE 15.5 Different kinds of scanning strategies.

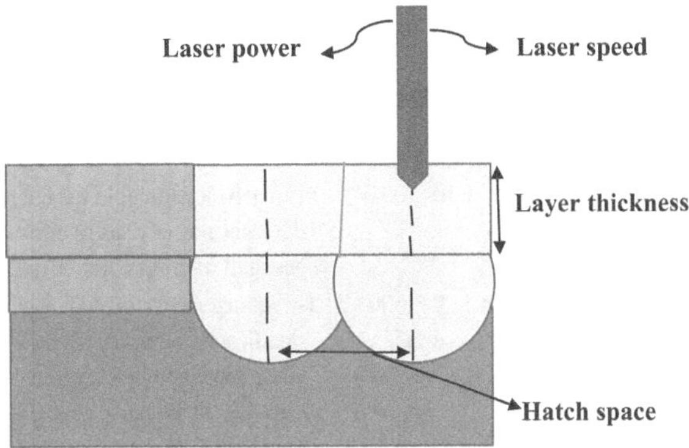

FIGURE 15.6 Shows diagram of laser power, scan speed, and hatch spacing.

15.3.4 Atmospheric Condition

The LPBF process atmosphere must be carefully controlled to prevent undesirable reactions and to initiate some desired responses. If there is presence of oxygen during LPBF process then oxidation of material takes place and will also lead to balling formation. The oxygen content in the argon-protected room was less than five parts per million. Argon was utilized as the shielding gas. The microstructure of the sections that were treated in the air had larger, coarser dendritic-like patterns, according to the data. Simultaneously, the portion created within the protected environment included evenly dispersed finer particles [28, 47]. Table 15.4 discusses the various process parameters of LPBF. Table 15.5 discusses laser power, scan speed, and hatch spacing. Table 15.6 explains layer thickness, scan strategy, and density effect.

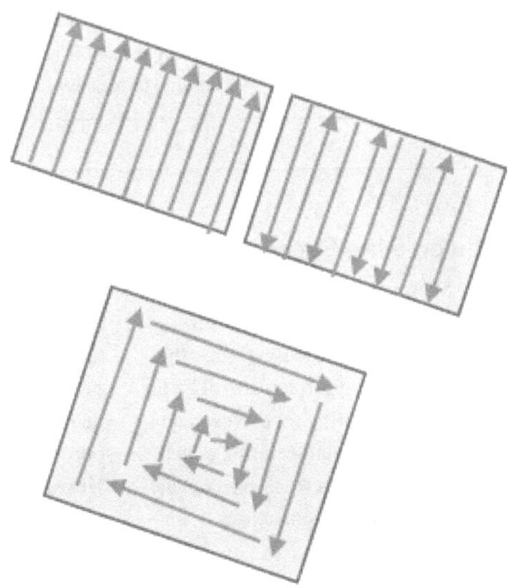

FIGURE 15.7 Shows scan strategy.

TABLE 15.4
Shows Different Process Parameter Range and Its Effect

Sl No.	Parameter	Sub Parameter	Range	Effect
1	Energy	Laser power	10–100W	Primarily accountable for the process of melting, the laser power rises in correspondence with the elevated melting point of the powder employed.
2	Scan related parameters	Scan speed Scan spacing Scan pattern	100–600 mm/s generally	The arrangement and method of scanning can significantly influence the build-up of residual stress in a component. Relocating a part within a machine can result in alterations to the specific laser paths used for building the part. These modifications in laser paths may lead to uneven distortion across different areas of the part. Consequently, a part may be successfully constructed in one area but encounter issues in another area within the same machine due to variations in how the scanning strategy is implemented.
3	Powder-related parameters	Particle size Particle shape Particle size Distribution width Layer thickness	15–50 μm spherical 1.38 μm 20–100 μm	The morphology, dimensions, and distribution of particles in powder greatly impact the way in which laser energy is absorbed. Powder size play a vital role in determining the density of components, the conductivity of materials and how the powder is spreadable. Smaller particles offer a larger surface area and are more effective in absorbing laser energy as compared to larger particles.
4	Temperature-related parameters	Powder bed temperature Powder feeder temperature		The temperature of the powder bed depends upon the absorptivity properties of the powder bed, which are affected by various factors such as powder size, shape material used and its composition, and packing density. To regulate the temperature infrared or resistive heater are employed.

TABLE 15.5
Definition of Laser Power, Scan Speed, and Hatch Spacing

Laser power is the prime focus of this process. It shows how much amounts of energy inputs into the powder bed during laser powder bed fusion. Laser power directly effects depths of molten pools as, laser power increases the penetration power also increases.	**Scan speed** is the speed at which laser beam move across powder bed during process. Higher the scanning speed lower will be the manufacturing time. Now a day low time manufacturing play a vital role as a result the demands of this process increases.	**Hatch spacing** refers to the distance between two parallel scan lines during laser powder bed fusion process. It play vital role in determining the density and porosity of final printed materials. Hexagonal hatching have superior properties as compared to the others hatching methods.

TABLE 15.6
Definition of Layer Thickness, Scan Strategy, and Density

Layer thickness is defined as the vertical distance between two adjacent layers which deposit during laser powder bed fusion techniques. This parameter is mainly play important role in determine surface finish of the final specimen, density, and dimensional accuracy unlike other manufacturing process.	**Scan strategy** is used to determine the order and direction in which laser is moving across powder bed. It pay important role in determining mechanical and physical properties of the materials.	**Density** in laser powder bed fusion technique refers to how the compact or concentrated powder is fused to obtained final product. It directly effects the quality and mechanical properties of the materials. Higher the density greater will be the durability and strength. The density is given by following formula, $ED = P / (v \times h \times t)$ where P – laser power (W) V – laser scan speed (mm/s) H – hatch distance (mm), the distance between adjacent scan tracks T – layer thickness (mm)

15.4 PROPERTIES OF LPBF PARTS

15.4.1 Microstructures

Microstructure of materials processed by LPBF is mainly controlled by cooling rate, higher the cooling rate or we can say smaller the solidification time finer the microstructure (Brittle the materials is), lower the cooling rate or we can say that larger the solidification time coarse the microstructure (Ductile the materials is). By controlling the process parameter we can control the phase percentage, grain size, and phase compositions. The microstructure depends on process parameter such as layer thickness, hatch spacing, scan speed, and laser power [29, 48]. Figure 15.8 highlights SEM imaging of Ti fabricated by LPBF.

Various parameter that effect the parameter of microstructures during LPBF

- Various process parameter.
- Solidification time.
- Heat treatment/heat loss.

15.4.2 Mechanical Properties

15.4.2.1 Tensile Strength

The components made by using LPBF technique have better properties as compared to the material manufactured by basic traditional methods such as casting. Better superior properties because layer by layer addition approach is the key points of the LPBF manufacturing and direction in which layers are built significantly effects the tensile strength.

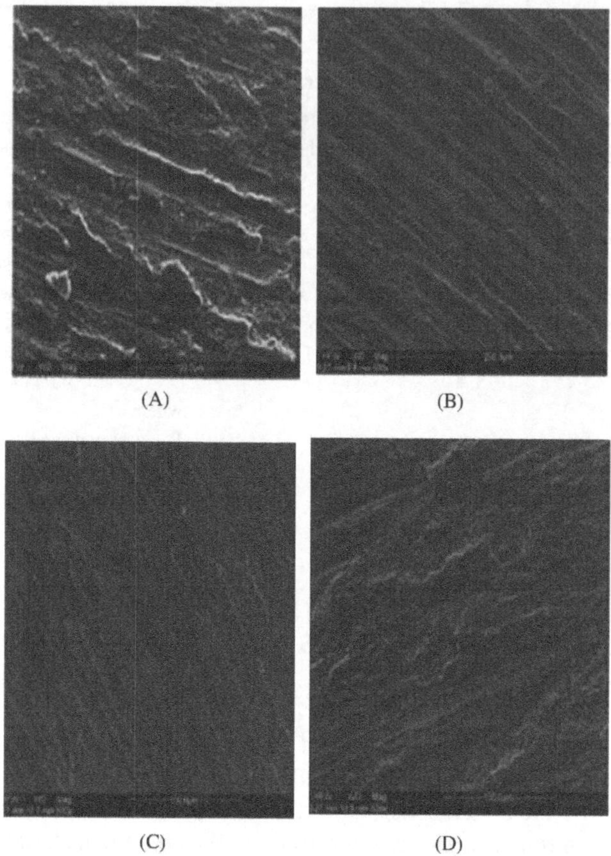

FIGURE 15.8 SEM imaging of Ti fabricated by LPBF (A) 900 J/m and 100 mm/s, (B) 450 J/m and 200 mm/s, (C) 300 J/m and 300 mm/s, and (D) 225 J/m and 400 mm/s [49].

15.4.2.2 Wear Resistance and Hardness

Initially it was claimed that scanning speed and laser power have no effect on hardness but that was not true in recent scenario as scanning speed increases the hardness of materials also increases. Hardness also depends on cooling rate, higher the cooling rate greater will be the hardness or we can say that the fine particles have good hardness as compared to the coarse particles. Specimen made by this technique have low coefficient of friction (COF) thus it improved the wear resistances of the portion [50–54].

15.4.2.3 Ductility

Ductility is usually induced in the materials on the expense of strength. Fine grains materials have high strength as compared to coarse grains materials but ductility of coarse grains materials is high as compared to fine grains; this is all controlled by solidification rate and temperature gradients. To increases the ductility of the materials sufficient amount of heat is always mentioned [55, 56].

15.4.2.4 Fatigue

One of the major failures of materials made by this technique is fatigues failures which arise when cyclic load is applied on the materials. Materials fail because there are pores defects present in the materials; this leads to the stress concentration and decreased load carrying capacity and thus decreases dynamic strength of the materials. First cracks start from pores and further it communicates to rest of the specimen [57, 58].

15.5 DEFECTS IN LPBF PROCESS

While making the materials with the help of LPBF there are many defects arises during the process. So a lot of work still have to done to overcome these defects or we can say in other words we have to regulate various process parameter.

15.5.1 Balling

This defects arises due to presence of loose powder presence in powder bed thus result in poor microstructure [59] (Figure 15.9).

15.5.2 Porosity

During LPBF process all powder particles melted completely producing unstable melt pools. But suppose a conditions arises where all the powder particles not reached, then fault's arise that is porosity. The different types of pores arise during this process are due to lack of shrinkage, gas, and lack of fusion [60] (Figure 15.10).

15.5.3 Surface Roughness

One of the major defects produced during this process is poor surface finish which is one of the major concern it arises due to oxide formation on the surface, partial melting of powder particles, due to presence of coarse powder size. It is also main disadvantages of additive manufacturing [62].

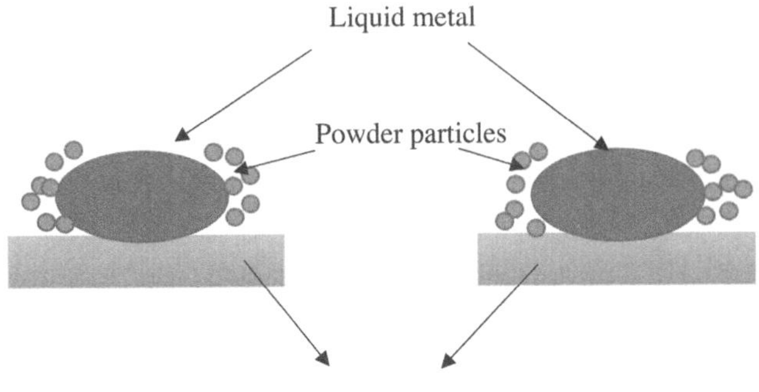

FIGURE 15.9 Show balling effect.

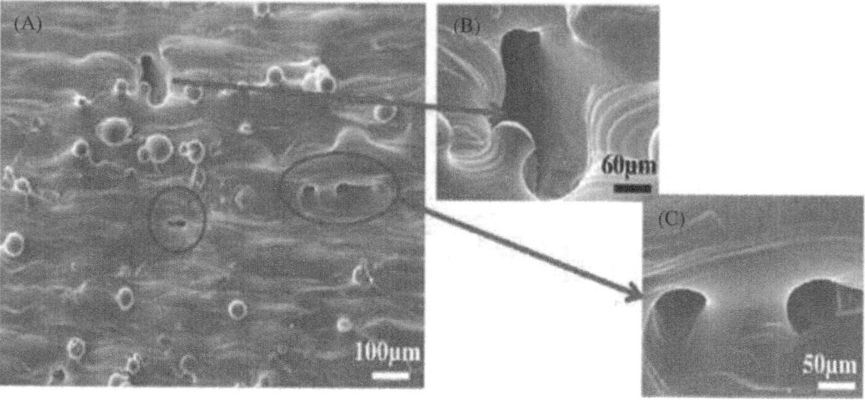

FIGURE 15.10 Show porosity [61].

15.5.4 Cracks and Residual Stresses

The LPBF technique identifies two main types of cracks. These are the two types of cracks: Cold and hot. The solidification cracks, also known as hot cracks, often produce the last step of solidification. The primary cause of hot fractures is distortion in the part's solid structure during solidification. Moreover, inadequate convection in the liquid zone might result in hot fractures. In the LPBF procedures, cold fractures brought on by residual strains are more frequent [5].

15.5.5 Loss of Alloying Elements

Because of the extremely high temperatures, volatile metals – mostly magnesium, zinc, aluminum, etc. – mainly evaporate from the molten pool. These material vaporized due to low boiling point and low vapor pressures, thus it produce poor alloying elements. These metal vaporized changing both the mechanical and compositional characteristics of resultant materials. Thus we can say that it effects the microstructure, strength, resistance to corrosion creep, elongation [63, 64].

15.5.6 Oxide Inclusion

Oxide formation is one of the major defects of this process, it came into picture when process is carry out in the presence of oxygen thus oxygen diffuses as a result oxide formation take place it is also the major cause of balling. To overcome such type of defects we have to carry out process in inert atmosphere [65]. Table 15.7 discusses defects in LPBF process.

15.6 CONCLUSION

Due to its wide range application and it is suitable for many metals and its alloys, LPBF is drawing a lot of interest. A comprehensive after the LPBF process was

TABLE 15.7
Show Various Defects Due Laser Powder Bed Fusion

Flaws	Reason	Remedy
Variation in composition	Due to excess heating vapors formation takes place.	By increasing layer thickness, by reducing laser power.
Oxide formation	Presence of atmospheric oxygen during processing.	Create a atmosphere of inert gas like argon, helium.
Cracks formation due to residual stress	Due to cooling rate is more than critical rate of cooling.	Control cooling rate by use of heated chamber.
Surface roughness	Due to oxide formation, due to coarse powder particles, partial melting of powder.	Use fine grain, complete melting, increase laser power.
Vacancy in specimen	Due to insufficient molten pool as a result low penetration take place.	Reheating/remelting.
Balling	Due to the presence of loose powder in powder bed.	Increase laser power, low scan speed, reheating.

reviewed, a few crucial issues surfaced that are extremely important. But it is also necessary to address various process parameter in order to reduce defects/flaws.

- In order to apply the LPBF process to any material and obtain the LPBF part's desired mechanical characteristics and maximum density, it is the most crucial to pay close attention to the processing parameter.
- The characteristics induced by the LPBF specimen demonstrate that LPBF may generate specimen with better characteristics than those generated by the traditional methods.
- The primary purpose of this method is densification of specimen formed which mainly depends on process parameter such as laser beam density, laser speed, scanning, and several other process parameter.
- As we know that in LPBF technique all the powder particles melt completely so the impact of powder size and shape is less significant. But in SLS process where melting happen partially their powder shape and size play a vital role.
- After study we come to know about the mechanical property of material made by LPBF technique mainly depends upon thermal history such as cooling rates (heat looses in the form of conduction, convection, radiation), the solidification rate, and thermal/temperature gradient.
- The key factor of this technique is the melt pool instability because it play crucial role in their key characteristics. Microstructural defects may also arise due to false selection of process parameter as we know defects negatively effects the parts characteristics.
- Moreover melting bigger powder particles is also challenges so when we melt coarse powder particles we are unable to get good surface finish.

REFERENCES

1. A. Singh, H. Parveen, and B. AlMangour (Eds.), *Handbook of Smart Manufacturing: Forecasting the Future of Industry 4.0* (1st ed.). CRC Press, 2023. doi: 10.1201/9781003333760.
2. A. Kumar, P. Kumar, A. K. Srivastava, and V. Goyat (Eds.), *Modeling, Characterization, and Processing of Smart Materials.* IGI Global, 2023. doi: 10.4018/978-1-6684-9224-6.
3. P. Kumar, S. S. Hussain, A. Kumar, A. K. Srivastava, M. Hussain, and P. K. Singh, '10 Finite element method investigation on delamination of 3D printed hybrid composites during the drilling operation', in *3D Printing Technologies: Digital Manufacturing, Artificial Intelligence, Industry 4.0*, 2024, De Gruyter, p. 223.
4. G. Goyal, A. Kumar, and D. Sharma, '12 recent applications of rapid prototyping with 3D printing: A review. 3D printing technologies: Digital manufacturing', in *Artificial Intelligence, Industry 4.0*, 2024, De Gruyter, p. 245.
5. Y. Zheng, W. Zhang, D. M. Baca Lopez, and R. Ahmad, 'Scientometric analysis and systematic review of multi-material additive manufacturing of Polymers', *Polymers*, vol. 13, no. 12, p. 1957, Jun. 2021, doi: 10.3390/polym13121957.
6. K. S. Munir, Y. Li, and C. Wen, 'Metallic scaffolds manufactured by selective laser melting for biomedical applications', in *Metallic Foam Bone*, Elsevier, 2017, pp. 1–23. doi: 10.1016/B978-0-08-101289-5.00001-9.
7. X. Gan, G. Fei, J. Wang, Z. Wang, M. Lavorgna, and H. Xia, 'Powder quality and electrical conductivity of selective laser sintered polymer composite components', in *Structure and Properties of Additive Manufactured Polymer Components*, Elsevier, 2020, pp. 149–185. doi: 10.1016/B978-0-12-819535-2.00006-5.
8. J. C. Najmon, S. Raeisi, and A. Tovar, 'Review of additive manufacturing technologies and applications in the aerospace industry', in *Additive Manufacturing for the Aerospace Industry*, Elsevier, 2019, pp. 7–31. doi: 10.1016/B978-0-12-814062-8.00002-9.
9. R. J. N. Joshua *et al.*, 'Powder bed fusion 3D printing in precision manufacturing for biomedical applications: A comprehensive review', *Materials*, vol. 17, no. 3, p. 769, Feb. 2024, doi: 10.3390/ma17030769.
10. W. Shi *et al.*, 'Simulation and experimental study of the hole-making process of Ti-6Al-4V Titanium alloy for selective laser melting', *J. Manuf. Process.*, vol. 106, pp. 223–239, Nov. 2023, doi: 10.1016/j.jmapro.2023.10.004.
11. A. K. Kushwaha *et al.*, 'Powder bed fusion–based additive manufacturing: SLS, SLM, SHS, and DMLS', in *Tribology of Additively Manufactured Materials*, Elsevier, 2022, pp. 1–37. doi: 10.1016/B978-0-12-821328-5.00001-9.
12. L. Jiao, Z. Chua, S. Moon, J. Song, G. Bi, and H. Zheng, 'Femtosecond laser produced hydrophobic hierarchical structures on additive manufacturing parts', *Nanomaterials*, vol. 8, no. 8, p. 601, Aug. 2018, doi: 10.3390/nano8080601.
13. S. Franchitti *et al.*, 'Investigation on electron beam melting: Dimensional accuracy and process repeatability', *Vacuum*, vol. 157, pp. 340–348, Nov. 2018, doi: 10.1016/j.vacuum.2018.09.007.
14. W. Toh, P. Wang, X. Tan, M. Nai, E. Liu, and S. Tor, 'Microstructure and wear properties of electron beam melted Ti-6Al-4V parts: A comparison study against as-cast form', *Metals*, vol. 6, no. 11, p. 284, Nov. 2016, doi: 10.3390/met6110284.
15. M. Galati, and L. Iuliano, 'A literature review of powder-based electron beam melting focusing on numerical simulations', *Addit. Manuf.*, vol. 19, pp. 1–20, Jan. 2018, doi: 10.1016/j.addma.2017.11.001.
16. K. P. M. Lee, and M. Kajtaz, 'Experimental characterisation and finite element modelling of polyamide-12 fabricated via multi jet fusion', *Polymers*, vol. 14, no. 23, p. 5258, Dec. 2022, doi: 10.3390/polym14235258.

17. T. M. Wischeropp, H. Tarhini, and C. Emmelmann, 'Influence of laser beam profile on the selective laser melting process of AlSi10Mg', *J. Laser Appl.*, vol. 32, no. 2, p. 022059, May 2020, doi: 10.2351/7.0000100.
18. J. Kruth, P. Mercelis, J. Van Vaerenbergh, L. Froyen, and M. Rombouts, 'Binding mechanisms in selective laser sintering and selective laser melting', *Rapid Prototyp. J.*, vol. 11, no. 1, pp. 26–36, Feb. 2005, doi: 10.1108/13552540510573365.
19. S. H. Riza, S. H. Masood, and C. Wen, 'Laser-assisted additive manufacturing for metallic biomedical Scaffolds', in *Comprehensive Materials Processing*, Elsevier, 2014, pp. 285–301. doi: 10.1016/B978-0-08-096532-1.01017-7.
20. I. Yadroitsev, and I. Smurov, 'Surface morphology in selective laser melting of metal Powders', *Phys. Procedia*, vol. 12, pp. 264–270, 2011, doi: 10.1016/j.phpro.2011.03.034.
21. C. Y. Yap *et al.*, 'Review of selective laser melting: Materials and applications', *Appl. Phys. Rev.*, vol. 2, no. 4, p. 041101, Dec. 2015, doi: 10.1063/1.4935926.
22. I. Gurrappa, 'Characterization of Titanium alloy Ti-6Al-4V for chemical, marine and industrial applications', *Mater. Charact.*, vol. 51, no. 2–3, pp. 131–139, Oct. 2003, doi: 10.1016/j.matchar.2003.10.006.
23. S. Q. Wang, J. H. Liu, and D. L. Chen, 'Effect of strain rate and temperature on strain hardening behavior of a dissimilar joint between Ti–6Al–4V and Ti17 alloys', *Mater. Des. 1980–2015*, vol. 56, pp. 174–184, Apr. 2014, doi: 10.1016/j.matdes.2013.11.003.
24. D. Banerjee, and J. C. Williams, 'Perspectives on Titanium science and technology', *Acta Mater.*, vol. 61, no. 3, pp. 844–879, Feb. 2013, doi: 10.1016/j.actamat.2012.10.043.
25. H. Attar, 'Manufacturing and properties of titanium-based materials produced by selective laser melting', *Theses Dr. Masters*, Jan. 2015 [Online]. Available: https://ro.ecu.edu.au/theses/1596
26. V. Cain, L. Thijs, J. Van Humbeeck, B. Van Hooreweder, and R. Knutsen, 'Crack propagation and fracture toughness of Ti6Al4V alloy produced by selective laser melting', *Addit. Manuf.*, vol. 5, pp. 68–76, Jan. 2015, doi: 10.1016/j.addma.2014.12.006.
27. S. Das, "Physical aspects of process control in selective laser sintering of metals." *Adv. Eng. Mater.*, vol. 5, no. 10, pp. 701–711, 2003.
28. G. B. Schaffer, B. J. Hall, S. J. Bonner, S. H. Huo, and T. B. Sercombe, "The effect of the atmosphere and the role of pore filling on the sintering of aluminium." *Acta Materialia, vol.* 54, no. 1, pp. 131–138, 2006.
29. J. M. Martín, and F. Castro, 'Liquid phase sintering of P/M aluminium alloys: Effect of processing conditions', *J. Mater. Process. Technol.*, vol. 143–144, pp. 814–821, Dec. 2003, doi: 10.1016/S0924-0136(03)00335-2.
30. A. Kumar, P. Kumar, and Y. Liu, *Industry 4.0 Driven Manufacturing Technologies.* Springer Nature. doi: 10.1007/978-3-031-68271-1.
31. S. Saedi, N. Shayesteh Moghaddam, A. Amerinatanzi, M. Elahinia, and H. E. Karaca, "On the effects of selective laser melting process parameters on microstructure and thermomechanical response of Ni-rich NiTi", *Acta Mater.*, vol. 144, pp. 552–560, Feb. 2018, doi: 10.1016/j.actamat.2017.10.072.
32. J. P. Oliveira, A. D. LaLonde, and J. Ma, "Processing parameters in laser powder bed fusion metal additive manufacturing", *Mater. Des.*, vol. 193, p. 108762, Aug. 2020, doi: 10.1016/j.matdes.2020.108762.
33. A. K. Srivastava, A. Kumar, P. Kumar, P. Gautam, and N. Dogra, "Research progress in metal additive manufacturing: Challenges and opportunities", *Int. J. Interact. Des. Manuf.*, pp. 1–17, 2023.
34. Y. M. Ren, X. Lin, X. Fu, H. Tan, J. Chen, and W. D. Huang, "Microstructure and deformation behavior of Ti-6Al-4V alloy by high-power laser solid forming", *Acta Mater.*, vol. 132, pp. 82–95, Jun. 2017, doi: 10.1016/j.actamat.2017.04.026.
35. G. Jian, and Z. Zhao, "Micro Machines | Free Full-Text | Study of Surface Integrity of Titanium Alloy (TC4) by Belt Grinding to Achieve the Same Surface Roughness

Range", 2022. Accessed: May 21, 2024 [Online]. https://www.mdpi.com/2072-666X/13/11/1950
36. T. Odaira, S. Xu, K. Hirata, X. Xu, T. Omori, K. Ueki, K. Ueda et al., "Flexible and tough superelastic Co–Cr alloys for biomedical applications", *Adv. Mater.*, 34, no. 27, p. 2202305, 2022.
37. Ó. Barro, F. Arias-González, F. Lusquiños, R. Comesaña, J. Del Val, A. Riveiro, A. Badaoui, F. Gómez-Baño, and J. Pou, "Characterization of Co-Cr-W dental alloys with veneering materials manufactured via subtractive milling and additive manufacturing LDED methods." *Materials*, vol. 15, no. 13, 4624, 2022.
38. M. Sathishkumar *et al.*, "Possibilities, performance and challenges of nitinol alloy fabricated by directed energy deposition and powder bed fusion for biomedical implants", *J. Manuf. Process.*, vol. 102, pp. 885–909, Sep. 2023, doi: 10.1016/j.jmapro.2023.08.024.
39. S. Ali *et al.*, "Biocompatibility and corrosion resistance of metallic biomaterials", *Corros. Rev.*, vol. 38, no. 5, pp. 381–402, Oct. 2020, doi: 10.1515/corrrev-2020-0001.
40. A. K. Srivastava, A. Dubey, M. Kumar, S. P. Dwivedi, R. K. Sing, and S. Kumar, "Measurement of form errors and comparative cost analysis for the component developed by metal printing (DMLS) and stir casting", *Instrumentation Measure Metrologie*, vol. 19, no. 5, pp. 363–369, 2020.
41. C. Liu, Z. Ren, Y. Xu, S. Pang, X. Zhao, and Y. Zhao, "Biodegradable magnesium alloys developed as bone repair materials: A review", *Scanning, vol.* 2018, no. 1, p. 9216314, 2018.
42. R. Kumar, and P. Katyal, "Effects of alloying elements on performance of biodegradable magnesium alloy", *Mater. Today Proc.*, vol. 56, pp. 2443–2450, Jan. 2022, doi: 10.1016/j.matpr.2021.08.233.
43. A. K. Srivastava, N. Kumar, and A. Rai Dixit, "Friction stir additive manufacturing–An innovative tool to enhance mechanical and microstructural properties", *Mater. Sci. Eng. B*, vol. 263, p. 114832, 2021.
44. X. Lu, X. Yang, X. Zhao, H. Yang, and M. V. Li, "Additively manufactured AlSi10Mg ultrathin walls: Microstructure and nano-mechanical properties under different energy densities and interlayer cooling times", *Mater. Sci. Eng. A*, vol. 835, p. 142652, Feb. 2022, doi: 10.1016/j.msea.2022.142652.
45. E. Yasa, J. Deckers, and J. Kruth, 'The investigation of the influence of laser re-melting on density, surface quality and microstructure of selective laser melting parts', *Rapid Prototyp. J.*, vol. 17, no. 5, pp. 312–327, Jan. 2011, doi: 10.1108/13552541111156450.
46. A. Ozbilen, A. Unal, and T. Sheppard, "Influence of oxygen on morphology and oxide content of gas atomised aluminium powders", *Phys. Chem. Met. Powders*, pp. 489–505, 1989.
47. A. Srivastava, C. Navaneetha, N. K. Abed, N. Singh, R. Chandrashekar, and H. Singh, "Rapid solidification techniques for metal processing: Microstructure and properties", *E3S Web of Conferences*, vol. 505, p. 01020. EDP Sciences, 2024.
48. W. M. Steen, and J. Mazumder, *Laser Material Processing*, Springer Science & Business Media, 2010.
49. D. Gu *et al.*, "Densification behavior, microstructure evolution, and wear performance of selective laser melting processed commercially pure titanium", *Acta Mater.*, vol. 60, no. 9, pp. 3849–3860, May 2012, doi: 10.1016/j.actamat.2012.04.006.
50. T. Sathish, R. Saravanan, S. J. Arunachalam, A. Parthiban, and J. Giri, 'Comparing Kevlar Fiber influence on PP/sisal/SiO_2 nanoparticle fillers/Kevlar hybrid nanocomposites with plain nanocomposite for enhanced mechanical properties', *Interactions*, vol. 245, no. 1, 2024. doi: 10.1007/s10751-024-01935-9.
51. P. A. Thakare, N. Kumar, V. B. Ugale, J. Giri, N. Sunheriya, and H. A. Al-Lohedan, 'Effect of impact and flexural loading on hybrid composite made of Kevlar and natural fibers', *AIP Advances*, vol. 14, no. 4, 2024. doi: 10.1063/5.0195907.

52. M. Natarajan, T. Pasupuleti, J. Giri, H. A. Al-Lohedan, L. N. Katta, F. Mohammad, N. Sunheriya, R. Chadge, C. Mahatme, P. Giri, S. Mallik, and T. Sathish, 'Optimization of wire spark erosion machining of Grade 9 titanium alloy (Grade 9) using a hybrid learning algorithm', *AIP Advances*, vol. 14, no. 1, 2024. doi: 10.1063/5.0177658.
53. K. L. Narasimhamu, M. Natarajan, P. Thejasree, E. Makki, J. Giri, N. Sunheriya, R. Chadge, C. Mahatme, P. Giri, and T. Sathish, 'Development of hybrid optimization model using grey-ANFIS-Jaya algorithm for CNC drilling of aluminium alloy', *J. Eng.*, vol. 2024, pp. 1–12, 2024. doi: 10.1155/2024/1476770.
54. N. Praveen, S. K. NG, C. Prasad, J. Giri, I. Albaijan, U. Mallik, and T. Sathish, 'Effect of pulse time (Ton), pause time (Toff), peak current (Ip) on MRR and surface roughness of Cu–Al–Mn ternary shape memory alloy using wire EDM', *J. Mater. Res. Technol.*, 2024. doi: 10.1016/j.jmrt.2024.03.122.
55. H. D. Carlton, A. Haboub, G. F. Gallegos, D. Y. Parkinson, and A. A. MacDowell, 'Damage evolution and failure mechanisms in additively manufactured stainless steel', *Mater. Sci. Eng. A*, vol. 651, pp. 406–414, Jan. 2016, doi: 10.1016/j.msea.2015.10.073.
56. A. Verma, H. Yadav, K. Kumar, P. K. Singh, M. Sharma, V. Shankar Srivastava, and A. K. Srivastava, 'State of art on microstructural and mechanical characterization of wire and arc additive manufacturing (WAAM)', *Computational and Experimental Methods in Mechanical Engineering: Proceedings of ICCEMME 2021* (2022): 93–104.
57. Y. Wang, J. Bergström, and C. Burman, 'Characterization of an iron-based laser sintered material', *J. Mater. Process. Technol.*, vol. 172, no. 1, pp. 77–87, Feb. 2006, doi: 10.1016/j.jmatprotec.2005.09.004.
58. Y. Wang, J. Bergström, and C. Burman, 'Four-point bending fatigue behaviour of an iron-based laser sintered material', *Int. J. Fatigue*, vol. 28, no. 12, pp. 1705–1715, Dec. 2006, doi: 10.1016/j.ijfatigue.2006.01.007.
59. D. Gu, *Laser Additive Manufacturing of High-Performance Materials*, Springer, 2015.
60. Q. Jia, and D. Gu, 'Selective laser melting additive manufacturing of inconel 718 superalloy parts: Densification, microstructure and properties', *J. Alloys Compd.*, vol. 585, pp. 713–721, Feb. 2014, doi: 10.1016/j.jallcom.2013.09.171.
61. C. Qiu, C. Panwisawas, M. Ward, H. C. Basoalto, J. W. Brooks, and M. M. Attallah, 'On the role of melt flow into the surface structure and porosity development during selective laser melting', *Acta Mater.*, vol. 96, pp. 72–79, Sep. 2015, doi: 10.1016/j.actamat.2015.06.004.
62. Y. Li, H. Yang, X. Lin, W. Huang, J. Li, and Y. Zhou, 'The influences of processing parameters on forming characterizations during laser rapid forming', *Mater. Sci. Eng. A*, vol. 360, no. 1, pp. 18–25, Nov. 2003, doi: 10.1016/S0921-5093(03)00435-0.
63. X. Cao, M. Jahazi, J. P. Immarigeon, and W. Wallace, 'A review of laser welding techniques for magnesium alloys', *J. Mater. Process. Technol.*, vol. 171, no. 2, pp. 188–204, Jan. 2006, doi: 10.1016/j.jmatprotec.2005.06.068.
64. 'Kumar, A., Liu, Y., & Kumar, R. (Eds.). (2024). Handbook of Intelligent and Sustainable Manufacturing: Tools, Principles, and Strategies. CRC Press. https://doi.org/10.1201/9781003405870
65. J. Campbell, *Complete Casting Handbook: Metal Casting Processes, Metallurgy, Techniques and Design*, Butterworth-Heinemann, 2015.

16 Innovative Approaches to Metal Matrix Composite Fabrication via Friction Stir Additive Manufacturing

Sanjay Kumar, Pramod Kumar, Manoj Kumar, Ashish Kumar Srivastava, Ajay Kumar, and Anuj Gupta

16.1 INTRODUCTION

Friction stir welding (FSW) and friction stir processing (FSP) share the same mechanism for welding or composite fabrication as shown in Figure 16.1. Due to this, the two terms are sometimes used interchangeably in literature. They do, however, serve many functions in actual applications. While FSP attempts to change the microstructure of one or more workpieces, FSW intends to connect two plates together. Aerospace and automotive offer a huge potential for the development of metal additive manufacturing (MAM). Most of these techniques are melt-based, suffer from solidification problems, and are not applicable to all alloys. Friction stir additive manufacturing (FSAM) is a family of new technologies that use the principle of FSW for layer-by-layer additive manufacturing (AM) of materials. This chapter discusses major FSAM technologies, concentrates on recent advancements in the still largely unexplored field of FSAM, and lists the benefits of FSAM versus fusion-based alternatives. The study also discusses FSAM technology's prospects and potential in the context of industrial manufacturing. This chapter summarizes by highlighting some of the notable research in this field. This innovation has been at the forefront for the past 30 years, currently surpassing traditional methods for producing complex materials with minimal cost and effort [1–4].

In recent years, the field of AM, which includes a variety of activities, has emerged as one of the most active, everything from machine design to material adaptability. Engineers have been inspired and given new opportunities by unprecedented topology design and on-demand printing capabilities. Despite the size of the AM materials and technology environment, there are several classifications organizing concepts. Of course, the more general categories of polymers, metals, and ceramics can be used to classify materials. The material class of interest greatly influences the choice of AM techniques. Metals have been the main topic of this overview. The AM

DOI: 10.1201/9781032725086-18

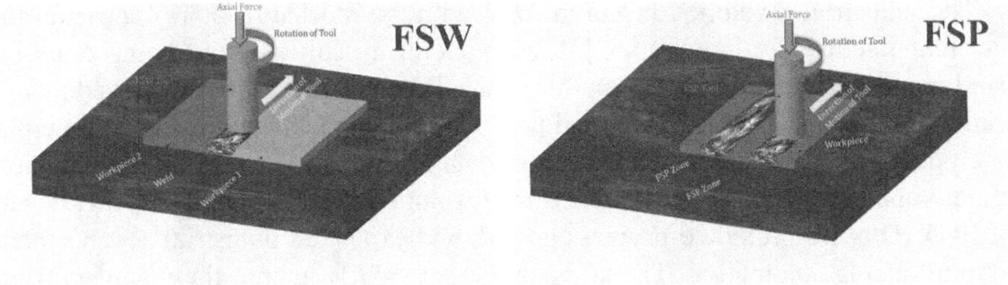

FIGURE 16.1 Schematic diagram of FSW and FSP.

method for metallic materials can be set up according to the alloy's state at the time of printing. A subgroup of solid AM methods is friction stir-based AM processes. AM processes using solid-state technology. In this review, we will look deeper into friction stir-based AM technologies and emphasize the significance of choosing the appropriate alloy for a given AM process. Insist that the greatest way to optimize the process potentially is through alloy and process co-design [5–8].

Although FSP is most frequently employed with aluminium [9–11], it is also used to process other alloys, such as tool steels [12], Mg alloys [13], intermetallic compound modified bronzes based on nickel [12], and Zr alloys. The benefits of the FSP method make it appropriate for a wide range of uses. Recent uses of FSP technologies include the railroad, automobile, aerospace, and marine industries. Friction stirs processing nanocomposites have been the subject of experimental [14–16], analytical [17–19], and computational research [10–13]. Investigations are made into the impact of processing factors on the development of microstructures, deformation behaviours, and mechanical characteristics [17–20].

The rapid industrial revolution is driving up the need for a new class of particular engineered materials with the required qualities [1]. The upshot is the development of new manufacturing techniques that are far quicker and more efficient than the conventional approaches created during the first industrial revolution as well as the second [2]. The history of the industrial revolution and the concurrent growth of the manufacturing sector are seen in Figure 16.2.

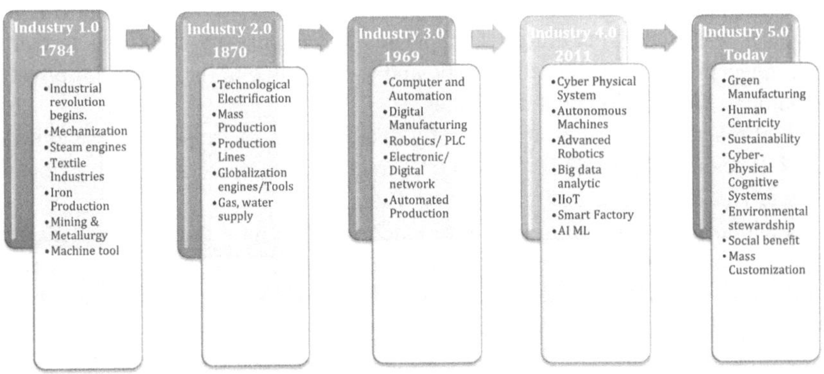

FIGURE 16.2 Industrialization and the development of industries.

The industrial revolution began in 1784 with the mechanization of agricultural and manufacturing communities [3]. Manufacturing shifted from being done by hand to utilizing machinery during this phase. Beginning in 1840, the second industrialization produced metal, electrical power, and oil and its byproducts. The output rate phases of the manufacturing sector advanced quickly to the mass production of auto and aeroplane parts [4]. The third revolution, digital manufacturing, began in 1945. This era promoted business growth and expedited industrial advancement by introducing automation. The key factors that swiftly altered the manufacturing industry's landscape were new 3D printers, inventive production methods, and state-of-the-art software such as CAM (Computer Aided Manufacturing), CIM (Computer Integrated Manufacturing), and CAD (Computer Aided Design) to connect the computer-controlled machinery to the bigger machinery [5]. With the help of machine learning and intelligent autonomous systems, Industry 4.0 will adapt automated and computer interfaces from the third revolution [6]. Industry 4.0, or I4.0, includes simulation, big data, augmented reality (AR), Internet of Things (IoT), and virtual systems as essential resources. Other resources include simulation and simulation-related technologies. It increases the viability of I4.0, which promotes intelligent manufacturing. One of the important resources created by Industry 4.0 is AM, which today functions as a quick-growing production method [7].

16.1.1 Metal Matrix Composite

Metal matrix composites (MMCs) have many special qualities that they can offer. Ample corrosion resistance, superior rigidity, enhanced damping capacity, and robust wear resistance are common combinations of these. Since MMCs are made of stiffeners incorporated into a metal matrix, they are better than base metal alloys. Rapid advances in technology have increased the need for better-quality materials. As a result, manufacturers are realizing that the performance of monolithic materials is inadequate for their production needs, and this is one of the main drivers for the continued development of composite materials [3].

The production method and characteristics of metal matrix composites reinforced with microparticles, platelets, and discontinuous (short) and continuous (long) fibres. The most popular procedures for creating composite materials and composite parts are based on powder metallurgy. Composites are extensively employed in electrical engineering, automotive technology, and aviation due to their superior physical, mechanical, and developmental qualities.

Metal matrix composites are currently one of the most frequently researched in materials science. This talk reviews improvements in the processing, utilization, and characteristics of light metal matrix composites.

As friction additive techniques have emerged, they have drawn the attention of numerous writers who have investigated them for engineered materials. As was previously described, the created component's microstructural and mechanical qualities can be improved using solid-state processing techniques.

Additionally, the tool shoulder or tool pin controls the material flow at several points during the production of the tool. As a result, throughout the FSAM and AFS (additive friction stir) processes, the microstructural characteristics vary at various layers. Material flow is also influenced by the geometry of the tool and important process variables, including tool rotation, traverse speed, and tilt angle.

16.1.2 Reinforcement

It involves blending a little quantity of an additional substance with the basic metal. It is done in order to provide the base metal with the qualities that are needed. Generally, weaker metals are strengthened by adding elements that increase their strength. Carbon nanotubes, titanium carbides, silicon carbide, and other materials are examples of reinforcements.

Reinforcement means to strengthen. Applications of the MMCs have a major role in reinforcement. Hybrid composites are made by combining reinforcement components with a matrix to further boost the strength of the composite. Additionally, hybrid MMCs could combine the same reinforcement in both micron and nm sizes or a variety of distinct reinforcement particles in both sizes.

16.1.3 Additive Manufacturing

In today's production environment, AM is a leading method [9]. It includes the layer-by-layer addition of material to create a variety of 3D objects, regardless of size and shape (complex objects), if the substance is formed of human tissue, concrete, metal, or plastic [10–12]. Technology has been at the cutting edge for the past 30 years; it has supplanted the traditional manufacturing method in terms of producing complex materials with the least amount of expense and work [13–15]. Modern tools like CAD, CNC, and simulation software are frequently used by AM to fabricate many types of materialistic items [16]. While creating the actual physical product with AM technology, there are several processes involved. Selecting raw materials, a particular AM technique, design parameters, and a layout section are the first steps in creating a realistic product. The prerequisites for post-processing are the next [17, 18]. When compared to conventional processes, the product created with the AM technology offers significant advantages, including little material and energy waste [19]. The method can produce products of higher quality with good dimensional precision and material characteristics [20].

16.1.4 Metal Additive Manufacturing (MAM)

MAM is a fully integrated area of AM that focuses on the layer-by-layer creation of metals [21–23]. Even if cold spraying AM is done at room temperature and with air conditioning, the deposited layers show noticeably inferior bulk physical attributes, such as strength, due to the unfused deposition [22, 25]. The highly accurate depositing of metal powder in binder jet AM is confronted with similar difficulties. However, this technology demonstrates that non-metals outperform metals in the essential field of strength [23, 24]. It restricts the freedom of sizes when the material is extruded using AM and a filament that is deposited on the substrate using a nozzle [24]. Contrary to popular belief, despite its subtractive nature, electrochemical AM is not used in complicated rapid constructions [25] and is unaffected by the features of substrates depositing faster in differences in materials and alloys [18]. PBF (Powder Bed Fusion) and direct laser sintering of metal are comparable processes that are used to make larger equipment parts, such as turbine blades, but have higher running costs and can produce materials that are thick and multi-layered [17]. Selective laser melting is more expensive to operate than other AM techniques, and similar electron beam melting requires a vacuum to generate functional parts [9, 11]. Material jetting (MJ) is another widely utilized AM method.

16.2 MATERIALS AND FABRICATION METHODS

To perform our experiment, we have to select the right kind of materials available in the market. Based on the literature reviews, we have selected the below-mentioned materials. Their compatibility is good, and they are widely used nowadays. They have better structural properties after the operation. In present scenarios, it is widely in demand in automobile industry.

After various reviews, we concluded that aluminium and magnesium are the best-suited materials for our experiment as their temperatures are compatible while heating. All these materials are better in strength and ductility, and they are easily processable.

Aluminium has a lower density as compared to other materials. The aluminium density is approximately one-third of the steel. Aluminium 6063 is used in this research work, Al 6063 has good mechanical properties. Various rotating and reciprocating components used in a wide range of household goods and businesses are made of aluminium. Due to its exceptional qualities, including its lightweight, high strength-to-weight ratio, easy fabrication and machinability, and high resistance to atmospheric corrosion. Al 6063 has a wide range of applications in the marine industry, the production of truck frames and bodies, the furniture industry, and automotive structures.

In this research work, HSS (high-speed steel) tool is used. First of all, design the tool as per requirement. Simple rod converted the tool; there are three types of tools in this work. First is triangular shown in Figure 16.3, second is circular shown in Figure 16.4, and third is in square shape shown in Figure 16.5. Every tool. Three tool profiles with different shapes and sizes were used in this research work. The first was tool triangular, 3 mm thick and 4 mm long. The second tool profile has a tip diameter of 2 mm and a tip length of 2 mm, and the last square tool is 4 mm thick and 7 mm long.

Reinforcement means to strengthen. A hybrid composite is made by combining the reinforcement with a matrix to further boost the strength of the composite. The same mix of reinforcements may be included in hybrid MMCs, but they may also be micron- and nanometre-sized or a combination of distinct micron- and nanometre-sized reinforcement particles. For this study, eggshell, B_4C (boron carbide), and SiC (silicon carbide) were used [26–30].

16.3 EXPERIMENTAL SETUP

Regarding the layer-by-layer connecting based on additive, FSAM used two techniques: powder base additive friction stir deposition (AFSD) and stacked base FSAM. The concepts behind both systems are the same; however, they are handled

FIGURE 16.3 Triangular shape.

differently. Tool pin length is greater than the created layer to ensure that two stacked layers can be connected simultaneously: The sequential layer bonding and schematic layout of the stacked type FSAM method. Using a pin-and-shoulder tool, the overlapping sheets' upper surface is penetrated in this manner. The sheets stick together as the tool advances because of its significant plastic deformation and frictional heat. The tool's movement causes the material to shift from the pin's front to its back, creating a junction. Consequently, a sandwich structure is formed by adhering the stacked sheets together.

FIGURE 16.4 Circular shape.

FIGURE 16.5 Square shape.

AFSD differs significantly from stacked type FSAM in terms of mechanism. It has a hollow cylindrical tool shoulder that delivers feed materials like solid rods or powder directly to the stirring region without the use of a tool pin. At the point where the tool shoulder and substrate plate meet in rapid rotation, friction heat is produced (base plate). Frictional heat causes the feed material to heat up, become softer, and undergo plastic deformation, binding it to the base plate at the contact. The continuous deposition of an additional layer with the necessary thickness is made possible by the bonding of feed material that has been plastically deformed to the substrate plate. Layers are then added one at a time to create a 3D object.

After selecting the machine, set all the parameters (fixer, tool angle, tool speed) and execute the operation. First, work on a single-layer aluminium plate with a thickness of 3 mm, fill the hole with reinforcing material, rotate the tool in the forward direction, and complete two passes. Another one-to-two aluminium plate is joined like a sandwich, and reinforcement is filled between the two plates to perform the operation. Finally, with a tool angle of 10, he joins the three aluminium layers at the same rate. After assembling all the plates in pairs two and three, set them aside for cooling. Please pay attention to the processing time, as it is 700–800°C. After some time, the sample is ready for sectioning, and the sample is prepared.

16.3.1 Process Parameter Selection

The process of layering several materials together to create a 3D item from digital data is used in the rapidly expanding AM technique known as FSAM. It functions using the FSW principle. However, the layer-by-layer joining that is connected to warming and re-sintering distinguishes FSAM from FSW. The friction stir lap welding (FSLW) concept is comparable to the preferred approach of the FSAM

TABLE 16.1
Processes Parameter during Experimentation

Serial Number	Rotational Speed	Feed	Tool Tilt Angle	Traverse Speed
1	1000, 1600 RPM	2.5 mm/min	1^0	1625 mm/min
2	1600 RPM	2.5 mm/min	1^0	1625 mm/min

technology. Unlike the FSLW process, the FSAM and AFS procedures are able to control the heat treatment and re-sintering duration, which allows them to attain the required mechanical qualities and control the microstructural properties. The welding interface at lower levels may also fail if multiple layers are deposited right away for FSLW. But in all honesty, FSAM isn't any slimier than the FSP. Differentiating the FSAM from the FSP are its spontaneous properties like warming and re-sintering. However, FSP allows for the scheduling of reheating and re-stirring. The melting point and the desired characteristics of the finished product influence the material choice. Table 16.1 lists the parameters that were followed during the experimental preformation.

16.4 COMPOSITION OF THE SAMPLES

In this study, the composition was chosen based on the weight percentage of the aluminium plate, and the reinforcement is also based on aluminium. In this work, the eggshell and SiC compositions are in the same proportions, but B_4C is different, i.e., 2% of the aluminium plate is shown in Table 16.2.

16.5 RESULTS AND DISCUSSION

FSAM was successfully implemented, and three specimens with the desired proportion of sample after being polished were created. Microstructural investigations were carried out on a polished sample of AA 6063; a sample of AA 6063 was prepared for FSAM. To further understand the micromechanical behaviour of the manufactured FSAM composite, microstructural research was conducted and reported.

TABLE 16.2
Composition of the Composite Material

Serial Number	Aluminium Plate (wt. in g)	Eggshell (wt.%)	SiC (wt.%)	B_4C (wt.%)
1	100	0.10	0.10	2
2	100	0.25	0.25	2
3	200	0.50	0.50	2
4	200	0.75	0.75	2
5	300	1	1	2
6	300	1.25	1.25	2

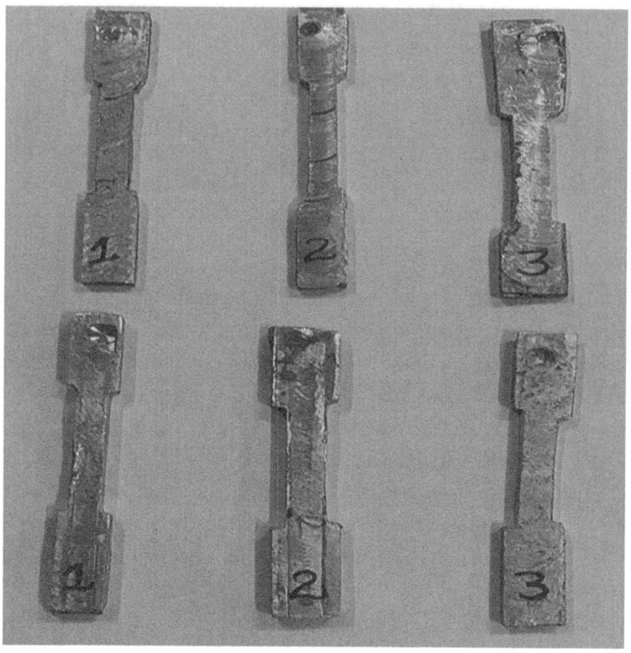

FIGURE 16.6 Tensile test specimen.

16.5.1 Tensile Test

Tensile tests are used to evaluate the safety and cohesiveness of materials. This guarantees the producer that the product will be of the best calibre. The greatest load that a material can bear is ascertained using a tensile test. This test is typically used to evaluate the quality of comparable manufactured product batches. Six samples are prepared for testing as shown in Figure 16.6. Determine the sample number 06's maximum tensile strength (198.3) after performing the tensile test. Test results are tabulated in Table 16.3 and shown in Figure 16.7.

TABLE 16.3
Tensile Test Reading

Sample Number	Tensile Reading (MPa)
Sample 1	160.56
Sample 2	167.93
Sample 3	176.35
Sample 4	184.46
Sample 5	191.24
Sample 6	187.3

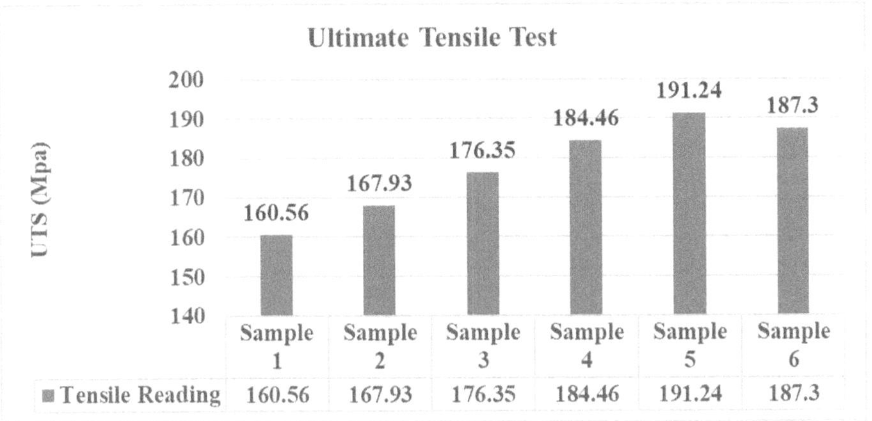

FIGURE 16.7 Tensile reading graph.

16.6 CONCLUSIONS

This research's major objectives were to create Al-B_4C MMCs using multiphase friction stir additive processing and analyse their attributes considering potential uses in the nuclear sector. It has been demonstrated that Al-B_4C MMCs with a thickness of up to 7 mm can be created using FSAM. The tensile strength and hardness of the Al 6063 with SiC/B_4C/eggshell HAMMC increase with the addition of B_4C. The tensile strength of the Al 6063 with SiC/B_4C/eggshell shows a maximum value of 191.24 MPa.

REFERENCES

1. Kumar Murmu, S., S. Chattopadhayaya, R. Cep, A. Kumar, A. Kumar, S. Kumar Mahato, A. Kumar, P. Ranjan Sethy, and K. Logesh. "Exploring tribological properties in the design and manufacturing of metal matrix composites: An investigation into the AL6061-SiC-fly ASH alloy fabricated via stir casting process." *Frontiers in Materials* 11, (2024): 1415907.
2. Srivastava, A. K., A. Kumar, P. Kumar, P. Gautam, and N. Dogra. "Research progress in metal additive manufacturing: Challenges and opportunities." *International Journal on Interactive Design and Manufacturing (IJIDeM)* (2023): 1–17.
3. Srivastava, A. K., S. Tiwari, P. Pachauri, N. Gupta, B. Sunil, and A. Kumar. "Bonding strength and microstructural features of Al5083-AZ31B alloys laminated sheet through friction stir additive manufacturing." *Journal of Adhesion Science and Technology* 38, no. 4 (2024): 583–596.
4. Kumar, A., R. K. Mittal, and A. Haleem. *Advances in Additive Manufacturing: Artificial Intelligence, Nature-Inspired, and Biomanufacturing*, 2023. https://doi.org/10.1016/c2020-0-03877-6
5. Srivastava, A. K., S. Dwivedi, A. Nag, D. Kumar, A. R. Dixit, and S. Hloch. "Microstructural, mechanical and tribological performance of a magnesium alloy AZ31B/Si_3N_4/eggshell surface composite produced by solid-state multi-pass friction stir processing." *Materials Chemistry and Physics* 301, (2023): 127694.

6. Ajay, Parveen, S. Ahmad, J. Sharma, and V. Gambhir (Eds.). *Handbook of Sustainable Materials: Modelling, Characterization, and Optimization* (1st ed.). CRC Press, 2023. https://doi.org/10.1201/9781003297772
7. Ajay, H. Singh, Parveen, and B. AlMangour (Eds.). *Handbook of Smart Manufacturing: Forecasting the Future of Industry 4.0* (1st ed.). CRC Press, 2023. https://doi.org/10.1201/9781003333760
8. Kumar, A., P. Kumar, A. K. Srivastava, and V. Goyat (Eds.). *Modeling, Characterization, and Processing of Smart Materials.* IGI Global, 2023. https://doi.org/10.4018/978-1-6684-9224-6
9. Dwivedi, R., R. K. Singh, A. Srivastava, A. Anand, S. Kumar, and A. Pal. "Statistical optimization of process parameters during the friction stir processing of Al7075/Al 2 O 3/waste eggshell surface composite." In *Recent Trends in Industrial and Production Engineering: Select Proceedings of ICCEMME 2021*, pp. 107–118. Springer, 2022.
10. Kumar, A., P. Kumar, N. Sharma, and A. K. Srivastava (Eds.). *3D Printing Technologies: Digital Manufacturing, Artificial Intelligence, Industry 4.0.* 2024. https://doi.org/10.1515/9783111215112
11. Goyal, G., A. Kumar, and A. Gupta. "Developments in 3D printing: A critical analysis and deep dive into innovative real-world applications." In *3D Printing Technologies: Digital Manufacturing, Artificial Intelligence, Industry 4.0*, p. 335. De Gruyter, 2024.
12. Goyal, G., A. Kumar, and D. Sharma. "Applications of rapid prototyping with 3D printing: A review." In *3D Printing Technologies: Digital Manufacturing, Artificial Intelligence, Industry 4.0*, p. 245. De Gruyter, 2024.
13. Kumar, P., S. S. Hussain, A. Kumar, A. K. Srivastava, M. Hussain, and P. K. Singh. "Element method investigation on delamination of 3D printed hybrid composites during the drilling operation." In *3D Printing Technologies: Digital Manufacturing, Artificial Intelligence, Industry 4.0*, p. 223. 2024.
14. Burande, D. V., K. Kalita, R. Gupta et al., "Machine learning metamodels for thermo-mechanical analysis of friction stir welding." *International Journal on Interactive Design and Manufacturing* (2024). https://doi.org/10.1007/s12008-024-01871-6
15. Jin, B. and B. Jin. "Industrial revolution: In full swing in both East and West." In *China's Path of Industrialization: Endeavors and Inclusiveness*, pp. 1–29. Springer, 2020.
16. Nahavandi, S. "Industry 5.0—A human-centric solution." *Sustainability* 11, no. 16 (2019): 4371.
17. Kumar, A., P. Kumar, R. K. Mittal, and V. Gambhir. "Materials processed by additive manufacturing techniques." In *Advances in Additive Manufacturing: Artificial Intelligence, Nature-Inspired, and Biomanufacturing*, pp. 217–233. Elsevier, 2023. https://doi.org/10.1016/B978-0-323-91834-3.00014-4
18. Grabowska, S., S. Saniuk, and B. Gajdzik. "Industry 5.0: Improving humanization and sustainability of industry 4.0." *Scientometrics* 127, no. 6 (2022): 3117–3144.
19. Srivastava, A. K., N. Kumar, and A. Rai Dixit. "Friction stir additive manufacturing–An innovative tool to enhance mechanical and microstructural properties." *Materials Science and Engineering: B* 263, (2021): 114832.
20. Kumar, S., S. Kumar, and A. Kumar. "Optimization of process parameters for friction stir welding of joining A6061 and A6082 alloys by Taguchi method." *Proceedings of the Institution of Mechanical Engineers, Part C: Journal of Mechanical Engineering Science* 227, no. 6 (2013): 1150–1163.
21. Dwivedi, R., R. K. Singh, S. Kumar, and A. K. Srivastava. "Parametric optimization of process parameters during the friction stir processing of Al7075/SiC/waste eggshell surface composite." *Materials Today: Proceedings* 47, (2021): 3884–3890.
22. Kumar, A., P. Kumar, R. K. Mittal, and H. Singh. "Preprocessing and postprocessing in additive manufacturing." In *Advances in Additive Manufacturing: Artificial*

Intelligence, Nature-Inspired, and Biomanufacturing, pp. 141–165. Elsevier, 2023. https://doi.org/10.1016/B978-0-323-91834-3.00005-3

23. Kumar, A., P. Kumar, R. K. Mittal, and H. Singh. "Printing file formats for additive manufacturing technologies." In *Advances in Additive Manufacturing: Artificial Intelligence, Nature-Inspired, and Biomanufacturing*, pp. 87–102. Elsevier, 2023. https://doi.org/10.1016/B978-0-323-91834-3.00006-5
24. Bandyopadhyay, A., and B. Heer. "Additive manufacturing of multi-material structures." *Materials Science and Engineering: R: Reports* 129, (2018): 1–16.
25. Kumar, A., P. Kumar, H. Singh, A. Haleem, and R. K Mittal. "Integration of reverse engineering with additive manufacturing." In *Additive Manufacturing Materials and Technologies, Advances in Additive Manufacturing*, edited by Kumar, A., Mittal, R. K. and Haleem, A., pp. 43–65. Elsevier, 2023. https://doi.org/10.1016/B978-0-323-91834-3.00028-4
26. Mahatme, C., J. Giri, F. Mohammad, M. S. Ali, T. Sathish, N. Sunheriya, and R. Chadge. "Experimental and numerical investigation of PLA based different lattice topologies and unit cell configurations for additive manufacturing." *The International Journal, Advanced Manufacturing Technology* (2024). https://doi.org/10.1007/s00170-024-13882-4
27. Giri, J., T. Sathish, T. Sheikh, N. Sunehriya, P. Giri, R. Chadge, C. Mahatme, and A. Parthiban. "Automatic liver segmentation using U-Net deep learning architecture for additive manufacturing. *Interactions* 245, no. 1 (2024). https://doi.org/10.1007/s10751-024-01927-9
28. Natarajan, M., T. Pasupuleti, M. M. S. Abdullah, F. Mohammad, J. Giri, R. Chadge, N. Sunheriya, C. Mahatme, P. Giri, and A. A. Soleiman. "Assessment of machining of Hastelloy using WEDM by a multi-objective approach." *Sustainability* 15, no. 13 (2023): 10105. https://doi.org/10.3390/su151310105
29. Kamble, P. D., J. Giri, E. Makki, N. Sunheriya, S. B. Sahare, R. Chadge, C. Mahatme, P. Giri, S. T., and H. Panchal. "An application of hybrid Taguchi-ANN to predict tool wear for turning EN24 material." *AIP Advances* 14, no. 1 (2024). https://doi.org/10.1063/5.0186432
30. Tufail, M. S., J. Giri, E. Makki, T. Sathish, R. Chadge, and N. Sunheriya. "Machinability of different cutting tool materials for electric discharge machining: A review and future prospects. *AIP Advances* 14, no. 4 (2024). https://doi.org/10.1063/5.0201614

Index

Note: **Bold** page numbers refer to **tables** and *italic* page numbers refer to figures.

L

M

T

V

W

Z

For Product Safety Concerns and Information please contact our EU representative GPSR@taylorandfrancis.com Taylor & Francis Verlag GmbH, Kaufingerstraße 24, 80331 München, Germany

Batch number: 10397790

Printed by Printforce, the Netherlands